The Big Bang and Georges Lemaître

GEORGES LEMAÎTRE

Congress of Zurich, August 1948.

The Big Bang and Georges Lemaître

Proceedings of a Symposium in honour of G. Lemaître fifty years after his initiation of Big-Bang Cosmology, Louvain-la-Neuve, Belgium, 10-13 October 1983

edited by

A. Berger

Institut d'Astronomie et de Géophysique Georges Lemaître,
Université Catholique de Louvain, Louvain-la-Neuve, Belgique

Under the High Patronage of His Majesty Baudouin the first, King of Belgium

D. Reidel Publishing Company

A MEMBER OF THE KLUWER ACADEMIC PUBLISHERS GROUP

Dordrecht / Boston / Lancaster

Library of Congress Cataloging in Publication Data

CIP

Main entry under title:

EURASIP Directory, 1983.

Includes index.
1. Signal processing–Research–Europe–Directories.
I. Gerbrands, Jan J., 1948- II. European Association for Signal Processing.
III. Title: E.U.R.A.S.I.P. Directory, 1983.
TK5102.5E88 1984 621.38'043'0254 84-15892
ISBN-13: 978-94-009-6489-1 e-ISBN-13: 978-94-009-6487-7
DOI: 10.1007/978-94-009-6487-7

Published by D. Reidel Publishing Company,
P.O. Box 17, 3300 AA Dordrecht, Holland.

Sold and distributed in the U.S.A. and Canada
by Kluwer Academic Publishers,
190 Old Derby Street, Hingham, MA 02043, U.S.A.

In all other countries, sold and distributed
by Kluwer Academic Publishers Group,
P.O. Box 322, 3300 AH Dordrecht, Holland.

TABLE OF CONTENTS

PART III - STRUCTURE OF THE UNIVERSE AND COSMIC RAYS

PART IV - GEORGES LEMAITRE: THE MAN AND HIS WORK

AN INTRODUCTION TO THE INTERNATIONAL SYMPOSIUM GEORGES LEMAITRE

A. Berger

Institut d'Astronomie et de Géophysique G. Lemaître
Université Catholique de Louvain-la-Neuve, Belgium

The world has proceeded from the condense to the diffuse... The atom-world was broken into fragments, each fragment into still smaller pieces... We can conceive of space beginning with the primeval atom and the beginning of space being marked by the beginning of time... But it is quite possible that the expansion has already passed the equilibrium radius, and will not be followed by a contraction. In this case, ... the suns will become colder, the nebulae will recede, the cinders and smoke of the original fireworks will cool off and disperse...

from "The Primeval Atom, An Essay on Cosmogony" by G. Lemaître, D. Van Nostrand Company, Inc. New York, 1950.

Georges Lemaître was born in Charleroi on the 17 July, 1894. It was in the thirties that the Abbé Lemaître proposed for the first time what became ultimately the theory of the Primeval Atom. In fact, the most important ideas in the work of Lemaître took shape between 1927 and 1933. Three fundamental publications during this period were followed by further works on the expansion of the Universe and on the primeval atom. The first of these seminal papers dates from 1927: "un Univers homogène de masse constante et de rayon croissant". The second paper - "The beginning of the world from the point of view of quantum theory" - was presented as a communication to the Royal Society and contained the seeds of a later publication having

A. Berger (ed.), The Big Bang and Georges Lemaitre, vii–xiv.

the title "Hypothèse de l'Atome Primitif". This theory is now known generally as the Big-Bang model.

Physical cosmology was born.

The ideas presented in these two publications of 1927 and 1931 were subsequently synthesized by Lemaître in his 1933 paper which indicated how the theory of the expanding universe relates to the idea of a primeval atom.

It thus seemed appropriate to celebrate in 1983 the fiftieth anniversary of this work. In fact, the idea of organizing a symposium was conceived in Belgrade in October 1979, during the international scientific assembly organized by the Serbian Academy of Sciences and Arts for the celebration of the hundredth anniversary of the birth of Milutin Milankovitch (1879-1958). This idea became more definite during the commemoration of the hundredth anniversary of the birth of Alfred Wegener (1880-1930). The organization of such a symposium became all the more urgent as we wanted not only to invite the scientific community working in the fields embraced by the work of Lemaître, but also to ask the participation of as many of his early collaborators and friends as possible: among others, Professors J. Oort, W.H. McCrea, O. Godart, L. Bouckaert and P. Ledoux.

Immediately on my return to Louvain-la-Neuve, the idea was put forward to those who would play a fundamental role in such a symposium. Their reaction was most encouraging and in 1981 an organizing committee was formed, which comprised former students of Lemaître : O. Godart, L. Bossy, P. Pâquet, J. Henrard, A. Berger together with J. Demaret. The organizing committee soon enlisted the help of an advisory panel of well-known Belgian cosmologists. An International Scientific Committee was also formed of scientists working in topics dear to Lemaître: cosmology, celestial mechanics and cosmic rays. It was also decided to invite André Deprit to occupy the Chair Georges Lemaître of the Sciences Faculty at the Catholic University of Louvain. Professor Deprit, first a student, then a collaborator of Lemaître, formerly a Belgian citizen, now a naturalized citizen of the U.S.A., whose wife, Andrée Bartholomé was an assistant of Lemaître, was indeed a most appropriate person to participate actively in the project.

This is how the International Symposium organized by the Institute of Astronomy and Geophysics Georges Lemaître from 10 to 13 October, 1983, in Louvain-la-Neuve was realised in order to commemorate its renowned student, professor, and eponym.

Fully aware of the work accomplished by Mgr. Lemaître, His Majesty King Baudouin enhanced this occasion by placing it under His High Patronage. His Holiness the Pope Jean-Paul II accepted to testify his paternel solicitude for the work of the scientists participating in the symposium. The President of the Pontifical Academy of Sciences and the Director of the Vatican Observatory transmitted their fervent wishes for the full success of the symposium. Numerous other eminent people graced the ceremony with their patronage.

The academic opening, the addresses of which are published by the Revue des Questions Scientifiques de Bruxelles (*), was presided over by Mgr. E. Massaux, Rector of the Catholic University of Louvain who spoke about Lemaître, the University professor. Professor Ch. de Duve, Nobel Prize winner in Medicine, called to mind the role of Lemaître as President of the Pontifical Academy of Sciences; the Emeritus Professor O. Godart, founder of the Institute, recalled the life and work of Mgr. Lemaître; Professor A. Deprit, Senior Mathematician at the National Bureau of Standards, spoke about Lemaître's work in celestial mechanics and his keen interest for computers; Professor J. Peebles, Professor of Physics at Princeton University, summarized the fundamental contributions of Lemaître to modern cosmology.

The attendance of more than three hundred people was enhanced by the presence of Mgr. A. Pedroni, Papal Nuncio, Mr Ph. Maystadt, Minister of Research Policy, Mr E. Knoops, Secretary of State, Mr Y. de Wasseige, Senator, Professor E. Boulpaep, President of the Belgian American Educational Foundation, Mr P. Lienardy, Principal Private Secretary to the Ministry of Education, His Lordship Y. du Monceau, Senator-Burgomaster. In addition, Professors M. Woitrin, General Administrator, H. Buyse, Scientific Advisor, P. Macq, Dean of the Sciences Faculty, G. de Ghellinck Vaernewycke, Dean of the Faculty of Applied Sciences, and A. Bruylants, Director of the Classe des Sciences de l'Académie Royale de Belgique, were accompanied by their University colleagues. The family of Lemaître and former students and collaborators of Mgr Lemaître were also present to witness this homage of the whole university community to the memory of G. Lemaître and in recognition of his work.

The symposium supported by the Commission of Celestial Mechanics and Cosmology of the International Astronomical Union,

---

(*) fasc. 2, 1984. Inquire to 61, rue de Bruxelles, B-5000 Namur.

has drawn attention to the contributions of Lemaître in cosmology, but also to celestial mechanics and numerical analysis, subjects in which he was passionately fond. Thirty-four papers, presented during plenary sessions or workshops recall the importance of Lemaître's works in the development of Astronomy and Geophysics. About a hundred scientists coming from fourteen different countries participated in the discussions and have contributed to this volume, which is intended to commemorate the Symposium.

Today, all facets of Lemaître's cosmology remain the preoccupation of active researchers. His discussion of the meaning of the zero-point of the radius of the Universe lies at the origin of modern, highly mathematicial, work on cosmological singularities. Also, his hypothesis of the primeval atom has stimulated further studies concerning cosmological nucleosynthesis and works on the physics of the primordial universe, which have experienced great development owing to the considerable progress of high energy physics and the recent ideas about the quantification of the gravitational field. The problems relating to the formation of Galaxies and to the possible existence of a cosmological constant were also part of Lemaître's work. These problems are still very much under active study and have possible bearings on recent gauge theories of physical interactions. These speculations, related, on the one hand, to the idea of an extremely early physical origin of the fluctuations leading to the birth of the galaxies and, on the other hand, to the development of inflationary models, promise to help considerably in the resolution of the enigmas related to the isotropic and quasi-flat character of the present universe.

This Symposium was organized to commemorate this crucial period of G. Lemaître's scientific career and also to give testimony to the results of his research in celestial mechanics and cosmic particles. Lemaître's works in celestial mechanics were mainly concerned with the Three Body Problem. He managed to regularize their equations in the case of binary encounters by a transformation of coordinates which maintained the Hamiltonian formalism. Moreover, he applied improved methods in celestial mechanics to problems in mechanics; in particular to the motion of charged particles in the field of a magnetic dipole in connection with the geophysical study of cosmic radiation.

ACKNOWLEDGMENTS

We are greatly indebted to the members of the following committees :

Patronage Committee

HIS MAJESTY KING BAUDOUIN THE FIRST

S. Em. le Cardinal G. Danneels, Grand Chancelier de l'Université Catholique de Louvain.
Mgr E. Massaux, Recteur de l'Université Catholique de Louvain.
S. Exc. Mgr A. Pedroni, Nonce Apostolique.
S. Exc. M. C.H. Price II, Ambassadeur des Etats-Unis d'Amérique
S. Exc. M.E. Jackson, Ambassadeur de Grande Bretagne.
M. H. De Croo, Ministre des Communications et des Postes, Télégraphes et Téléphones.
M. Ph. Maystadt, Ministre du Budget, de la Politique Scientifique et du Plan.
M. A. Bertouille, Ministre de l'Education Nationale.
M. Ph. Moureaux, Ministre-Président de l'Exécutif de la Communauté Française.
M. J.M. Dehousse, Ministre de l'Economie Wallonne, Président de l'Exécutif Régional Wallon.
M. E. Knoops, Secrétaire d'Etat à l'Energie.
Mme P. D'Hondt-van Opdenbosch, Secrétaire d'Etat aux Postes, Télégraphes et Téléphones.
M. F. Aerts, Secrétaire d'Etat à la Santé Publique et à l'Environnement.
M. F.X. de Donnéa, Secrétaire d'Etat à la Coopération au Développement.
M. M. Wathelet, Ministre des Techniques Nouvelles et des PME, de l'Aménagement du Territoire et de la Forêt pour la Région Wallonne.
M. V. Féaux, Ministre de la Région Wallonne pour l'Eau, l'Environnement et la Vie Rurale.
Mme J. Mayence-Goossens, Ministre du Logement pour la Région Wallonne.
Prof. A. Molitor, Président de la Fondation Roi Baudouin et Chef de Cabinet Honoraire du Roi.
M. I. Roggen, Gouverneur de la Province de Brabant.
M. M. Tromont, Gouverneur de la Province du Hainaut.

M. J. Hallet, Président du Conseil d'Administration de l'Université Catholique de Louvain.
Prof. M. Woitrin, Administrateur Général de l'Université Catholique de Louvain.
M. Y. du Monceau de Bergendael, Bourgmestre d'Ottignies-Louvain-la-Neuve.
M. G. Deprez, Président du Parti Social Chrétien.
M. L. Michel, Président du Parti du Renouveau et de la Liberté.
M. J.C. Van Cauwenberg, Bourgmestre de Charleroi.
M. A. Stenmans, Secrétaire Général du Conseil National de la Politique Scientifique.

M. Max Wasterlain, Secrétaire Général du Ministère de la Communauté Française.
Cdt M. Renson, Chef de Cabinet Adjoint du Secrétaire d'Etat à la Santé Publique et à l'Environnement.

Prof. Ch. de Duve, Prix Nobel, Université Catholique de Louvain
Prof. I. Prigogine, Prix Nobel, Université Libre de Bruxelles.
Prof. H. Buyse, Conseiller Scientifique, Université Catholique de Louvain.
Prof. P. Macq, Doyen de la Faculté des Sciences, Université Catholique de Louvain.
Prof. G. de Ghellinck Vaernewijck, Doyen de la Faculté des Sciences Appliquées, Université Catholique de Louvain.
Prof. F. Brouillard, Président du Département de Physique, Université Catholique de Louvain.
Prof. M. Leroy, Secrétaire Perpétuel de l'Académie Royale des Sciences, des Lettres et des Beaux-Arts de Belgique.
Prof. G. Verbeke, Vast Secretaris van de Koninklijke Academie voor Wetenschappen, Letteren en Schone Kunsten van België.
Prof. A. Bruylants, Directeur de la Classe des Sciences, de l'Académie Royale des Sciences, des Lettres et des Beaux-Arts de Belgique.
Prof. C. Chagas, Président de l'Académie Pontificale des Sciences.
M. P. Levaux, Secrétaire Général du Fonds National de la Recherche Scientifique.
Prof. P. Melchior, Directeur de l'Observatoire Royal de Belgique.
Prof. P. Pâquet, Président de la Société Belge d'Astronomie, de Météorologie et de Physique du Globe.
Prof. A.J. De Ruytter, Président de la Société Belge de Physique.

Prof. J. De Cuyper, Président de l'Association des Amis de l'Université de Louvain.
Prof. M. Clairbois, Secrétaire National de l'Association des Amis de l'Université de Louvain.
M. R. Novis, Président de l'Association des Ingénieurs Civils sortis de l'U.C.L.
Mgr L. Gillon, Président de la Société Scientifique de Bruxelles.
Prof. P. de Béthune, Secrétaire Général de la Société Scientifique de Bruxelles.
Prof. C. Courtoy, Secrétaire de la Société Scientifique de Bruxelles.
M. E.L. Boulpaep, Président de la Belgian American Educational Foundation.
Prof. F. Bingen, Président du Cercle des Alumni de la Fondation Universitaire.

Prof. G. Muraille, Président du Centre d'Histoire des Sciences et des Techniques, Université Catholique de Louvain.
Prof. A. D'Haenens, Président du Centre de Documentation sur l'Histoire de l'Université Catholique de Louvain.

Scientific International Committee

Prof. J. Audouze, CNRS, Institut d'Astrophysique, Paris, France.
Prof. A. Deprit, Center for Applied Mathematics, National Bureau of Standards, Washington, USA.
Prof. J. Kovalevsky, Directeur du CERGA, Grasse, France.
Prof. P. Ledoux, Institut d'Astrophysique, Université de Liège, Belgium.
Prof. J. Lequeux, Observatoire de Paris, Section d'Astrophysique, France.
Prof. M.S. Longair, Royal Observatory, Edinburgh, Scotland.
Prof. P. Melchior, Secrétaire Général de l'Union Géodésique et Géophysique Internationale, Directeur de l'Observatoire Royal de Belgique and Université Catholique de Louvain-la-Neuve.
Prof. P.J.E. Peebles, Institute for Advanced Study, Princeton University, USA.
Prof. D.W. Sciama, Dept. of Astrophysics, University of Oxford, U.K.
Prof. J.A. Wheeler, Center for Theoretical Physics, The University of Texas at Austin, USA.

Advisory Committee

Prof. Emérite L. Bouckaert, Université Catholique de Louvain-la-Neuve and Katholieke Universiteit te Leuven.
Prof. D. Callebaut, Dept. Natuurkunde, Universiteit Instelling Antwerpen.
Prof. R. Debever, Dept. de Mathématiques, Université Libre de Bruxelles.
Prof. C. De Loor, Astrophysisch Instituut, Vrije Universiteit Brussel.
Dr. J. Demaret, Institut d'Astrophysique, Université de Liège.
Prof. P. Dingens, Observatorium Meteorologie-Geofisica, Rijksuniversiteit te Gent.
Prof. P. Smeyers, Astronomisch Instituut, Katholieke Universiteit te Leuven.
Prof. D. Speiser, Institut de Physique Théorique, Université Catholique de Louvain-la-Neuve.

Our colleagues from the Department of Physics have supported our effort and, more specifically, all the members of the Institute of Astronomy and Geophysics Georges Lemaître, whose devotion was instrumental in assuring the success of this Symposium. We particularly wish to thank Professor Abbé M. Clairbois, Professor A. Deprit, Mrs A. Deprit-Bartholomé, Dr J. Steyaert and Mrs L. Moens for their very effective help.

We thank very much the Société Scientifique de Bruxelles for having agreed to publish all the Addresses delivered during the Opening Ceremony on 10 October, 1983, in commemoration of Lemaître who published a great number of his papers in the "Annales" and "Revue des Questions Scientifiques".

Many thanks also to Dr D. J. Larner, of the Reidel Publishing Company, who welcomed the idea of publishing these proceedings and helped us with his advice.

These academic manifestation could not be held without the generous help of the Association des Ingénieurs Civils Sortis de l'UCL, Banque Nationale de Belgique, Bell Telephon, British Council, Caisse Générale d'Epargne et de Retraite, Commissariat Général aux Relations Internationales de la Communauté Française, Commission des Communautés Européennes, Conseil de l'Europe, Digital Equipment, Fabrique Nationale de Herstal, Faculté des Sciences de l'UCL, Fonds National de la Recherche Scientifique, Kredietbank, Lion's Club d'Ottignies - Louvain-la-Neuve, Ministère de l'Education Nationale et de la Culture Française, Troisième Cycle en Physique de l'Université Catholique de Louvain, the members of the Support Committee and Les Amis de l'UCL to whom the Organizing Committee is particularly grateful for their unceasing help.

Some participants to the International Symposium Georges Lemaître, Université Catholique de Louvain-la-Neuve, October 1983.

Former students and/or collaborators of Lemaître, Institut d'Astronomie et de Géophysique Georges Lemaître, Louvain-la-Neuve, October 1983.
From left to right: First row: P. Pâquet, L. Bossy, O. Godart, A. Berger. Second row: J. Steyaert, R. Van der Borght, J. Henrard, R. De Vogelaere, A. Deprit, J. Roels, R. Broucke, S. Szücs.
(Photo G. Schayes)

LIST OF PARTICIPANTS

AUDOUZE J.
Institut d'Astrophysique
C.N.R.S. CP 165
Bd Arago 98 bis
F - 75014 Paris
France

BARBERIS B.
Istituto di Fisica Matematica
"J.L. Lagrande"
Via Carlo Alberto 10
I - 10123 Torino
Italy

BERGER A.
Univ. Catholique de Louvain
Institut d'Astronomie et de
Géophysique G. Lemaître
Chemin du Cyclotron 2
B - 1348 Louvain-la-Neuve
Belgium

BERNARD E.
I.R.M.
Av. Circulaire 3
B - 1180 Bruxelles
Belgium

BOSSY L.
Univ. Catholique de Louvain
Institut d'Astronomie et de
Géophysique G. Lemaître
Chemin du Cyclotron 2
B - 1348 Louvain-la-Neuve
Belgium

BRICMONT J.
Univ. Catholique de Louvain
Unité de Physique Théorique et
de Physique Mathématique
Chemin du Cyclotron 2
B - 1348 Louvain-la-Neuve
Belgium

BROUCKE R.
University of Texas
WRW - 414
78712 Austin (Texas)
U.S.A.

BRYANT J.G.
Université de Paris VI
Av. Félix Favre 47
F - 75015 Paris
France

CALLEBAUT D.K.
U.I.A.
Univesiteitsplein 1
B - 2610 Wilrijk (Antwerpen)
Belgium

CERULUS F.
K.U.L.
Departement Natuurkunde
Celestijnenlaan 200D
B - 3030 Heverlee
Belgium

CHALANTON A.

COLOMBO S.
Rue G. Roannet 6
F - 75018 Paris
France

COOPER A.
The Open University
Physics Discipline
Walton Hall
Milton Keynes MK7 6AA
England

CRILLY A.
The Open University
Physics Discipline
Walton Hall
Milton Keynes MK7 6AA
England

DE BECKER M.
Observatoire Royal de Belgique
Av. Circulaire 3
B - 1180 Bruxelles
Belgium

DEHANT V.
Univ. Catholique de Louvain
Institut d'Astronomie et de
Géophysique G. Lemaître
Chemin du Cyclotron 2
B - 1348 Louvain-la-Neuve
Belgium

DELVA M.A.I.
Institut für Astronomie
Universitatsplatz 5
A - 8010 Graz
Austria

DEMARET J.
Univ. de Liège
Institut d'Astrophysique
Parc de Cointe
Av. de Cointe 5
B - 4200 Cointe-Ougree
Belgium

DEPRIT A.
U.S. Dept of Commerce
National Bureau of Standards
Washington D.C. 20234
U.S.A.

DEPRIT-BARTHOLOME A.
19119 Roman Way
Gaithersburg, MD 20879
U.S.A.

DEVIRON Fr.
Univ. Catholique de Louvain
Unité de Physique Théorique et
de Physique Mathématique
Chemin du Cyclotron 2
B - 1348 Louvain-la-Neuve
Belgium

DE VOGELAERE R.
Dept of Mathematics
University of California
Evans Hall 970
Berkely, Ca. 94708
U.S.A.

DUCARME B.
Observatoire Royal de Belgique
Av. Circulaire 3
B - 1180 Bruxelles
Belgium

GALLEE H.
Institut d'Astronomie et de
Géophysique G. Lemaître
Chemin du Cyclotron 2
B - 1348 Louvain-la-Neuve
Belgium

GALLETTO D.
Istituto di Fisica Matematica
"J.L. Lagrange"
Via Carlo Alberto 10
I - 10123 Torino
Italy

GASPAR PH.
Institut d'Astronomie et de
Géophysique G. Lemaître
Chemin du Cyclotron 2
B - 1348 Louvain-la-Neuve
Belgium

GILSON H.
Univ. Catholique de Louvain
Unité de Physique Théorique et
de Physique Mathématique
Chemin du Cyclotron 2
B - 1348 Louvain-la-Neuve
Belgium

GODART O.
Institut d'Astronomie et de
Géophysique G. Lemaître
Chemin du Cyclotron 2
B - 1348 Louvain-la-Neuve
Belgium

GOOSSENS CHR.
Institut d'Astronomie et de
Géophysique G. Lemaître
Chemin du Cyclotron 2
B - 1348 Louvain-la-Neuve
Belgium

GREVESSE N.
Univ. de Liège
Institut d'Astrophysique
Parc de Cointe
Av. de Cointe 5
B - 4200 Cointe-Ougree
Belgium

GUSBIN D.
Univ. Catholique de Louvain
Unité de Physique Théorique et
de Physique Mathématique
Chemin du Cyclotron 2
B - 1348 Louvain-la-Neuve
Belgium

LAMBERT G.
Univ. Catholique de Louvain
Unité de Physique Théorique et
de Physique Mathématique
Chemin du Cyclotron 2
B - 1348 Louvain-la-Neuve
Belgium

HANQUIN J.L.
Univ. de Liège
Institut d'Astrophysique
Parc de Cointe
Av. de Cointe 5
B - 4200 Cointe-Ougree
Belgium

HENRARD J.
Faculté N.D. de la Paix
Dépt. de Mathématiques
Rempart de la Vierge 8
B - 5000 Namur
Belgium

HOUZIAUX L.
Univ. de Liège
Av. des Tilleuls 15
B - 4000 Liège
Belgium

HUT P.
Institute for Advanced Study
Building E
Olden Lane
Princeton, New Jersey 08540
U.S.A.

IMBRIE J.
Brown University
Dept of Geological Sciences
Providence
Rhode Islaand 02912
U.S.A.

IONNI G.
Univ. Catholique de Louvain
Unité de Physique Théorique et
de Physique Mathématique
Chemin du Cyclotron 2
B - 1348 Louvain-la-Neuve
Belgium

KOCH-MIRAMOND L.
CEN Saclay
Bât. 28
F - 91191 Gif-sur-Yvette
France

KOVALEVSKY J.
CERGA
Avenue Copernic
F - 06130 Grasse
France

KOZAI Y.
Tokyo Astronomical Observatory
Osawa Mitaka
J - 181 Tokyo
Japan

KRASINSKI A.P.
Fak. für Physik
Universitat Konstanz
Postfach 5560
D - 7750 Konstanz 1
R.F.A.

LAUSBERG A.
Univ. de Liège
Institut d'Astrophysique
Parc de Cointe
Av. de Cointe 5
B - 4200 Cointe-Ougree
Belgium

LEDOUX P.
Univ. de Liège
Institut d'Astrophysique
Parc de Cointe 5
B - 4200 Cointe-Ougree
Belgium

LEMAITRE A.
Fac. Univ. N.D. de la Paix
Rempart de la Vierge 8
B - 5000 Namur
Belgium

LEMAITRE J.
Rue R. Christiaens 20
B - 1160 Bruxelles
Belgium

MARTINEZ-BENJAMIN J.S.

Espagna

MASHHOON B.
Inst. für Theoretische Physik
Universität zu Köln
Zülpicher Strasse 7
D - 5000 Köln
R.F.A.

McCREA W.H.
Astronomy Centre
University of Sussex
Sussex BNI 9QH Brighton
England

MELCHIOR P.
Observatoire Royal de Belgique
Av.Circulaire 3
B - 1180 Bruxelles
Belgium

MEURICE Y.
Univ. Catholique de Louvain
Unité de Physique Théorique et
de Physique Mathématique
Chemin du Cyclotron 2
B - 1348 Louvain-la-Neuve
Belgium

MOONS M.
Fac. Univ. N.D. de la Paix
Dépt. de Mathématique
Rempart de la Vierge 8
B - 5000 Namur
Belgium

MURIGANDE Chr.
Fac. Univ. N.D. de la Paix
Dépt. de Mathématique
Rempart de la Vierge 8
B - 5000 Namur
Belgium

NGOIE R.
Univ. Catholique de Louvain
Unité de Physique Théorique et
de Physique Mathématique
Chemin du Cyclotron 2
B - 1348 Louvain-la-Neuve
Belgium

NOBILI A.M.
Department of Astronomy
The University of Glasgow
Glasgow G12 8QQ
U.K.

OORT J.H.
Sterrewacht
P.O. Box 9513
NL - 2300 RA Leiden
The Netherlands

PAQUET P.
Observatoire Royal de Belgique
Av. Circulaire 3
B - 1180 Bruxelles
Belgium

PAUWELS TH.
Sterrenkundig Observatorium
Rijksuniversiteit te Gent
Krijgslaan 281
B - 9000 Gent
Belgium

PEEBLES P.J.E.
Joseph Henry Labs.
Princeton University
Princeton N.J.08544
U.S.A.

PESTIAUX P
Institut d'Astronomie et de
Géophysique G. Lemaître
Chemin du Cyclotron 2
B - 1348 Louvain-la-Neuve
Belgium

PLETZER
C/O Dent
Rue Basse 84
B - 1180 Bruxelles
Belgium

RADELET P.
Univ. Catholique de Louvain
Unité de Physique Théorique et
de Physique Mathématique
Chemin du Cyclotron 2
B - 1348 Louvain-la-Neuve
Belgium

RONVEAUX A.
Fac. Univ. N.D. de la Paix
Rue de Bruxelles 61
B - 5000 Namur
Belgium

SALVADOR-DELBOURGO P.
Institut d'Astrophysique
C.N.R.S.
Bd Arago 98 bis
F - 75014 Paris
France

SATO K.
Dept. od Physics
Fac. Sc. Univ. of Tokyo
Hongo 7-3-1
J - 113 Tokyo
Japan

SAUVAL A.J.
Observatoire Royal de Belgique
Av. Circulaire 3
B - 1180 Bruxelles
Belgium

SCHAYES G.
Univ. Catholique de Louvain
Institut d'Astronomie et de
Géophysique G. Lemaître
Chemin du Cyclotron 2
B - 1348 Louvain-la-Neuve
Belgium

SCHMIDT M.
Dept. of Astronomy 105-24
California Institute of Technology
Pasadena, California 91125
U.S.A.

SCIAMA D.W.
Dept. of Astrophysics
University of Oxford
South Parks Road
Oxford OX1 3RQ
U.K.

SERGIJSELS R.
Univ. Libre de Bruxelles
Institut d'Astrophysique
Av. F.D. Roosevelt 50
B - 1050 Bruxelles
Belgium

SILK J.I.
Astronomy Dept.
University of California
Berkeley, Ca. 94720
U.S.A.

SIMPSON J.A.
The Enrico Fermi Institute
The University of Chicago
933 East 56th street
Chicago, Illinois 60637
U.S.A.

SMETS E.
Lange Lozanastraat 134
B - 2018 Antwerpen
Belgium

SMEYERS P.
Astronomisch Instituut
K.U.L.
Naamsestraat 61
B - 3000 Leuven
Belgium

SPEISER D.
Univ. Catholique de Louvain
Unité de Physique Théorique et
de Physique Mathématique
Chemin du Cyclotron 2
B - 1348 Louvain-la-Neuve
Belgium

STASSART M.
Univ. Catholique de Louvain
Unité de Physique Théorique et
de Physique Mathématique
Chemin du Cyclotron 2
B - 1348 Louvain-la-Neuve
Belgium

STEYAERT J.
Institut de Physique Nucleaire
Chemin du Cyclotron 2
B-1348 Louvain-la-Neuve
Belgium

SWINGS P.
Univ. de Liège
Parc de Cointe
Av. de Cointe 5
B - 4200 Cointe-Ougree
Belgium

SZEBEHELY V.
Dept. of Aerospace Engineer.
The University of Texas at Austin
Area code 512, 471-7593
Austin, Texas 78712-1085
U.S.A.

TASIAUX B.
Univ. Catholique de Louvain
Unité de Physique Nucléaire
Sc. 1
Chemin du Cyclotron 2
B - 1348 Louvain-la-Neuve
Belgium

THIRION D.
Univ. Catholique de Louvain
Institut d'Astronomie et de
Géophysique G. Lemaître
Chemin du Cyclotron 2
B - 1348 Louvain-la-Neuve
Belgium

TRICOT CHR.
Univ. Catholique de Louvain
Institut d'Astronomie et de
Géophysique G. Lemaître
Chemin du Cyclotron 2
B - 1348 Louvain-la-Neuve
Belgium

VAN DEN BERGH N.
Physics Dept.
U.I.A.
Universiteitsplein 1
B - 2610 Antwerpen
Belgium

VAN DEN BORGHT R.
Dept. of Mathematics
Monash University
Wellington Road
Clayton (Vic.) 3168
Australia

VERSTEGEN D.
Univ. Catholique de Louvain
Unité de Physique Théorique et
de Physique Mathématique
Chemin du Cyclotron 2
B - 1348 Louvain-la-Neuve
Belgium

VERVIER J.
Univ. Catholique de Louvain
Unité de Physique Nucléaire
Sc. 1
Chemin du Cyclotron 2
B - 1348 Louvain-la-Neuve
Belgium

VIGNERON G.
Koninklijke Baan 252
B - 8460 Kosijde
Belgium

WELSCH J.P.
Univ. Catholique de Louvain
Fac. de Théologie et de Droit canonique
Collège A. Descamps
Grand'Place 45
B - 1348 Louvain-la-Neuve
Belgium

WEYVERBERGH R.
Rue de Gaesbecq 4
B - 1460 Ittre
Belgium

Part I
Cosmology

PHYSICS AND COSMOLOGY: SOME INTERACTIONS

W.H. McCrea
Astronomy Centre, University of Sussex, U.K.

Introduction

Georges Lemaître (1894-1966) was one of the founders of modern cosmology - expanding universe cosmology, as it may be called. He was *the* founder of modern physical cosmology - big bang cosmology, as it has come to be called. His ideas in this field seem to have become well-defined by 1933, although any date for their inception is harder to identify, and now, 50 years later, we are invited to commemorate this historic scientific adventure. Particularly for those of us who knew Lemaître, it is a high privilege to participate and to do so in Lemaitre's own University in the Institute that bears his name.

After half-a-century of enormous developments in physics and astronomy, most of the particulars of Lemaître's model have been superceded. Probably he expected this to happen, and he did not in fact pursue them in much detail. Nevertheless the clarity and sureness with which he recognized the basic problems and the general lines along which they should be approached remain astonishing. The purpose of this paper is to sketch in some of the background to Lemaître's cosmology, to recall its main features, briefly to review the development of observational cosmology since the time when Lemaître proposed his model, and then to note some sequels to his ideas in some of the most recent models. Finally, since Lemaître sought to relate the physics and the cosmology of his day, it seems appropriate to end with some attempt to assess the present-day state of the relationship.

A. Berger (ed.), The Big Bang and Georges Lemaitre, 3–22.

Lemaître's lifetime

Lemaître published his now famous first paper on the expanding universe in 1927 in Belgium. At the time he did not know that the Russian mathematician and meteorologist Alexander A. Friedman (1888-1925) had published similar work in 1922 in Germany. The names of these two men will evermore be together linked with one of the most audacious developments in physical thought. They were near contemporaries, but each lived as though the other had never been.

To notice when that was, it may help if we remember that one of the great founders of astrophysics - who must seem to most people a figure in the distant past - E. Arthur Milne (1896-1950) was actually about two years *younger* than Lemaître. By contrast, one of the great founders of geophysics, Harold Jeffreys (b.1891), was three years *older* than Lemaitre, and he is still an active scientist!

Natural philosophy

The general procedure of natural philosophy seems inevitable. *Observations* of something recognized as being observable suggest a *mathematical model* of that something; the model serves to predict the outcome of *further observations*; the actual outcome suggests an *improved model*, and so forth.

In the Newtonian approach, a model consists of the (model) system being studied + a reference frame (which models the rest of the Universe) + universal time + laws (of motion, of electromagnetism, ...) obeyed by the (model) system and regarded as unchanging with time.

Cosmology is the study of the Universe as a whole. It is therefore not amenable to the Newtonian approach. The aim of cosmology must be to construct cosmological models, not to 'discover' laws. This is the Einsteinian approach, as realized in general relativity (GR). Every GR model is a universe of its own; there is no 'rest of the universe'.

In GR any completely defined Riemann 4-space (of suitable signature) is a universe. It can be interpreted as a conceivable system of mass and stress under self-gravitation, again with no 'rest of the Universe'. This is what Einstein himself appears first to have appreciated when he wrote his paper 'Cosmological considerations on the general theory of relativity' (Einstein 1917). Of course, in general the mass and stress in such a model could not be reproduced by any real matter. There is no way of ensuring *a priori* that the contents are real in this sense.

Status of GR in cosmology

The comparison between the Newtonian and Einsteinian approaches in the preceding section shows that the latter must be preferred for use in cosmology. But GR presents problems and limitations that have to be recognized. To start with, a GR model is the whole history of the 'universe' concerned all laid out before us. It is a frozen picture; nothing happens in four dimensions; an observer in the model gets the illusion of things happening because he is supposed to experience a succession of spatial sections in a certain sequence. Such a model cannot, in particular, depict itself coming into existence; that would require another time-dimension, and so on.

If a model has simple topology, it is possible self-consistently to admit an *arrow of time* and an associated *causality* concept. But it is difficult to see how it can admit thermodynamic *irreversibility* or quantum theory *uncertainty*.

It appears to be a recommendation for GR that according to the well-known work of Hawking and Penrose (1970) (See also Hawking and Ellis 1973), every GR spacetime of physical interest has at least one singularity. The case of one singularity is that of a big-bang cosmological model. Penrose (1982) quotes an example for which the big-bang singularity has 'degree of specialness' of general order one part in $10^{10^{123}}$, suggesting, as he says, "very precise physical laws in operation at the big-bang itself. The new physics involved is necessarily time-asymmetric." This is a difficult concept since any such laws could themselves have originated only from the big-bang along with whatever is assumed to obey them.

We must in fact think of there coming into existence from the big-bang

the content of the Universe
physics
mathematics and logic
existence itself

but, if we do entertain the notion of existence coming into existence, we seem to be embarking upon an infinite regress.

It is at any rate the plain fact that current cosmological models are in general based upon GR.

GR and cosmology

It is interesting to examine the extent to which cosmology has tested specifically Einstein's theory of gravitation. Some predictions of relativistic cosmology depend only upon the postulation of a Robertson-Walker metric without saying anything about gravitation. This is involved only if the predictions concern energy and stress in the cosmological model. In that case relations of these to the expansion factor R(t) of the metric are needed. If the relations are those given by Einstein's theory, the Friedman-Lemaître cosmological models result. The simplest of these is the well-known Einstein-de Sitter (ES) model. This is commonly employed as a standard of comparison. In particular, for any other model the density parameter $\Omega(t)$ is defined as the ratio of the density of that model at cosmic epoch t to the density at the same epoch in an ES universe having the same Hubble constant at that epoch.

Barrow & Ottewill (1983) have shown that Friedman-Lemaître type universes exist for gravitation theories derived from a Lagrangian of a form more general than Einstein's. This may be significant because, if we do not regard Einstein's form of general relativity as the only one to be considered, then we need not assign special status to the ES model, i.e. that having $\Omega = 1$ for all t.

It is known that, on Einstein's theory, unless in the very early big-bang universe the value of Ω is unity to fantastic accuracy, the model would explode or collapse within the 'very early' time and never reach the state that we observe. This is the same as saying that the spatial section of the very early Universe must be *flat* to fantastic accuracy. The problem of how this comes about is the well-known 'flatness' problem. The solution is generally sought in a combination of particle physics and Einsteinian gravitation. But maybe it is the use of Einsteinian gravitation that creates the problem.

One recent suggestion is Adler's (1983) that Einstein's theory should be regarded as a 'long-wavelength effective field theory' arising from a 'fundamental theory' more like other quantum field theories. The difference from Einstein would be significant only in the very early universe. It is not yet known, so far as I am aware, whether this would have any immediate bearing upon the flatness problem. But it certainly has bearing upon the fundamental problem of gravity in the very early Universe - that of quantization. Physicists conclude that quantization must occur then, even if it is significant only before cosmic time of the order of the Planck time, that is t $\approx 10^{-43}$s. There is no accepted scheme for this. At any rate in part this must be owing to the basic feature of relativistic treatments of gravitation that it and space-time itself are inextricably interrelated. So quantization of gravitation presumably requires quantization of space-time. This has often been mentioned, but never achieved.

To return to the question at the beginning of this section: Going back to the work of Friedman and of Lemaître, it was a tremendous triumph for GR to predict the expansion of the Universe. But the success was and remains essentially qualitative. No relativistic cosmological model has ever been tested in a way that a physicst could regard as quantitatively crucial. Also for the reasons mentioned we expect GR to demand modification sufficiently near to the big-bang singularity. There are, too, the conceptual difficulties to which allusion has been made. Most of these perplexities should be resolved before long; none of them calls in question any of the 'confirmation' of quantitative predictions of GR on the scale of, say, the Solar System or a binary pulsar.

Cosmology of G. Lemaître

In the paper already quoted Einstein (1917) introduced his cosmical constant Λ that enabled him to formulate his static model universe (assuming $\Lambda > 0$). In the same year de Sitter (1917) produced his model, which is properly regarded as the first non-static model. Then Friedman (1922) and Lemaître (1927) produced their more general non-static models. Friedman pointed out that if a non-static model be regarded as acceptable, the need for a non-zero Λ has disappeared; in due course Einstein agreed, and thenceforth dropped Λ from his theory. Using Lemaître's treatment, Eddington showed that the original Einstein model is unstable; if disturbed so that expansion commences, it goes on expanding forever, and this was the model adopted by Eddington. Lemaître took the commonsense attitude for a mathematical physicist; in effect, he said, keep Λ in the equations until we find observations that contradict some two of the hypotheses $\Lambda < 0$, $\Lambda = 0$, $\Lambda > 0$.

Lemaître identified three basic problems for the expanding universe which he discussed for homogeneous, isotropic relativistic models:

A. *Age of the Universe*

Let t_0 be cosmic time at the observer, i.e. the age of the universe at the observer; let T_0 be the Hubble time as measured by the observer at t_0. If $\Lambda = 0$ then for the model $t_0 < T_0$. The value of T_0 inferred by Hubble was smaller than current values of geological ages. So the model would imply that the age of the universe is less than the age of the Earth. Therefore Lemaître rejected $\Lambda = 0$. Other arguments led him to reject also $0 < \Lambda \lesssim \Lambda_E$ where Λ_E corresponds to an Einstein static universe of 'radius' R_E. If $R(t)$ is the Robertson-Walker expansion factor normalized to $R = R_E$ for the Einstein model, then $R(t)$ for a Lemaître model having $\Lambda > \Lambda_E$ has a graph as shown

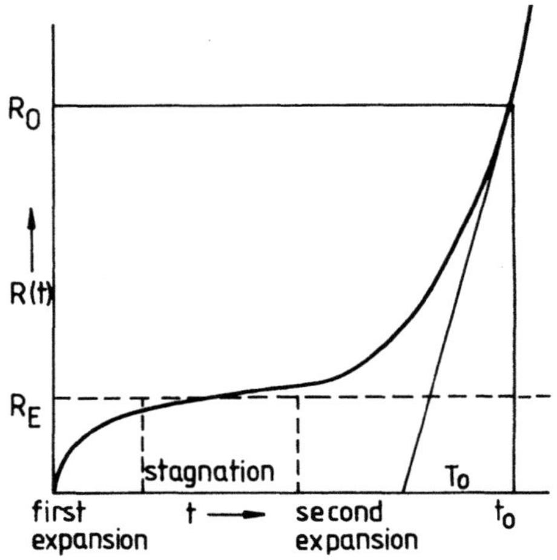

Figure 1 Lemaître cosmological model (schematic diagram).

qualitatively in Figure 1. It is drawn for Λ relatively little more than Λ_E. It is seen that $R(t) \to 0$, $dR/dt \to \infty$, as $t \to 0$ so that $t = 0$ is a singularity in the density and in dR/dt. Three phases of the expansion may be recognized the 'first expansion' from $R = 0$ to R only a little less than R_E, an interval of near-'stagnation' in which R increases to only a little more than R_E, the 'second expansion' in which R moves increasingly rapidly away from R_E. Lemaître showed that he could find a case for which $t_o \simeq 10T_o$ and $R(T_o) \approx 10R_E$, and these appeared plausible values, i.e. giving a plausible age and a plausible mean density of the Universe.

In this way Lemaître was the first to propose a resolution of the *age problem* in cosmology.

B. *Galaxy-formation*

Lemaître was also the first explicitly to recognize that the culminating problem of cosmology is the origin of the structure of the Universe,as composed of galaxies and clusters of galaxies,within the time available. In Lemaître's model this last meant the time allowed under A.

He presented a rather qualitative scheme starting apparently early in the 'stagnation' phase with 'small accidental fluctuations in the original distribution' of matter. These he saw as producing clouds which by processes of agglomeration, collision and merging would lead to concentrations of material sufficient to produce galaxies or clusters of galaxies. Stars would result by gravitational contraction of portions of the material of a proto-galaxy. He estimated that all this could take place within 'a few' Hubble times. Such an inadequate summary makes it appear even more speculative than in Lemaître's own presentation. Even so it does read much like a summary of the modern theory of 'isothermal fluctuations' (see below). Speculative it undoubtedly was, but it had all the right ingredients, and not all modern attempts take account of the time available, as it did seek to do.

C. *Interpretation of the big-bang; origin of the raw material for galaxy-formation*

Lemaître was the first to appreciate the possibility of a singularity as $t \to o$ and to attempt to assign physical significance to this. He postulated that the Universe began as a single 'primeval atom' - which he supposed to undergo disintegration by *cosmic radioactivity*. In fact Lemaître (1946) entitled his small volume of essays on the subject *L'Hypothèse de l'atome primitif* and the English translation (1950) was called *The primeval atom*. It should be scarcely necessary to remark that the picture is of the *entire* Universe being initially (whatever that may mean) this one 'atom', not of an atom existing somewhere in space; so the disintegration is to be pictured as a fragmentation accompanying the initial expansion. Lemaître wrote, "if matter existed as a single atomic nucleus, it makes no sense to speak of space and time in connexion with this atom. Space and time are statistical notions which apply to an assembly of a great number of individual elements; they were meaningless notions, therefore, at the instant of first disintegration of the primeval atom".

As must be remarked, there may be some inconsistency in speaking of "the instant of the first disintegration" after asserting that "it makes no sense to speak of space and time..." That apart, Lemaître must be credited with the first attempt to contend with the notion of a singularity in space-time. Our reference to the consequences of any quantization of gravity for the meaning of space-time in the very early Universe shows that this was another instance of Lemaître recognizing a basic problem that is still with us.

The picture that he proceeded to develop was the disintegration of his primeval atom - which he described as an 'isotope of the neutron' - first into supermassive nuclei, the further disintegration of which resulted in both the *cosmic-ray background*, that in his picture is still with us, and the normal atoms that then constituted the gas which provided the *raw material* for the processes in B.

All this was Lemaître's invention of the big-bang, which we are now celebrating. The details are greatly different from those that are now generally accepted. Nevertheless, yet again he produced features of broadly the right character - a present background surviving from the early Universe and a process in the early Universe that yielded the raw material for the present galaxies. As we shall see the most essential change since Lemaître's work is that cosmologists now contemplate a *hot* big-bang; his picture assigned no particular significance to any cosmic temperature.

The observed cosmos

At this point it is necessary briefly to review the changes in empirical knowledge of the cosmos between the time when Lemaître developed his cosmology and the time of this commemoration.

About 1933 such knowledge was much what Hubble (1936) described in his book *The realm of the nebulae*. This gave for the Hubble time $T_o \lesssim 2 \times 10^9$ years, whereas it is now almost certain that $10^{10} \lesssim T_o \lesssim 2 \times 10^{10}$ years. The then current estimate of the mean density of galactic matter was quite reasonable. The age of the Earth was inferred to be more than 2×10^9 years, but by how much was not known. Compared with more recent times, knowledge about cosmic rays was rudimentary. As regards the structure of the Universe, the hypotheses of large-scale homogeneity and isotropy were not contradicted by observation, while on a smaller scale the clustering of galaxies was well recognized although there was not much systematic information.

It has to be appreciated that Hubble had started publishing his observations of the 'expanding Universe' only in 1929 and that hitherto there had been little systematic work in extragalactic astronomy. So we are in fact looking back to the very early days of such astronomy. Two things now may strike us as surprising: (a) that nobody raised an insistent call for 'more observations', (b) that everybody in the business seemed to accept Hubble's measurements quite uncritically. The reason for both of these was that Hubble had the use of the Mt. Wilson 100-inch telescope, and no other existing telescope could compete.

Moving on to the time of this celebration in 1983, there is vastly more information than there was 50 years earlier, and most of it - like results from radioastronomy - is of sorts that were unknown around 1933. From the standpoint of cosmology it is in the following categories, as compared with information accessible to Lemaître:

1. Improvements upon old results
2. New results having cosmological applicability
3. New results without present cosmological applicability
4. Results awaited

To take these briefly in turn:

1. As we mentioned above, Hubble's value for T_o was too small by a factor of order 10, but the actual value is still uncertain to within a factor about 2.

Various estimates of *mean densities* in the Universe at the present cosmic epoch are now available; they include that for the galactic matter, baryonic matter, total energy (including rest mass). Some of the values are independent of the Hubble time T_o and some are proportional to T_o^{-2}. Comparisons may therefore set bounds upon the value of T_o. For some purposes it is more convenient to express results in terms of the density-parameter Ω_o rather than in mass per unit volume.

Particularly in the last few years there have been extensive studies of the *large-scale structure of the Universe*. Statistical studies employing 2-point or higher order correlation functions, particularly those of Peebles (1980) and his school have yielded much more systematic quantitative knowledge of the clustering of galaxies. The work of Abell (1958) had earlier provided far more descriptive knowledge than had been available in Lemaître's day. All such work supports the early hypothesis of the large-scale homogeneity and isotropy of the Universe.

Special studies strongly indicate, however, that there exists a detailed structure more complex than had ever before been envisaged. If they are broadly correct galaxies and clusters of galaxies are arranged in the form of a rough network that outlines great voids each of the order of a million cubic megaparsecs in which there are effectively no bright galaxies. Some workers are still inclined to doubt whether the 'strings' of galaxies and clusters are significantly different from features that occur fortuitously in any random distribution. Others seem to be so convinced of the non-random character of the structure that they wish to regard it as the 'fossil' of some structure in the early Universe.

2. All observations using electromagnetic radiation outside the optical and near infrared wavelength-range have come since Lemaître's time, as well as the bulk of cosmic-ray observations. Some of these observations have assisted in improving results in category 1. But others apply to new

discoveries. Of these probably the most important is the *microwave background radiation.* It provides the only known means of observing the Universe before any galaxies had been formed - if the standard interpretation is correct. In that case it shows that the Universe at that epoch was isotropic to an exceedingly high degree. As we have seen, in a general sense it plays the role envisaged by Lemaître for a cosmic-ray background.

Another empirical parameter of cosmological significance is the baryon: photon ratio η, estimated to be about 10^{-9} and believed to have remained effectively constant since the beginning of the 'radiation era' of the Universe.

Quantities also of cosmological importance are the relative abundances of the atomic nuclei ^{1}H, ^{2}D, ^{3}He, ^{4}He that are inferred to have been 'frozen in' to the cosmos from the end of about the first 3 minutes until the first stars were formed. Significant empirical values of these primordial abundances are now claimed. They form the best basis we have for estimating the present mean density of baryonic matter.

3. Radio-galaxy and quasar number-counts have been expected to yield important cosmological information particularly with the object of selecting a cosmological model. It seems now that their usefulness from that aspect is obscured by what are classed as 'evolutionary effects'. Sooner or later the information will have to be adequately analysed.

In the same general category, but far more pressing and important, is the evidence that has been discussed now over many years regarding the existence and quantity of 'dark matter' in the Universe. If the amount is near the upper bound that has been considered, then the Universe is an almost totally different place from what astronomers had hitherto thought. All their past endeavours would have been concentrated upon less than 1 per cent of its content. The other 99 per cent of the mass could almost certainly not mainly be ordinary (baryonic) matter; it might be 'massive' neutrinos or more exotic particles. On the other hand, if the amount of dark matter is near the lower bound considered, then it need imply nothing more alarming than that some galaxies may be surrounded by rather many faint stars or 'jupiters', and some clusters may contain rather more intergalactic matter than had been thought. The resolution of this uncertainty has become surely the central problem for present-day astronomy.

To quote an example of a discovery which, when correctly interpreted, must be a clue to the evolution of the Universe, we cite that of the so-called "Lyman-alpha clouds". These were evidently scattered through intergalactic space before a few 10^9

years ago and they produced most of the absorption lines in the spectra of quasars. They seem not to contain an important mass of the matter in the cosmos, but since their material was apparently left over after the formation of galaxies they should help to reveal the nature of the formation process.

4. It is well known that a very small positive rest mass of the neutrino, no more than the energy of a few electron volts, would suffice to ensure that at the present cosmic epoch the neutrinos in the Universe should furnish most of its mass. [Different neutrino species might have different positive rest masses, unless all have zero mass] . It is therefore of the utmost importance to know if the rest mass of any neutrino is non-zero. The experimental evidence seems still to be inconclusive.

Cosmology since Lemaître

Lemaître himself after about 1933 worked mainly in fields other than cosmology. Although he was always generous about responding to invitations to expound in lectures and essays his views on the subject, he did not develop them much further during the rest of his life.

In the 1930s Eddington was developing his ideas regarding the constants of physics; his scheme demanded a positive Λ . Lemaître was practically the only other worker to retain Λ . Almost everyone else at the time regarded an isolated constant of this sort as being out of keeping with the spirit of GR. It has then to be asked why they were not concerned as much as Lemaître was about the age paradox. Strangely enough for most astronomers at the time the paradox worked the other way. I think that because the Hubble time was so short they took the view that not much more could be inferred from Hubble's result than that the Universe had been in a rather highly congested state at a time about T_o before the present. T_o being so much less than the ages assigned to the stars and galaxies, they had to suppose that these would have retained their identities while experiencing that state. A Friedman-Lemaître model as they saw it, was a grossly simplified representation of the actual Universe, in which all the elaborate system of stars, galaxies and clusters was replaced by a uniform stress-free dust. So the model need not be taken seriously anywhere near its singularity.

Other aspects of relativistic cosmology and alternatives to it continued to be studied until after World War II. Then in 1948 Bondi and Gold, and to some extent independently, Hoyle propounded *steady-state cosmology,* necessarily implying *continual creation.* While it would be incorrect to say that this was ever widely *accepted,* it was certainly the case that its concepts continued to have a dominating influence upon cosmological thinking until about 1965. This is not an occasion to attempt to recount the history of those years. For one thing, steady-state concepts seem never to have had much impact upon Lemaître. Historically what for most astronomers was the strongest reason for rejecting steady-state cosmology in the form in which it had been presented was the discovery in 1965 of the microwave background radiation. This was taken as evidence of an explosive start for the Universe. It is recorded that Lemaître expressed satisfaction about this feature a short while before he died in 1966. It is an irony of history that the general acceptance of big-bang cosmology is to be dated from the year of the death of its inventor. However, two comments must be made: When big-bang cosmology regained favour, for most cosmologists this meant a *hot* big-bang. The current version cannot be final, and it is conceivable that whatever succeeds it will contrive to combine some of its concepts with some of the more attractive concepts of steady-state theory.

Meanwhile hot big-bang cosmology has furnished a history of the cosmos that in general terms seem to be acceptable on all the available evidence. Briefly it is:

Early Universe - from say 10^{-43}s to 10^{-3}s, forming the 'particle era', beginning with a quark-gas, followed by hadrons.

Radiation era - about 10^{-2}s to 10^{13}s (about 4×10^5 years), up to about 3 minutes in regard to nuclear reactions there is effective thermodynamic equilibrium at each instant, but approaching about 3 minutes nuclear abundances are determined by reaction rates after which they become "frozen in"; effectively only ^{1}H, ^{2}D, ^{3}He, ^{4}He remain. At about 4×10^5 years matter and radiation decouple.

Matter era - after about 4×10^5 years the energy-density is predominantly from the rest-mass of matter. The fact that decoupling works out to occur about the end of the radiation era appears to be an arithmetical coincidence brought about by the property $\eta \approx 10^{-9}$.

All this gives a self-consistent picture using a Robertson-Walker metric with expansion factor R(t) satisfying the Friedman-Lemaître equations. These follow from GR and they may first be derived with the retention of Λ. They may then be solved for Λ, which is thus expressed in terms of the Hubble constant, the acceleration parameter, the function R(t), the mean density and pressure at epoch t. For the actual Universe at the present epoch bounds may be set to all these quantities and they are found to imply $|\Lambda| \lesssim 10^{-120}$ in absolute units. But since Λ is by definition a universal constant, this result must hold good at all epochs. Λ is thus the quantity in physics most accurately measured to be zero (Hawking 1983).

Inflation

Several properties of vacuum (quantum) states have closely the same effect as non-zero values of Λ. They are significant only at very high energies. There has been a suggestion that a phase transition occurred when an original unified 'electroweak' force split into electromagnetic and nuclear-weak constituents. Times of order 10^{-35}s from the big-bang have been mentioned for this. During the transition a vacuum effect of the sort mentioned is inferred to have produced an enormous 'cosmic repulsion' that caused the Universe to inflate by a factor estimated at 10^{20}. When the transition was complete and the two kinds of force had been 'frozen out' with their familiar characters, the repulsion would vanish. This would, of course, be consistent with using $\Lambda = 0$ for the subsequent normal expansion. Consistently with GR, the repulsion cannot then be exactly equivalent to having a non-zero value of Λ for part of the time; even so Guth (1981) noted that it is hard to represent a smooth return to non-inflation.

One important consequence appeared to be that the huge inflationary expansion would smooth away any initial irregularities in the universe and so produce the high degrees of homogeneity and isotropy which are inferred to have existed at an early stage of the normal expansion. Another would be that it would explain why the homogeneity can hold good between regions that otherwise could not have been in causal contact when their contents were determined.

Unfortunately it now appears that this original inflationary model has to be rejected as depending upon a too naive interpretation of the particle physics. Physicists seem now to favour a 'bubbly' early Universe. One version envisages the observable Universe as arising from the inflation of one small bubble of an early state. Another regards even the present Universe as 'bubbly' on a micro-scale, but very smooth on the scale on which we observe it.

Even the latest models thus invoke an essential role for cosmical repulsion - as did Lemaître's model half-a-century ago - but now only for the very early Universe, whereas Lemaître had an effect of his Λ that became relatively more important as the expansion proceeds.

Galaxy formation

As already mentioned, Lemaître very properly saw the formation of galaxies - or maybe clusters of galaxies as the culminating problem of cosmology. Here we shall briefly review the current approach to this problem.

A well known argument shows that since there are now fluctuations in the density of matter in the Universe, there must always have been fluctuations. More specifically, if what is basically a Friedman-Lemaître model possesses galaxies at some epoch after decoupling there must have been fluctuations of density at any epoch before decoupling. So the 'modern' approach to the problem of the origin of galaxies is to consider arbitrary fluctuations $\delta\rho/\rho$ before decoupling, and to enquire how they develop as the model expands into the matter era. If some such fluctuations are inferred in due course to produce galaxies, then we can ask what fluctuations in some earlier era could lead to these fluctuations before decoupling. The aim is then to discover what were the most primitive significant fluctuations.

This approach implicitly supposes that a FL model was a better match to the actual Universe in the past than it is in our era. In particular, it is assumed that in the radiation era the matter and radiation were almost uniformly distributed in space. So far as the actual Universe is concerned this is strongly supported by the high degree of isotropy of the microwave background radiation.

Two sorts of fluctuations are studied; the names they have acquired should not be taken literally: -

1. 'Adiabatic' fluctuations The initial fluctuation is taken to apply to both the matter and the radiation. So long as the matter is to a considerable extent ionized - that is, until decoupling is largely complete - *radiation damping* is strong for condensations of relatively small mass. This leads to the conclusion that condensations surviving recombination are mostly in the range 10^{12} to 10^{14} solar masses. It is inferred that such a condensation then collapses first as a 'pancake', which proceeds to fragment into clusters of galaxies. Peebles (1980) identifies three characteristic lengths associated with the process.

2. 'Isothermal' fluctuations These are taken to involve the matter alone. Radiation causes less damping in this case. No characteristic lengths emerge; some astronomers consider that this is in better agreement with observation. The first condensations, after decoupling, may be on the scale of globular clusters; if so, these would merge to form galaxies.

Neither picture leads to a quite convincing account of how a condensation of the raw material is transformed into a real galaxy as it is seen in the sky. Some phenomenon besides gravitational instability seems to be required to play some crucial part. This could be the occurrence of *shocks* either between condensations or within a collapsing condensation (McCrea 1982, 1983).

Primeval fluctuations

The work that has been done on adiabatic and isothermal fluctuations, whatever may be its inconclusiveness in detail, is almost certainly sufficient to show that the existence of galaxies in the matter era implies the existence in the preceding era of fluctuations $\delta\rho/\rho$ in a certain range of size and amplitude. As regards amplitude appeal may then be made to the observed absence of anisotropy, exceeding a certain very small amount, in the observed microwave background radiation. This leads to the inference that, in the region in the radiation era in which most of this radiation last interacted significantly with matter, that matter must have been of uniform density ρ to within fluctuations not exceeding $\delta\rho/\rho \approx 10^{-4}$. On the other hand, fluctuations weaker than this would not be expected to lead to galaxy formation. It is therefore generally inferred that fluctuations of this amplitude existed in the cosmos at an epoch of order 10^5 years after the big bang.

Astronomers ask, Is this a fundamental property of the Universe that, at any rate in our present state of knowledge, has simply to be accepted as such ? Or can it be traced to something more primitive ?

As regards the latter question, among possibilities contemplated are:

Quantum fluctuations as an inherent element in the concept of the very early Universe. Some cosmologists have discussed how these might leave an imprint that could survive through all subsequent phases.

Primeval chaos, part of which somehow achieved considerable homogeneity at an early epoch but never without some irregularities.

Primeval turbulence as a possibly more comprehensible version of 'chaos'.

Astronomers also ask, Do we learn anything about primeval fluctuations from the present large-scale structure of the Universe as observed ? If significant, is the 'cellular' or 'network' structure that is claimed to exist a fossil of the early Universe ? It seems unlikely that this can be true in any simple way, for I am told that what evidence there is from numerical simulations shows that such structure would be unlikely to survive from an early stage. Nevertheless, in a very general sense it seems that it *must* be true. For the existence of condensations at any stage depends upon the existence of condensations at an earlier stage, and in the same way the existence of any general structure at any stage would depend upon the existence of structure at an earlier stage. But there remains the question as to whether significant general structure does actually exist.

In summary, the whole problem of condensations in the cosmos is still beset by uncertainties, the most serious being at the two ends, the one concerning the nature of the most primitive condensations, the other concerning the process by which a condensation of the raw material is converted into stars, stellar clusters, nebulae ... to make a galaxy.

Physics

Cosmology, observational and theoretical, and particle and high energy physics, experimental and theoretical, all seem at the present time to have arrived at a peak of activity and discovery. This is partly a cause and partly a result of the interaction of all these elements. It is resulting in a review of the foundations of physics that is more profound than any previously possible. It would have been highly desirable that this essay should have dealt with the most profound aspects of all these developments. But anyone attempting to do this would need to understand much more about modern physics than the writer. He can only mention a few of the aspects that have immediate significance for cosmology.

Here we mention a few cosmological considerations specially concerned with the *constants of physics*. It is the existence of these that makes physics what it is. They arise basically because everything in physics is quantized, so that the physical world itself provides *natural units* ('Planck units') in which it can be described. If our physical concepts are valid, this would in principle permit us to exchange precise physical information with physicists anywhere in the Universe. Consistently with this, it should be noted that a constant of physics has an *operational existence* that transcends any particular theoretical model. The experience that there exist operations that always yield the same outcome is a way of defining the 'external' world of physics. This is a paraphrase of the remark that the constants of physics make physics. Not surprisingly, therefore, it can be seen that properties of the world of astrophysics depend upon the values of a few constants. For example it can be shown that the mass of an asteroid, of a planet, of a star each lie within a particular interval dictated by these constants - the same constants that determine, say, the range of possible physical capabilities of the human animal.

What is at first surprising about the resulting situation is the sensitivity of its features to the values of the constants. The whole world of experience could be made so different by relatively small changes in one or two constants that we could not have evolved to observe it (Carr & Rees 1979; Press & Lightman 1983).

Such considerations are embodied in what have been called *anthropic principles*. The 'weak' principle asserts that man's experience of the Universe depends upon the circumstance that he can exist only within a restricted region of space-time. In itself this is self-evident; but it is obviously necessary for the cosmologist to appreciate that, when he thinks that he is discovering an important property of the cosmos, he may be doing no more than noticing a feature that happens to be present when he himself happens to be around to oberve it. Thus the weak principle may issue useful cautions, but in the form stated it cannot serve as a basis for making predictions about the cosmos. Also it is to be noted that it assigns no properties to the observer other than the existence of an ability to receive and record signals.

The 'strong' principle, on the other hand, takes note of the properties of the observers that actually exist, and it asserts that the constants of physics have to possess values such that the cosmos must cause these beings to exist. Again the assertion is self-evident, but now it is one that may lead to predictions. It seems almost certain that, if we suppose the familiar constants of physics simply to exist, then such a principle should impose *bounds* upon their values. But it is hard to see how it can be inferred that such constants *must* exist and that they must possess certain *precise* values.

It is interesting to remark that inferences of this last sort were something of what Eddington (1936, 1946) was trying to achieve in the work described in his last two books. Nowadays it has become more fashionable to ask, Do there in any sense 'exist' other universes in which the constants of physics have values different from those in 'our' Universe ? This appears to be broadly the same problem.

There are two other problems related to all this. One is, in a big-bang model universe, how and when do the constants of physics come into existence ? So far as one knows, nobody has made any useful approach to a solution.

The other is, Are there constants of cosmical physics that are not related to those of microphysics - at least in any way that we can discover at present. It is to this that we finally turn.

Cosmical numbers

According to the usual view of the constants of *physics* they concern, of course, entities that exist. But they tell us nothing about the amounts of these entities that exist and that would thus serve to specify the Universe that exists. We have remarked that the constants of physics make *physics* what it is. Are there additionally *cosmical numbers* that make the *Universe* what it is ?

The number of dimensions of space-time seems to be a 'given' constant of the Universe. The Universe would be fundamentally different were the number other than 4, so that an observer experiences one time dimension and three space dimensions (Barrow 1983). Actually some recent unified theories employ space-times with dimensions up to 11. But whatever the number it may be best to regard it as both a constant of physics and a cosmical number.

Rees (1983), who has given most explicit consideration to the question, has indicated 'three basic numbers that characterize our Universe'. In the terminology used here, these are: (i) The Robertson-Walker curvature radius at our cosmic epoch, $R(t_0) \approx 10^{60}$ Planck lengths. (ii) The baryon: photon ratio, $\eta \approx 10^{-9}$. (iii) The amplitude of the fluctuations that triggered galaxy formation $\delta\rho/\rho \approx 10^{-4}$. We do not know why they should have these values, or whether to expect any discovery of any dependence upon the values of the constants of physics.

If indeed, as mentioned above, 'our Universe' resulted from the inflation of one small bubble in a very early Universe, then we might conclude that the constants of physics arose from the latter, and that the cosmical numbers were determined by what happened to be the content of that one particular bubble.

This paper is largely a catalogue of unsolved problems. The present developments in physics seem to promise some imminent further progress. We may hazard the view that progress as a whole is likely to be gradual. For example, as regards the constants of physics the trend naturally seems to be to seek some new theory such that the constants of existing theory become expressible in terms of a smaller number of constants in the new theory (Weinberg 1983). It may be a long time before the number has been reduced to zero.

References

Abell, G.O. 1958 *The distribution of rich clusters of galaxies* Astrophys. J. Suppl. 3 no. 31

Adler, S.L. 1983 Phil. Trans. R. Soc. Lond. A 310 pp. 273-278

Barrow, J.D. 1983 Phil. Trans. R. Soc. Lond. A 310 pp. 337-346

Barrow J.D. & Ottewill, A.C. 1983 J. Phys. A Math. Gen 16 pp 2757-76

Carr, B.J. & Rees, M.J. 1979 Nature Lond. 278 pp 605-612

Eddington, A.S. 1936 *Relativity theory of protons and electrons*

Eddington, A.S. 1946 *Fundamental theory*

Einstein, A. 1917 trans. in *The principle of relativity* H. A. Lorentz et al 1923 pp 177-188

Friedman, A.A. 1922 Zeit.f. Physik 10 pp 377-386

Guth, H.A. 1981 Phys. Rev. D23 pp 347-356

Hawking, S.W. 1983 Phil. Trans. R. Soc. Lond. A 310 pp 303-310

Hawking, S.W. & Ellis, G.F.R. 1973 *The large-scale structure of space-time*

Hawking, S.W. & Penrose, R. 1970 Proc. R. Soc. Lond. A 314 pp 529-48

Helfand, D.J. 1983 Physics today 36 no. 10 pp 17-20

Hubble, E.P. 1936 *The realm of the nebulae*

Kibble, T.W.B. 1983 Phil. Trans. R. Soc. Lond. A 310 pp 293-302

Lemaitre, G. 1927 Ann. Soc. Scient. Brux., 47A pp 49-59, also 1931 trans. in Mon. Not. R. Ast. Soc. 91 pp 483-490

Lemaitre, G. 1950 *The primeval atom* French edition republished 1972 *L'hypothèse de l'atome primitif*

McCrea, W.H. 1982 *Progress in cosmology* (A.W.Wolfendale ed) pp 239-257

McCrea, W.H. 1983 *Early evolution of the Universe and its present structure* (G.O. Abell & C. Chincarini eds.) pp 383-85

Peebles, P.J.E. 1980 *The large-scale structure of the Universe*

Penrose, R. 1982 *Progress in cosmology* (A.W. Wolfendale ed.) pp 87-88

Press, W.H. & Lightman, A.P. 1983 Phil. Trans. R. Soc. Lond. A 310 pp 326-336

Rees, M.J. 1983 Phil. Trans. R. Soc. Lond. A 310 pp 311-322

Sitter, W. de 1917 Mon. Not. R. Astr. Soc. 78 pp 3-28

Weinberg, S. 1983 Phil. Trans. R. Soc. Lond. A310 pp 249-252

IMPACT OF LEMAITRE'S IDEAS ON MODERN COSMOLOGY

P. J. E. Peebles

Joseph Henry Laboratories
Princeton University

Physical scientists have a healthy attitude toward the history of their subject: by and large we ignore it. But it is good to pause now and then and consider the careers of those who through a combination of the right talent at the propitious time have had an exceptional influence on the progress of science. As I have noted on several occasions it seems to me that Georges Lemaître played a unique and remarkable role in setting out the program of research we now call physical cosmology (1,2).

In the next section I recall some of the history of the discovery of the expansion of the universe. In the following section I present my assessment of the present status of some of Lemaître's main ideas in physical cosmology.

1. THE EXPANDING UNIVERSE

Modern cosmology can be traced to Einstein's demonstration in 1917 that general relativity can describe an unbounded homogeneous mass distribution. At the time there was not much reason to think the universe really is homogeneous but by 1926 Hubble had shown that one could use galaxy counts as a probe of the large-scale distribution of galaxies and he had found that the realm of the nebulae is at least roughly uniform (3). The best modern evidence is indirect, from the accurate isotropy of deep galaxy and radio source counts and of the radiation backgrounds (4). As our galaxy seems no better a home for observers than many others it seems absurd to think that the universe might be inhomogeneous but isotropic about us, so we conclude that the universe is accurately homogeneous in the

A. Berger (ed.), The Big Bang and Georges Lemaitre, 23–30.

large-scale average.

Einstein's 1917 world model is static, the cosmological constant balancing gravity. During the 1920's people came to see that this model has some problems. One is the Olbers' paradox, that if stars had shone forever in an Einstein model starlight would accumulate indefinitely. The earliest reference I have found to this in connection with Einstein's model is in Lemaître's 1927 paper (5). Another problem noted by Weyl and Eddington is that a variable, the mass density, is set equal to a physical constant, $\Lambda/(4\pi G)$ (6,7). What would happen if the mass were rearranged? Only after the publication of Lemaître's 1927 paper did people see the answer: the universe is unstable, the perturbation tending to grow. Yet another problem was Slipher's discovery that galaxy spectra tend to be shifted toward the red. Already in 1917 deSitter had noted that Slipher's effect might be expected in his solution because of the g_{44} term in the time-independent form of the line element. The situation is complicated by the freedom of choice of galaxy orbits, but Weyl noted that if one demanded that an observer on any galaxy would see the same pattern of motions of neighbors, as befits a homogeneous universe, then one would find that at small distances the redshift is proportional to distance (8). This was independently discovered in 1925 in a little noted paper by Lemaître that is a fascinating step toward his famous 1927 paper (9). It was discussed again by Robertson (10), who also had the temerity to suggest that Hubble's distances and Slipher's redshifts for the nebulae were not inconsistent with a linear relation. Unfortunately Robertson offered no details on how he came to these conclusions.

The expanding matter-filled world model was discovered in 1922 by Friedmann (11). At the time it was thought that galaxy redshifts tend to increase with increasing distance but distance estimates were too crude to reveal the relation. In any event the possible connection between Friedmann's solution and galaxy redshifts was not discussed, and, although Friedmann's work was mentioned (in a somewhat negative way) by Einstein (12) it unfortunately dropped out of sight until about 1929. Lemaître had the good fortune to hit upon the matter-filled solution when the redshift-distance relation was in the wings if not already known (5).

What did Lemaître in 1927 know of the redshift-distance relation? As we have noted, in the following year Robertson stated (in connection with the deSitter model) that the data seemed to him to fit a linear relation (10). In his 1927 paper Lemaître estimated what we now call Hubble's constant H (he got 630 km s^{-1} Mpc^{-1}, close to Hubble's 1929 result $H \sim 500$) by using Hubble's value of the characteristic absolute magnitude of a galaxy and assuming a linear relation. He indicated that the accuracy of the available distance estimates seemed to him to be inadequate for an actual test of linearity. (It is curious that the crucial paragraphs describing how Lemaître estimated H and assessed the evidence for linearity were dropped from the 1931 English translation (13)). Hubble used distances based on the detection of stars with calibrated absolute magnitudes, and, at greater depths, distances to clusters based on mean magnitudes of several members. This reduced the scatter enough to reveal the linear relation (14). Hubble's use of clusters as standard candles was taken up by Sandage and more recently others, and Hubble's law now has been tested to an accuracy of 10% or so out to redshifts approaching unity (15).

In the three decades following the burst of discoveries in the early 1930's much of the discussion in cosmology centered on alternatives to the standard relativistic model. The main candidates were Milne's model and then the Steady State cosmology, but also important were Zwicky's tired light idea and the infinite clustering hierarchy picture of Charlier and others. This was healthy, because the empirical basis for the standard model was not all that strong, and a good way to assess the observational evidence and hit on ways to improve it is to compare alternative models. This phase ended quite abruptly with the discovery of the microwave background, which seems to offer almost tangible evidence that the universe really has expanded from a considerably denser state because no one has found a reasonable way to produce this radiation in the universe as it is now. Opinion rapidly crystallized around the standard relativistic model. My impression is that the case for this model is now strong, though certainly not definitive.

2. THE PRIMEVAL ATOM

Many were involved in the discovery of the connection between galaxy redshifts and the relativistic cosmological model: Weyl and Friedmann were on the track before Lemaître and before the observational situation was ripe; Robertson had all the pieces a year or so after Lemaître and Eddington and Tolman were close behind him (10,16,17). But in the recognition and exploration of the new vistas in physics opened up by the discovery of the expanding universe Lemaître was distinctly the pioneer, without equal until Gamow came on the scene a decade later. Lemaître's main early results were collected in a review published just fifty years ago, in 1933 (18). This paper is remarkable for the freshness and clarity and depth of the ideas. Except where noted all the following discussion of Lemaître's work refers to this paper and the references to be found therein.

We are used to linking the concepts of the expanding universe and the singular origin of classical space-time, but, as Godart and Turek point out, the connection was a daring step (19). In his 1927 paper Lemaître listed the possible courses of evolution of a closed relativistic model universe with non-zero cosmological constant, but he discussed most fully what has since come to be called the Eddington model, where the expansion asymptotically traces back to the static Einstein model. That was at least partly because this is the model that can accomodate the old stellar evolution ages, $\gtrsim 10^{13}$ y >> H^{-1}. However, Eddington and Lemaître noted that that requires an exceedingly delicate (and unlikely) balance in the quasi-static phase (16, 20). It was Lemaître who took the bold step: if the universe cannot have existed into the indefinite past in a quasi-static phase then let us consider the possibility that space expanded from a singularly dense state, what Lemaître came to call the Primeval Atom (and Gamow later termed the Big Bang).

In a Big Bang cosmology the ages of things are limited to a modest multiple of the Hubble time H^{-1}. Lemaître strongly felt that the only sensible model universes have space sections with finite volume (21). Many of us share that prejudice despite the continued lack of encouragement from the observations. If $\Lambda = 0$ this limits the age of the universe to $t < 2/3$ H. By 1933 Lemaître did not take seriously the very large stellar evolution ages in the old theory (in which energy is derived from annihilation of matter; the evolution ages came down considerably with the recognition that the energy supply is the much smaller nuclear binding energy) but he did have an important constraint from the radioactive decay ages of terrestrial minerals. With the then current estimate of H these ages exceeded $2/(3\ H)$, and he concluded that "from a purely aesthetic point of view that perhaps is regrettable. The solutions where the universe alternatively expands and contracts to an atomic state with the dimensions of the solar system have an incontestably poetic charm, bringing to mind the legendary phoenix" (my non-poetical translation). Some of us still find this aesthetically attractive.

Lemaître avoided the time-scale problem by adopting a closed model with positive cosmological constant, where the competition between gravitational attraction and cosmic repulsion of the Λ-term increases the time since the Big Bang. As it happens the modern estimates of H have gone down a factor of 5 to 10 from what Lemaître used, removing any problem with terrestrial ages. However, globular cluster star evolution ages now seem well established at $\gtrsim$ 16 billion years (22,23), and in a closed model with $\Lambda = 0$ that would require $H \lesssim 40$ km s^{-1} Mpc^{-1}, which is well below any of the current estimates. If this bind persists we certainly will have to consider Lemaître's cosmological model as one way out. (And of course an open universe with $\Lambda = 0$, where ages can approach H^{-1}, another possibility.)

In the Big Bang cosmology we encounter an end to classical spacetime at a finite time in the past. Lemaître expressed for us how spectacular that concept is: "The evolution of the universe can be compared to a display of fireworks that has just ended: some few wisps, ashes and smoke. Standing on a well-chilled cinder, we see the slow fading of the suns, and we try to recall the vanished brilliance of the origin of the worlds" (20). He remarked that the precursor of classical space-time must be a fully quantum phenomenon, what we would now call the Planck epoch (24). Einstein proposed that the singularity in classical theory might be eliminated by departures from the simplified homogeneous and isotropic world model. It was at Einstein's suggestion that Lemaître analyzed the behavior of a homogeneous anisotropic model with line element

$$ds^2 = dt^2 - b_1{}^2dx_1{}^2 - b_2{}^2dx_2{}^2 - b_3{}^2dx_3{}^2 \tag{1}$$

where the b_α are the functions of t alone. He concluded, as have many others since, that this is not a promising way out. Lemaître's suggestion that subatomic forces must stop the contraction of the universe is not now widely accepted: it is thought that the evolution can be traced back all the way to the quantum phase he had envisioned some years before. He noted that the entropy of the universe can only increase, so it is not unreasonable to suppose that the expansion commenced at zero entropy with the irreversible decay of some initial quantum state (20,24). We see this vision reflected in the exit (whether graceful or otherwise) from the inflationary phase of the Guth cosmology (25). It is still a vision, though perhaps nearer reality.

Lemaître emphasized that if the universe did expand from a dense state then we ought to be able to find some evidence of it, debris from the fireworks, and of course the nature of the debris would be an invaluable clue to the physics of the early universe. It was natural to guess that cosmic rays might be such remnants (26). That no longer seems likely because radiation would be a strong drag on energetic protons and photons. His early idea that stars are fragments from the decay of the initial quantum state was later abandoned (26,27), and indeed still seems unpromising. We do have a very strong candidate for a remnant in the microwave background radiation mentioned in the last section, and we are heavily involved with other possible remnants: quarks, magnetic monopoles, massive neutrinos, axions, supersymmetric partners, magnetic fields, strings and so on that may or may not be essential to our understanding of why the universe is the way it is (28).

Let us turn finally to the puzzle of the origin of galaxies and clusters of galaxies. In the early 1930's people saw full well that the homogeneous and isotropic cosmological model could only be a first approximation, and that the phenomenon of mass clustering must be telling us something important about the nature of the universe. It was Lemaître who laid out the research program that I think has the best promise of untangling the puzzle: consider scenarios for the evolution of structure that start at high redshift with initial conditions that do not seem unduly contrived, evolve according to accepted (or specifically conjectured) laws of physics, and end up looking more or less like the universe we observe (18,27). We may hope that as our understanding of physical processes and the physical universe improves we will come to see that some scenarios may be rejected, and that in the fullness of time we will be led to a useful approximation to the truth. And that may be the key to a deeper understanding of the physics of the Primeval Atom.

An essential element of the physics of evolution is the fitting of a mass concentration like a galaxy into an otherwise homogeneous cosmological model. In the early 1930's it was recognized that a spherically symmetric model for a mass concentration would be mathematically convenient and a sensible if rough approximation to a real object. Lemaître discovered the solution for this spherical model: when pressure may be neglected each mass shell evolves like a separate homogeneous world model. Since different models expand at different rates we arrive at the exceedingly important conclusion that the universe is gravitationally unstable. By considering the limiting case of high redshift Lemaître found the growth law $\delta\rho/\rho \propto t^{2/3}$ for linear perturbations to an Einstein-deSitter model (29). Tolman (30) and Bondi (31) enlarged on the analysis of the spherical model, but as they both referred to Lemaître's prior discovery I find it curious that this often is called the Bondi-Tolman solution.

The scenario that Lemaître analyzed in detail is based on a closed cosmological model with $\Lambda > 0$ where initial conditions can be chosen so that the universe expands from great density, passes through a quasi-static phase when gravity is very nearly balanced by the cosmological term, and then expands toward deSitter's limiting case. For a "reasonable" value of the density parameter Ω_0 the quasi-static phase would be at a redshift on the order of ten. Because the quasi-static phase depends on a close balance of Λ and gravity any small density fluctuations would be strongly amplified during this phase, and Lemaître accordingly proposed that the quasi-static phase triggered the fragmentation of the initially nearly smooth distribution of gas into protogalaxies and clusters. He pointed out that the collapse of a protogalaxy would be highly dissipative until the gas had fragmented into stars, and that the different galaxy morphological types might result from spin-up during collapse of the accidental initial angular momenta (27). A cluster of galaxies, being dissipationless, would remain at about the mean density $\Lambda/(4\pi G)$ of the quasi-static phase, and Lemaître was encouraged by the fact that estimates of $\Lambda/(4\pi G)$ and of cluster densities were quite similar (18,27). It now appears that the clustering pattern is more complicated than that, approximating a scale-invariant clustering hierarchy, so a scenario with a fixed characteristic density no longer seems to be indicated (though it still may be possible). The discussion of galaxy formation commencing with collapse of a gaseous protogalaxy seems quite familiar today, though it should be emphasized that there is no general agreement on how galaxies formed, whether by coalescence of "pre-galaxies," or by fragmentation of protoclusters, or as debris from stellar or relativistic explosions, or yet some other process. This has become a lively subject because the observational situation has been rapidly improving (32), and we may well see a crystallization of opinion around some picture or another in the next few years.

Lemaître set forth a program for the study of the mass clustering phenomenon that has become one of the mainstreams of modern research in cosmology, but it was not taken up in a systematic way for several decades. There were several reasons for the delay. Debate on the validity of the relativistic model had to take precedence until the observational situation had improved. While that debate was in progress the opinion developed that because the gravitational instability of the relativistic model is not exponential galaxies could not have developed out of reasonable initial conditions, such as thermal fluctuations. That has been resolved by observing that since

$$\delta\rho/\rho = (\delta\rho/\rho)_i \, (t/t_i)^n \quad , \tag{2}$$

with n on the order of unity, we can get all the amplification we need by taking t_i small enough. Of course, that is Lemaître's program: fix $(\delta\rho/\rho)_i$ ad hoc and then ask whether we can puzzle out a consistent scenario, leaving ultimate origins to a deeper future theory. Another barrier was the fact that the scenario Lemaître favored placed emphasis on the cosmological constant Λ at a time when Λ was becoming unpopular. If Lemaître had wanted to rally support to his ideas he would have been well advised to drop Λ. But I can no more imagine him following that advice than I can Einstein heeding the admonition to give up the search for a classical unified field theory. And it must be recorded that recent developments in elementary particle physics have very distinctly brought Λ to our attention once more (25).

REFERENCES

1. Peebles, P. J. E. 1971, "Physical Cosmology" (Princeton: Princeton University Press), Chapt. 1.
2. Peebles, P. J. E. 1980, "The Large-Scale Structure of the Universe" (Princeton: Princeton University Press), Chapt. 1.
3. Hubble, E. 1926, Ap. J. 64, p. 321.
4. For a review see Ref. 2, § 2.
5. Lemaître, G. 1927, Ann. Soc. Sci. Brux. 47A, p. 49.
6. Weyl, H. 1922, "Space Time Matter," tr. H. L. Brose, 4th ed. (Reprint, New York: Dover Publications).
7. Eddington, A. S. 1923, "The Mathematical Theory of Relativity" (Cambridge: Cambridge University Press).
8. Weyl, H. 1923, Phys. Zeits. 24, p. 230.
9. Lemaître, G. 1925, J. Math. Phys. 4, p. 188.
10. Robertson, H. P. 1928, Phil. Mag. 5, p. 835.
11. Friedmann, A. 1922, Z. Phys. 10, p. 377.
12. Einstein, A. 1922, Z. Phys. 11, p. 326; 1923, Z. Phys. 16, p. 228.
13. Lemaître, G. 1931, M.N.R.A.S. 91, p. 483.
14. Hubble, E. P. 1929, Proc. N.A.S. (U.S.A.) 15, p. 168.
15. **Oke, J. B. 1984, Clusters and Groups of Galaxies, F. Mardirossian et al.(eds.), D. Reidel, Holland, p. 99.**
16. Eddington, A. S. 1930, M.N.R.A.S. 90, p. 668.
17. Tolman, R. C. 1930, Proc. N.A.S. (U.S.A.) 16, p. 409.
18. Lemaître, G. 1933, Ann. Soc. Sci. Brux. 53A, p. 51.
19. Godart, O. and Turek, J. 1982, Rev. des Quest, Sci. 153, p. 145; 153, p. 311.
20. Lemaître, G. 1931, Rev. des Quest. Sci., Nov., p. 391.
21. Lemaître, G. 1929, Rev. des Quest. Sci., March, p. 189.
22. Harris, W. E., Hesser, J. E. and Atwood, B. 1983, Ap. J. Lett. 268, p. L111.
23. Sandage, A. 1983, A. J. 88, p. 1159.
24. Lemaître, G. 1931, Nature 127, p. 706.
25. Guth, A. 1982, paper presented at the Eleventh Texas Symposium on Relativistic Astrophysics, Austin, Dec. 13-17.
26. Lemaître, G. 1931, Nature 128, p. 704.
27. Lemaître, G. 1934, Proc. N.A.S. (U.S.A.) 20, p. 12.
28. Pagels, H. R. 1982, paper presented at the Eleventh Texas Symposium on Relativistic Astrophysics, Austin, Dec. 13-17.
29. Lemaître, G. 1933, C. R. Acad. Sci. 196, p. 1085.
30. Tolman, R. C. 1934, Proc. N.A.S. (U.S.A.) 20, p. 169.
31. Bondi, H. 1947, M.N.R.A.S. 107, p. 410.
32. Scripta Varia 48, "Astrophysical Cosmology," 1982, eds. H. A. Bruck, G. V. Coyne and M. S. Longair.
33. This research was supported in part by the U.S. National Science Foundation.

MASSIVE NEUTRINOS AND PHOTINOS IN COSMOLOGY AND GALACTIC ASTRONOMY

D.W. Sciama

International School for Advanced Studies, Trieste
International Centre for Theoretical Physics, Trieste
Department of Astrophysics, Oxford

ABSTRACT
The hot Primeval Atom would have produced pairs of neutrinos and of photinos (if they exist), many of which would have survived to the present day. If these particles have non-zero rest-mass they might dominate the universe, providing it with the critical density, and also individual galaxies, providing them with their missing mass. This hypothesis might be tested by searching for the photons which these particles would be expected to emit.

1. INTRODUCTION

While preparing my talk for this commemmorative symposium my mind went back thirty years, to the time when I met Georges Lemaitre. He was in England to receive the first Eddington Medal to be awarded by the Royal Astronomical Society. During his visit he came up to Cambridge, where I was then working. I can vividly recall meeting him in the Great Court of Trinity College. His infectious laughter shook, or so it seemed to me, the whole Court. I felt very privileged to meet him.

We are here during these days to celebrate an event that took place twenty years earlier still - the publication of his paper on the Primeval Atom. In considering what subject I should talk about, I thought that I could do no better than to choose a topic which would relate major astronomical features of the universe observable today to fundamental aspects of the Primeval Atom. I have chosen to discuss the possibility that massive neutrinos or the photinos of supersymmetry theory, formed by pair production in the high temperatures of the Primeval Atom, have survived in sufficient abundance till today to provide the critical density for the Universe, and the dark matter in galactic halos.

A. Berger (ed.), The Big Bang and Georges Lemaitre, 31–41.

2. COSMOLOGICAL CONSIDERATIONS

It is not yet known whether the universe possesses the critical density ρ_c, but it would be expected to do so to a good approximation if the inflationary theory of the universe is correct (Guth 1983). For the purposes of the present argument we shall accept this conclusion, which implies that the universe today very nearly conforms to the Einstein-de Sitter model. The value of ρ_c is determined by general relativity (for zero cosmical constant) to be given by

$$\frac{8\pi}{3} G\rho_c = H_0^2$$

where H_0 is the present value of the Hubble constant. There is considerable uncertainty in the observed value of H_0, but it is generally agreed that

$$50 \leqslant H_0 \leqslant 100 \text{ km.sec}^{-1}\text{Mpc}^{-1}.$$

Accordingly

$$5 \times 10^{-30} \leqslant \rho_c \leqslant 2 \times 10^{-29} \text{ gm.cm}^{-3}.$$

A related quantity is the age of the universe t_0 which is directly related to H_0 in the Einstein-de Sitter model by $H_0 t_0 = 2/3$. Thus we would expect that

$$3 \geqslant t_0 \geqslant 6.6 \text{ billion years.}$$

There is also considerable uncertainty in the observed value of t_0. The best value probably comes from observations of stars in globular clusters. A recent re-discussion of this evidence (Flannery and Johnson 1982) suggests that

$$3 \geqslant t_0 \geqslant 9 \text{ billion years.}$$

Accordingly we shall assume that

$$50 \leqslant H_0 \leqslant 75 \text{ km.sec}^{-1}\text{Mpc}^{-1},$$

and that

$$5 \times 10^{-30} \leqslant \rho_c \leqslant 10^{-29} \text{gm.sec}^{-3}.$$

Now it is unlikely that a density as large as ρ_c could be entirely due to baryons. This follows both from the direct observation of baryons in stars and interstellar material, and from considerations of the synthesis of the light elements D, He^3, He^4 and Li^7 in the hot big bang. Each of these arguments suggests that $\rho_b \sim 10^{-31}$gm cm^{-3}, which is only about one per cent of the critical density (Pagel 1982).

We wish to discuss the possibility that the critical density is mainly due to massive neutrinos or photinos. Let us first consider the possibility that it is mainly due to massive neutrinos. For this purpose we need to compute their present concentration n_ν for each neutrino flavour. This problem has been well understood for a long time (e.g. Weinberg 1972). The essential point is that at very early times neutrino pairs would have been thermally excited e.g. by the reaction

$$e^- + e^+ \leftrightarrow \nu + \bar{\nu}. \qquad (1)$$

Eventually the rates for these reactions become lower than the expansion rate of the universe, and the neutrinos decouple from the general heat bath. According to the Salam-Weinberg theory this decoupling occurs when the universe had a temperature $T_d \sim 2$ Mev. The present value of n_ν depends critically on whether the neutrinos were still relativistic at T_d, that is, on whether $m_\nu < 2$Mev. If this condition holds (as we shall assume) then at decoupling the neutrinos would have been as numerous as the photons in the heat bath (apart from a factor of 3/4 arising from Fermi-Dirac statistics). They would still be as numerous as the photons today were it not for the permanent annihilation of electron pairs at $T \sim \frac{1}{2}$Mev. The decay products of this annihilation would have boosted the photons without boosting the (decoupled) neutrinos. Because of this suppression effect one finds that today

$$n_\nu \sim \tfrac{1}{4} n_\gamma.$$

The photon heat bath is now at 2.7°K (microwave background) and so one obtains

$$n_\nu \sim 100 \text{ cm}^{-3}.$$

Thus for one neutrino flavour to provide the critical density, one would require that

$$25 \leqslant m_\nu \leqslant 50\text{ev}.$$

Since there are presumably at least three neutrino flavours, one must sum over their rest-masses. The question here arises of possible further neutrino flavours contributing to the critical density. Primordial nucleosynthesis of the light elements again plays a role here. Each neutrino flavour which was relativistic at nucleosynthesis helps to increase the expansion rate of the universe and so changes the resulting abundances of the light elements, thereby putting at risk the agreement with the observed abundances. This argument (especially as applied to He^4) is now believed to restrict the number of allowed neutrino flavours to less than 4 (e.g. Barrow and Morgan 1983, Yang et al 1984), in remarkable agreement with the presently accepted number of neutrino flavours. Independent evidence on the number of neutrino flavours will be forthcoming when the width of the intermediate Z boson state is measured by LEP.

At this point one might regard the neutrino hypothesis for the critical density as rather appealing. However, we will meet another argument in the next section according to which, if neutrinos were also to dominate individual galaxies and account for their missing mass, then the data on dwarf galaxies, if confirmed, would indicate that $m_\nu \geqslant 250$ev, in contradiction to the cosmological upper limit on m_ν. We therefore consider here the alternative possibility that the critical density is provided by massive photinos. Of course we would still have to solve the problem posed by the large mass of the particle.

Indeed we immediately face a further problem, since we are apparently not allowed one full further particle type which was relativistic at nucleosynthesis. We can however, solve both problems if photinos decoupled much earlier than neutrinos, so that particle species other than electrons which annihilated permanently would have boosted the neutrinos (and photons) without boosting the photinos (Dimopoulos and Turner 1982, Fayet 1982, Sciama 1982a). Now the lowest mass particle species available are muons and pions which annihilated when $T \sim 200$ Mev. Thus we would require for photinos that $T_d \geqslant 200$Mev.

To explore this possibility further we need to carry out two types of calculation. The first is of the amount of suppression as a function of T_d, resulting from the annihilation of all the relevant particle species. The second is the dependence of T_d on the coupling constants of the photino (whose numerical values have not yet been determined by supersymmetry theory). Both these calculations have already been carried out. The first calculation (Olive et al 1981) is simplified by the expectation that at $T \geqslant 200$ Mev the hadrons in the universe were broken down into quarks and gluons. If we define the suppression factor f by

$$n_{\tilde{\gamma}} = f n_\nu$$

then one finds that at temperatures above 200 Mev, f rapidly assumes a plateau value ~ 0.1, with an asymptotic value ~ 0.05. More specifically, one has $f \sim 0.53$ at $T_d \sim 150$ Mev, $f \sim 0.12$ at $T_d \sim 200$ Mev, while $f \sim 0.08$ for $T_d \sim 1$ Gev, and $f \sim 0.05$ for $T_d \geqslant 20$ Gev. Thus if one tentatively neglects the short stretch of T_d between 150 Mev and 200 Mev, one finds that for photinos to supply the critical density one must have

$$250 \leqslant m_{\tilde{\gamma}} \leqslant 1000 \text{ ev.}$$

The requirement that $T_d \geqslant 200$ Mev has implications for the coupling constants of the photino. Photino-photino processes analogous to (1) have a cross-section $\propto d^{-2}$, where d is the supersymmetry-breaking parameter E^2_0 (Fayet 1979a), and E_0 is the energy at which supersymmetry is broken. One must consider also photino-goldstino processes. (The goldstino is the spin ½ supersymmetry partner of the Goldstone boson which must accompany the breaking of supersymmetry.) These processes have a cross-section $\alpha\ (e/d_g)^2\ d^{-2}$, where e_g is the goldstino coupling constant (Fayet 1979a).

The rate at which these interactions tend to maintain thermal equilibrium is $\langle n\sigma v\rangle$, where n is the number of interacting particles, $\alpha\ T^3$ in thermal equilibrium, $\sigma\ \alpha\ E^2\ \alpha\ T^2$ and $v\sim c$. The expansion rate of the universe $\alpha\ T^2$, according to general relativity. Hence T_d is determined by the smaller of $d^{2/3}$ and $(e_g d/e)^{2/3}$. Inserting all the numerical coefficients one finds that $T_d \geqslant 200$ Mev implies that

$$e_g d > 10^7\ \text{Gev}^2$$
$$d > 10^8\ \text{Gev}^2$$

Are these restrictions reasonable? Neither e_g nor d is determined by supersymmetry theory, but there are weak lower limits imposed on d by experiment (Fayet 1979b) and by considerations of stellar evolution (Fukugita and Sakai 1982), namely, $d > 2.6 \times 10^3$ Gev^{2}and $e_g d > 2.6 \times 10^2$ Gev^2. An upper limit on $e_g d$ would follow from the requirement that supersymmetry should solve the hierarchy problem (e.g. Llewellyn-Smith 1982). This problem arises as follows. The Higgs potential for the electroweak theory has the form

$$V = -\mu_0{}^2\phi^2 + \lambda\phi^4,$$

with $\mu_0 \sim 300$ Gev. However, μ_0 is altered by radiative corrections, and one might expect μ_0 to be increased to a typical grand unified value $\sim 10^{14}$ Gev. This need not happen in a supersymmetric theory, however (Llewellyn-Smith 1982). There one finds that

$$\mu^2(p^2) = \mu^2(\Lambda^2) + O\left(\int_{M^2_{min}}^{M^2_{max}} dk^2\right),$$

where p represents a low energy, Λ an energy much greater than that at which supersymmetry breaking sets in, and $M_{max,min}$ is the maximum/minimum mass of a supermultiplet.

If we require that $\delta\mu^2 < \mu^2$, it follows that

$$\Delta M^2 < 10^7\ \text{Gev}^2,$$

say. Now $\Delta M^2 = e_g d$ (Fayet 1977).
Hence

$$e_g d < 10^7\ \text{Gev}^2,$$

which is the opposite of our previous condition. If both arguments are correct we accordingly require (Sciama 1982b) that

$$e_g d \sim 10^7 \text{ Gev}^2.$$

It may be significant that our value for ΔM^2 implies that $\delta\mu^2 \sim \mu^2$. This suggests that $\mu^2(\infty) = \mu_0{}^2 = 0$, so that both the symmetry breaking of the electroweak theory and the masses of the Higgs particles would be radiatively induced (Coleman and Weinberg 1973). Our lower limit on d would then imply that

$$e_g \leqslant 0.1.$$

One could have e_g very small (and d correspondingly large) as in some variants of supersymmetry theory, or one could have the other extreme of $e_g \sim 0.1$ (and so $\sim e$, which may be significant) and $d \sim 10^8$ Gev^2. We will comment further on this latter possibility in the last section.

Finally we note that with $e_g d$ closely determined, we would have $T_d \sim 200$ Mev, and so $f \sim 0.1$. Thus if photinos provide the critical density and the hierarchy argument is correct we would have

$$250 \leqslant m_\gamma \leqslant 500 \text{ ev}.$$

3. GALACTIC CONSIDERATIONS

Evidence has recently accumulated that galaxies possess massive halos which extend far beyond their optical boundaries. This follows both from a stability argument (Ostriker and Peebles 1973) and from observations of their rotation curves (Rubin 1979, Bosma and Van der Kruit 1979). This conclusion complements one known since 1933, namely, that galaxy clusters also possess considerable amounts of hidden mass, as judged by their large velocity dispersions. If neutrinos and/or photinos possess rest-masses of the right order, the hot big-bang would clearly supply them in adequate numbers for them to be possible candidates for the hidden mass in galaxy clusters and in individual galaxies. We therefore consider this question here.

The problem has two aspects. The first, more difficult one, is to understand the processes of galaxy formation in the presence of non-interacting massive particles. The second is to study the present properties of a galaxy or cluster which is dominated by such particles. The first problem is not well understood, and we consider here only the second one. We shall find that the observed properties of galaxies lead to a significant lower limit on the mass of the dominating particle.

This limit arises as a consequence of the Liouville theorem, which applies to the particles after they decouple (Tremaine and

Gunn 1979). Strictly speaking the phase space density of the particles remains constant along their world-lines, but in practice the discreteness of their distribution may lead to intricate phase-mixing, if they suffer violent relaxation during their collapse to form a galaxy (Lynden-Bell 1967). It would then become necessary to coarse-grain, which would lead to a reduction in their phase-space density. By comparing their present phase space density in a galaxy with their phase-space density at decoupling, one obtains the limit (Peebles 1980)

$$m_p^4 \geqslant \frac{1}{6\pi}\left(\frac{3}{2\pi}\right)^{3/2}\frac{h^3}{G\,v_0\,a^2},$$

where m_p is the mass of the particle, h is Planck's constant, v_0 is their three-dimensional velocity dispersion (assumed to be Gaussian and independent of position in the galaxy) and a is the core-radius of the particle-distribution. Let us first apply this limit to our own Galaxy, the Milky Way. In this case we would have $v_0 \sim 300$ km sec^{-1} and a~8 kpc (Caldwell and Ostriker 1981) and so

$$m_p \geqslant 25\text{ev}.$$

(Galaxy clusters would give a weaker limit). This argument is quite independent of the one leading to the critical density but is in remarkable agreement with it if the particles are neutrinos and the phase-space mixing is relatively unimportant (as it may be according to Melott's (1982) N-body simulations of the relevant gravitational collapse processes).

However, recently it has been discovered that dwarf galaxies like Draco may possess much hidden matter (Aaronson 1983, Faber and Lin 1983, Lin and Faber 1983, Faber 1984). In such cases both v_0 and a would be much smaller, and m_p correspondingly larger. For example, for Draco one has $v_0 \sim 10$ km sec^{-1} and a ~0.5 kpc, leading to

$$m_p \geqslant 250\text{ev}$$

As pointed out by Aaronson and by Faber and Lin, if this inequality is correct it would conclusively rule out neutrinos. However, we notice that it would fit in well with photinos, for which we deduced that $250 \leqslant m_{\tilde\gamma} \leqslant 500$ ev, if the phase-mixing is small only for dwarf galaxies. While many problems clearly remain to be solved, this numerical agreement is suggestive. We note in particular that systems smaller than dwarf galaxies are unlikely to lead to still larger estimates of the particle mass. For example, globular clusters have a mass to light ratio only about one tenth of that of dwarf galaxies (Innanen et al 1983), which would mean that they contain relatively much less missing mass and so would not lead to a further increase in the mass of the dominating particle.

We conclude from this discussion that dwarf galaxies may provide the first observational evidence in favour of broken supersymmetry.

4. ULTRA-VIOLET ASTRONOMY CONSIDERATIONS

Further evidence for the existence of massive photinos may come from the photons which they would be expected to emit (Cabibbo, Farrar and Maiani 1981). This idea is an extension of an earlier proposal by de Rujula and Glashow (1980) that one should look for photons emitted by massive neutrinos which might dominate our Galaxy and the universe. In the photino case of interest to us here the process envisaged is

$$\tilde{\gamma} \rightarrow \gamma + \tilde{g},$$

where $\tilde{g}$ is a goldstino.

If the parent particle is at rest relative to the observer, then conservation of energy and momentum in the decay process results in a photon energy E_γ given by

$$E_\gamma = (m^2_{\tilde{\gamma}} - m^2_{\tilde{g}})\ 2m_{\tilde{\gamma}}$$

If $m_{\tilde{g}} \ll m_{\tilde{\gamma}}$ we can simplify this to

$$E_\gamma \sim \tfrac{1}{2}\, m_{\tilde{\gamma}}.$$

Thus if $m_{\tilde{\gamma}}$ lies in the range 250-500 ev we would have

$$E_\gamma \sim 125\text{-}250 \text{ ev}.$$

(unless $m_{\tilde{g}} \approx m_{\tilde{\gamma}}$), so that the decay photons would lie in the extreme ultra-violet or soft X-ray part of the spectrum.

The radiative lifetime $\tau_{\tilde{\gamma}}$ of the photino has been calculated by Cabibbo, Farrar and Maiani. They obtain

$$\tau_{\tilde{\gamma}} \sim 1.7 \times 10^{26} \left(\frac{d}{10^8 \text{Gev}}\right)^2 \left(\frac{250 \text{ ev}}{m_{\tilde{\gamma}}}\right)^5 \text{ sec}$$

If $d > 10^8$ Gev^2, as we discussed above, we would have

$$\tau_{\tilde{\gamma}} > 1.7 \times 10^{26} \text{ sec} \quad \text{for } E_\gamma \sim 125 \text{ ev}$$
$$\tau_{\tilde{\gamma}} > 5 \times 10^{24} \text{ sec} \quad \text{for } E_\gamma \sim 250 \text{ ev}$$

We now consider whether the resulting photon flux would have observational consequences. To calculate the flux from photinos dominating the Galaxy we note that the surface density of dark matter at the sun required in addition to known stars and gas to account for the observed rotation velocity of the Galaxy is about

300 $M_\odot$ pc^{-2} (Caldwell and Ostriker 1981). If this is due to photinos of mass ~250 ev their surface density would then be of order 1.6 x 10^{29} cm^{-2}. Accordingly the photon flux at ~125 ev (neglecting absorption for the moment) for $\tau_{\tilde{\gamma}} \sim 1.7 \times 10^{26}$ sec would be ~1000 cm^{-2} sec^{-1}. Similarly, if $m_{\tilde{\gamma}}$ ~500 ev and $\tau_{\tilde{\gamma}} \sim 5 \times 10^{24}$ sec, we would have a flux ~ 1.6 x 10^4 cm^{-2} sec^{-1}.

By an interesting coincidence, the column density of cosmological photinos out to a Hubble radius ($n_{\tilde{\gamma}}c/H_0$) would also be of order 10^{29} cm^{-2}(10 x 10^{28}), so that the resulting cosmological photon flux would be of the same order as the galactic flux. The main difference is that the galactic flux would be nearly monochromatic (since the velocity dispersion of the galactic photinos ~300 km $sec^{-1} \ll c$), whereas the cosmological flux would be drawn out into a continuous spectrum by the differential red shift associated with the expansion of the Universe. For an Einstein-de Sitter model this spectrum would have the unabsorbed form

$$\underset{(\lambda \geq \lambda_c)}{I_\lambda} = \frac{c n_{\tilde{\gamma}}(z=0)}{H_o \tau_{\tilde{\gamma}}} \frac{\lambda_o^{3/2}}{\lambda^{5/2}},$$

where λ_0 is the rest wavelength of the decay photon and λ the received wavelength.

The observed photon background in the energy range 125-250 ev has been discussed recently by Fried et al (1980) and by Paresce and Stern (1981). The observations are broadband, but any line in the background in this energy range must have a limiting flux ~500 cm^{-2} sec^{-1}. The observed spectrum increases with increasing wavelength, and is believed to be mainly due to two components, namely the long wavelength end of the isotropic extragalactic X-ray background, and thermal emission from hot gas ($T \sim 10^6$ °K) within one hundred parsecs of the sun. The observed fluxes at 250 ev and 125 ev are about 2 and 8 photons cm^{-2} sec^{-1} $ster^{-1}$ ev^{-1} respectively.

The implied limits on the photon flux from photinos depend on estimates of galactic absorption. In directions at right angles to the galactic plane, the optical depth is about 0.5 at 250 ev and about 4 at 125 ev. Thus when we compare our estimated line and continuum photon fluxes from photinos with the observational upper limits, allowing for absorption, we find a discrepancy of about one order of magnitude at 250 ev, and no discrepancy at 125 ev. We could remove the discrepancy in the first case by increasing d by a factor 3 to ~3 x 10^8 Gev^2. In view of the various numerical uncertainties in our discussion such a change would still be consistent with our suggestion that $e_g \sim e$. Similar estimates for the line flux from the Andomeda galaxy M31 lead to a value which could be as large as 1 photon cm^{-2} sec^{-1}. This is 10^5 times greater than the limiting sensitivity of the various proposed X-ray satellites such as ROSAT, XMM, HTS and AXAF. Thus even if the true value of d were larger than our most optimistic estimate, many extragalactic sources might be readily detectable in the photino decay line. If this possibility is realised, a new branch of astronomy would be born.

We conclude that the photon fluxes implied by our choice of parameters are compatible with present observations. Future observations, particularly those designed to look for a narrow line, should clarify this situation. It would be a happy circumstance, and one which I believe would particularly have pleased Lemaitre, if X-ray astronomers and observers of galaxies could thus use the Primeval Atom hypothesis to demonstrate the correctness of the broken supersymmetry theory.

REFERENCES

1. Aaronson, M. 1983, Ap.J.Lett. 266, L11.
2. Barrow, J.D. and Morgan, J. 1983, Mon.Not.Roy.Astr.Soc. 203, 393.
3. Bosma, A. and van der Kruit, P.C. 1979, Astron.and Astrophys, 79, 281.
4. Cabibbo, N., Farrar, G.R. and Maiani, L. 1981, Phys.Lett.105B, 155.
5. Caldwell, J.A.R. and Ostriker, J.P. 1981, Ap.J. 251, 61.
6. Coleman, S. and Weinberg, E.J. 1973, Phys.Rev.D7, 188.
7. de Rujula, A. and Glashow, S. 1980, Phys.Rev.Lett. 45, 942.
8. Dimopoulos, S. and Turner, M. 1982, Proc.Moriond Meeting. ed.J. Tran Thanh Van.
9. Faber, S. and Lin, D.N.C. 1983, Ap.J.Lett. 266, L17.
10. Faber, S. 1984, in Large Scale Structure of the Universe, Cosmology and Fundamental Physics ESO-CERN Symposium, ed L. van Hove and G. Setti.
11. Fayet, P. 1977, Phys.Lett. 70B, 461.
12. Fayet, P. 1979a, Phys.Lett.84B, 272.
13. Fayet, P. 1979b, Phys.Lett.84B, 421.
14. Fayet, P. 1982, Proc. Moriond Meeting. ed.J. Tran Thanh Van.
15. Flannery, B.P. and Johnson, D.R. 1982, Ap.J. 263, 166.
16. Fried, P.M., Nousek, J.A., Sanders, W.T. and Kraushaar, W.L. 1980, Ap.J.242, 987.
17. Fukugita, M. and Sakai, N. 1982, Phys.Lett.114B, 23.
18. Golfand, Y.A. and Likhtman, E.P. 1974, JETP Letters 13, 454.
19. Guth, A.H. 1982. in The Very Early Universe, Nuffield Workshop. ed. G.W. Gibbons. S.W. Hawking and S. Siklos Cambridge University Press. To be published.
20. Innanen, K.A., Harris, W.E. and Webbink, R.F. 1983, Ann.Rev.Astron. and Astrophys.
21. Lin, D.N.C. and Faber, S. 1983, Ap.J.Lett. 266, L21.
22. Llewellyn-Smith, C.H. 1982. Proc.Stockholm Conference on Physics in Collision.
23. Lynden-Bell, D. 1967, Mon.Not.Roy.Astr.Soc. 136, 101.
24. Melott, A.L. 1982, Phys.Rev.Lett. 48, 894.

25. Olive, K.A., Schramm, D.N. and Steigman, G. 1981, Nucl.Phys.B. 180, 497.
26. Ostriker, J.P. and Peebles, P.J.E. 1973, Ap.J. 186, 467.
27. Pagel, B.E.J. 1982, Phil.Trans.Roy.Soc. 307, 19.
28. Paresce, F. and Stern, R. 1981, Ap.J. 247, 91.
29. Peebles, P.J.E. 1980. in Physical Cosmology ed. R. Balian, J. Audouze and D.N. Schramm, North-Holland, Amsterdam.
30. Rubin, V.C. 1979, Comments Astrophys. and Space Science 8, 79.
31. Sciama, D.W. 1982a, Phys.Lett. 114B, 19.
32. Sciama, D.W. 1982b, Phys.Lett. 118B, 327.
33. Tremaine, S. and Gunn, J.E. 1979, Phys.Rev.Lett. 42, 407.
34. Weinberg, S. 1972, General Relativity and Cosmology (Wiley, New York).
35. Wess, J. and Zumino, B. 1974, Nucl.Phys.B70, 39.
36. Yang, J., Turner, M.S., Steigman, G., Schramm, D.N. and Olive, K.A 1984, Ap.J. in press.

THE PRIMORDIAL NUCLEOSYNTHESIS

Jean Audouze

Institut d'Astrophysique du CNRS Paris France
and Laboratoire René Bernas Orsay France

Abstract

The review of the primordial nucleosynthesis presented in the honor of the memory of Mgr Georges Lemaitre is divided in three chapters : in the first one attempts to determine the primordial abundances of the lightest elements which can be formed by the Big Bang nucleosynthesis. This analysis leads to fairly large uncertainty ranges due to dispersions in the observations and also in the case of Deuterium to the still arbitrary choice between different models of galactic evolution.

Chapter 2 is a summary of the Standard Big Bang nucleosynthesis where it is recalled that (i) the abundances of D, ^{3}He and ^{7}Li allow to predict that the baryonic cosmological parameter is $0.10^{+0.11}_{-0.09}$ (i.e leading to a open universe if baryons are the major constituants of the matter of the Universe ; and (ii) that the He primordial abundance is consistent with the three different types of neutrinos which are presently observed. This simple and attractive model might be found in difficulty in the case of a primordial abundance of He $\leqslant$ 0.24 and/or in the case of models of galactic evolution allowing infall of external matter having a primordial composition.

A. Berger (ed.), The Big Bang and Georges Lemaitre, 43–61.

Finally chapter 3 summarizes two alternative proposals to the Standard Big Bang nucleosynthesis : the possible production of D by partial photodisintegration of He induced by energetic photons coming from the decay of massive and unstable neutrinos or some (at present quite unlikely !) spallation mechanisms induced by pregalactic cosmic rays on a pure hydrogen primitive interstellar/pregalactic gas.

1- Introduction

At epochs where the theory of a dense and hot primordial phase for the Universe (designated now as the Big Bang) was not as accepted as it is today, Mrg Georges Lemaitre (with George Gamow whose the memory should be associated to this ceremony) was may be the only one to push the idea of a singular primeval atom. In my own country because of the strong influence of Henri Poincaré, mathematical cosmology has been considered for a very long period more fashionable than physical cosmology. This attitude is changing slowly but definitely : the french research agencies like the Centre National de la Recherche Scientifique are now convinced that an important effort should be made in that field. This is why I feel much honored to participate to this celebration in honor of the most prominent french speaking physical cosmologist.

The topic that I am going to review here is certainly one of those which should have pleased most Mgr Lemaitre. While the recession of the galaxies and the cosmic background radiation allow to describe the early phases of the Universe but distant by about 10^5-10^6 years from the origin, the primordial nucleosynthesis, responsible for the formation of the lightest elements (D, ^{3}He, ^{4}He and ^{7}Li) has occured at times as early of 1-3 minutes after the primordial explosion. This is why the processes which are going to be presented here are so important with respect to the theory of the Big Bang itself.

Before a review of the Standard Big Bang nucleosynthesis and its consequences on the present density of the Universe and the maximum number of neutrino families, a quick survey of the primordial abundances of the light elements D, ^{3}He, ^{4}He and ^{7}Li has to presented here. The assets and liabilities of the Standard Big Bang nucleosynthesis are also discussed such as two proposals made by Joseph Silk (Berkeley) and myself as possible alternative solutions to the Standard theory.

Since I have written another review paper on the same topic (Audouze 1984) this presentation will be much shorter and will concentrate on the basic features of the nucleosynthesis. The reader interested in a more thorough review on this subject might consult the above reference.

2- A survey of the primordial abundances of the light elements

The elements of interest are D, ^3He, ^4He and ^7Li

2-1 Deuterium abundances

There are two sites where the D/H ratio can be observed (i) the Solar system : the best determination comes from the Solar Wind $^3He/^4He$ ratio given some assumptions on the presolar $^3He/^4He$ ratio (D is easily destroyed into ^3He). From Geiss and Reeves (1972), $(D/H)_{Sol\ Syst} = (2 \pm 1)10^{-5}$ (ii) The interstellar D/H ratio is measured by its UV absorption line at 910 Å (see Laurent 1983 for a review of the current interstellar measurements).

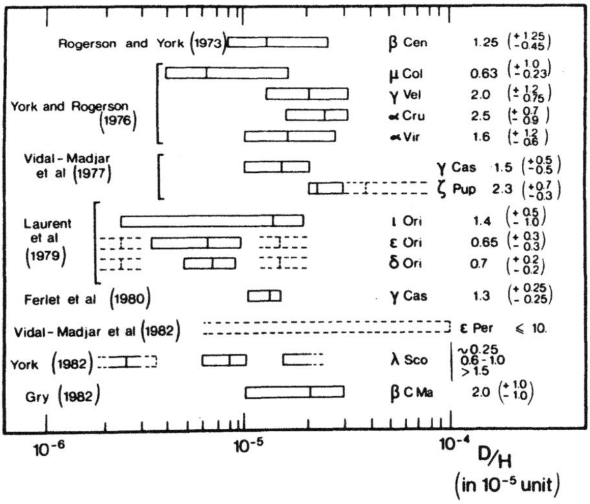

Figure 1 : Compilation of the available interstellar D/H determinations which shows a large dispersion between the different lines of sights (Vidal-Madjar 1983)

Figure 1 shows how dispersed are the present interstellar D/H determinations. It is argued by Vidal-Madjar and Gry (1983) that $(D/H)_{interst.}$ can be as low as $5\ 10^{-6}$.

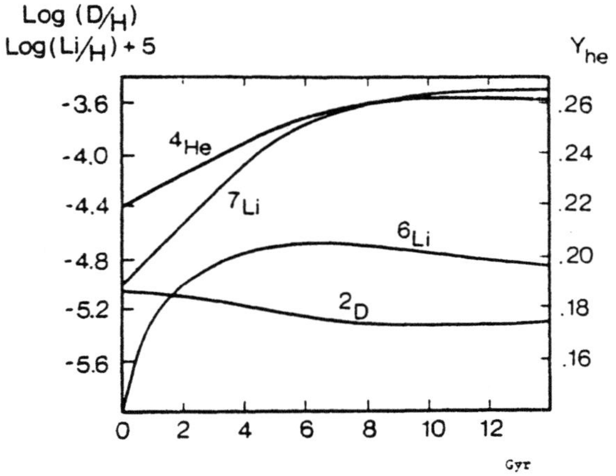

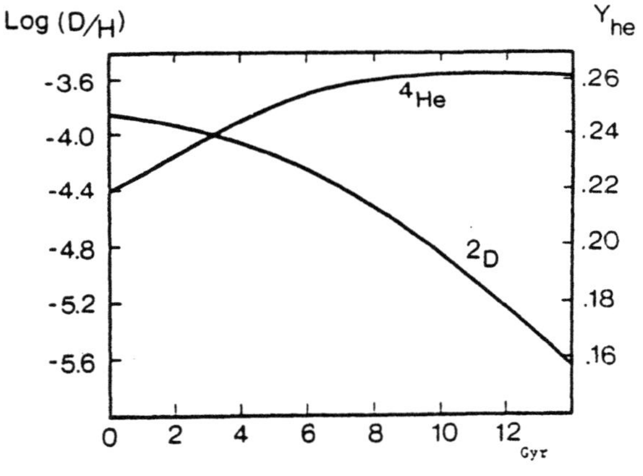

Figures 2a and 2b : These two figures show the evolution of some light elements with time in two models of chemical evolution of galaxy. In fig. 2a the considered zone is submitted to infall of primordial gas and in this case the D/H abundance does not vary much with time. By contrast in fig. 2b which represents the galactic evolution with inflow of processed material the D/H depletion factor can be as high as about 40 (Gry *et al* 1983)

From these two sets of values one cannot derive in a simple fashion the primordial (D/H) abundance. One has to take into account the astration processes which took place during the galactic evolution and which destroy D. According to Gry et al (1983), the D depletion during the galactic evolution can be as low as a factor 2 if the considered zone is submitted to infall of primordial material (fig. 2a) while it can be as high as ~40 if the considered zone is submitted to inflow of processed material (fig. 2b)

From the large uncertainties coming from the mixing of presolar ^{3}He with the by-product of the D destruction, the large dispersion in the interstellar D abundance and the large variations of the D depletion factor depending on the chosen galactic model the primordial D abundance range by mass is

$$7\ 10^{-6} \leq X(D) \leq 3\ 10^{-4}$$

It should be noted that $X(D) \geq 6\ 10^{-5}$ in the primordial phases only if a chemical evolution model with inflow is adopted.

2-2 The ^{3}He abundance

As for D, the ^{3}He abundance is only measured (i) in the Solar System (ii) in the interstellar medium. (i) For the Solar System ^{3}He/H = 1-2 10^{-5}. This result comes from the Solar wind measurements (Geiss and Reeves 1972) or from the study of gas rich meteorites (Black 1972). (ii) Wilson et al (1983) have reported a series of radio measurements of ^{3}He concerning different HII regions. They found values ranging from $^3\text{He/H} < 2\ 10^{-5}$ up to $2\ 10^{-4}$. Given these results it is impossible to-day to tell if stars enrich the interstellar medium into ^{3}He or destroy it (may be they do both !). This is why the range of possible primordial ^{3}He/H abundances (by mass) is

$$2\ 10^{-5} < {}^3\text{He/H} < 3\ 10^{-4}.$$

The combination of the D and ^{3}He primordial ranges leads to

$$3\ 10^{-5} < \frac{{}^3\text{He+D}}{\text{H}} < 3\ 10^{-4}$$

2-3 The ^{4}He abundance

The ^{4}He abundance has been looked for in many different astrophysical sites (see e.g. the book "Primordial Helium" edited by Shaver et al 1983). From the thorough observations of blue compact (metal poor) galaxies performed by Kunth and Sargent 1983, the range of primordial He abundance (by mass Y) is $0.22 < Y \leq 0.25$ while Gautier and Owen (1983) argue that Y can be as low as $0.15 \leq Y \leq 0.24$ based on IR spectrograms of the Jupiter surface.

We will adopt here $0.22 \leq Y \leq 0.25$ keeping in mind that there are no observational techniques regarding ^{4}He which are free from difficulties (ionization of He in blue compact galaxies which may represent material which is far to be primordial, chemical fractionation effects in Jupiter etc).

2-4 The ^{7}Li abundance

From the recent work of Spite and Spite (1982) who observed Li in spectra of halo stars, the primordial Li/H abundance (by number) seems to be Li/H $\sim 10^{-10}$ (figure 3). By taking into account possible Li destruction effects in the convective zones of these old stars (which should be less important than in disk stars) the primordial Li abundance (by mass) is

$5\ 10^{-10} < X(^7Li) < 1.5\ 10^{-9}$

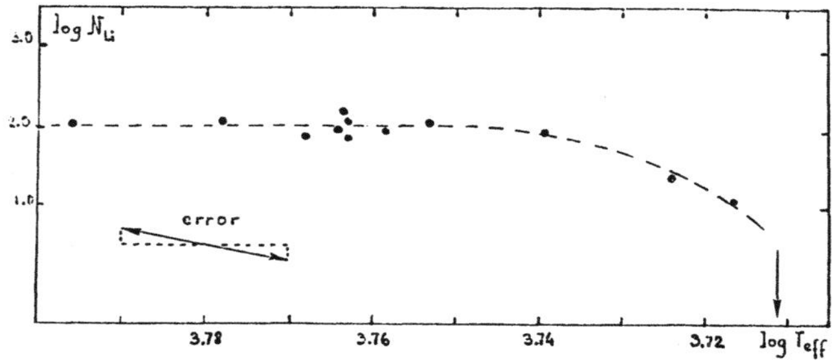

Figure 3 : Lithium abundances for some halo stars as a function of the effective temperature Teff. The corresponding Li/H ratio is 10^{-10} (log N_{Li} = 2 for log n H = 12) for stars such that Teff > 5500 K (Spite and Spite 1982)

The higher Li abundance observed in the Solar System should be due to a galactic enrichment induced either by red giants or by novae (Audouze et al 1983).

To sum up this discussion, table 1 presents the ranges for the primordial abundances of D, ^{3}He, ^{4}He and ^{7}Li which are considered here.

It is expected of course that future observations will be planned to attempt to obtain better determinations of these primordial abundances. In this respect the decision taken by ESA not to select the Magellan UV project intending to look for the D/H ratio in Magellanic Clouds and different interstellar regions of our Galaxy is quite unfortunate. Nevertheless, there are many difficulties (galactic evolution, chemical fractionation, ionization problems, convective zone depths, etc.) which will be hard to solve within a near future.

Table 1

Range of possible values for the primordial abundances of the light elements

elements	Range (abundance by mass)
D	$7\ 10^{-6} - 3\ 10^{-4}$
^{3}He	$3\ 10^{-5} - 3\ 10^{-4}$
D+^{3}He	$2\ 10^{-5} - 3\ 10^{-4}$
^{4}He	0.22 - 0.25
^{7}Li	$5\ 10^{-10} - 1.5\ 10^{-9}$

3 - The Standard Big Bang nucleosynthesis

In the Standard Big Bang model, six different hypotheses are made :

1 - The interacting particles (nucleons, electrons and positrons, neutrinos and photons) have been in statistical equilibrium which means that $T_{Big\ Bang} \gtrsim 10^{11}$ K at the beginning

2 - The laws of physics do not depend on time and location (Principle of Equivalence)

3 - The Universe is homogeneous and isotropic (Cosmological Principle)

4 - The rate of expansion of the Universe is fixed by the General Relativity theory, the characteristic expansion time τ is

$$\tau \sim \lfloor 24\ \pi\ G\ \rho(t) \rfloor^{-1/2} \qquad (1)$$

where G is the gravity constant and $\rho(t)$ the total density given as a function of the Universe age t.

5 - The expansion of the Universe is assumed to be adiabatic i.e. during the whole evolution of the Universe one has

$$\rho(t) = h\ T_9{}^3(t) \qquad (2)$$

where $T_9(t)$ is the temperature in 10^9 K units and h the so called baryon density parameter which remains constant and equal to $3.3\ 10^{-4}\ \eta$ all throughout the Universe history.

6 - The Universe is asymetric : namely the density of antimatter is negligible compared to that of matter. This is why the

baryon-photon density ratio η has values as large as 10^{-10} - 10^{-9}.

Within this framework, the nucleosynthesis starts when $T_9 \sim 1$ when the photodesintegration of D is slower than the neutron absorption by photons and a few minutes after the Big Bang.

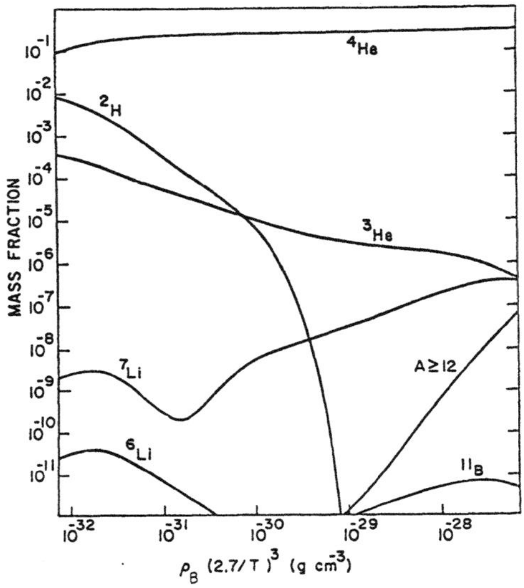

Figure 4 : Calculated primordial abundances of the light elements D, ^{3}He, ^{4}He and ^{7}Li as a function of the present baryonic-density of the Universe (after Wagoner 1973). The D, ^{3}He and ^{7}Li abundances favour values of the present baryonic density of the Universe lower than the critical density which means that the baryons by themselves do not close the Universe

Figure 4 is the very classical presentation of the calculated D, ^{3}He, ^{4}He and ^{7}Li abundances made after Wagoner (1973) with respect to the present density of the Universe. Figure 5 is an updated presentation made by Yang et al 1984 of these calculated abundances as a function of η the baryon-photon density ratio which is strictly proportional to the present density of the Universe.

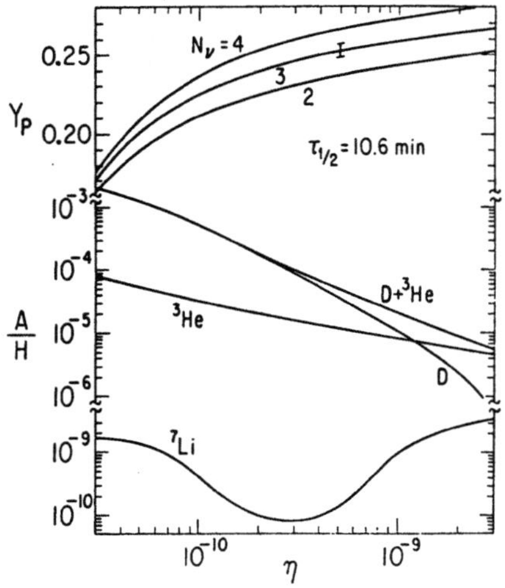

Figure 5 : calculated primordial abundances of ^4He (by mass i.e. Y), D, ^3He and ^7Li (by number) as a function of η the baryon over photon density for a neutron life-time of 10.6 minutes. The He abundance Y has been calculated for three values of the number of neutrino types N_ν = 2,3 and 4. The error bar on the $Y(N_\nu = 3)$ curve shows the range in Y corresponding to $10.4 < \tau\ 1/2 < 10.8$ minutes (After Yang et al 1984)

Two important predictions can be made from the Standard Big Bang nucleosynthesis :

1 - From D, ^3He and ^7Li abundances one can deduce both a lower limit and an upper limit of the present baryonic density of the Universe :

the baryon cosmological parameter Ω_B can be defined as

$$\Omega_B = \frac{3.5\ 10^{-3}}{h_o^2} \left(\frac{T_o}{2.7}\right)^3 \eta_{10} \qquad (3)$$

where h_o = (H/100), H being the Hubble constant in km s^{-1} Mpc^{-1}, T_o the cosmic background temperature and η_{10} = 10^{10} n_B/n_γ i.e. 10^{10} times η the baryon-photon density ratio.

When $\Omega_B > 1$ the Universe is closed (only by the baryons) while $\Omega_B \leqslant 1$ corresponds to a open Universe (unless neutrinos have masses $\gtrsim$ 30 ev). From the D and (D+He^3) primordial abundance dispersions, one deduces the following range

with $h_o = 1/2$ and $X(D) > 7\ 10^{-6}$ $\Omega_B \leq 0.21$
with $h_o = 1$ and $X(D+{}^3He) < 3\ 10^{-4}$ $\Omega_B \geq 0.01$
$0.01 \leq \Omega_B \leq 0.21$ corresponds to $3 \leq \eta_{10} \leq 15$

This means that by themselves the baryons cannot close the Universe. Apparently now there does not seem to be any compelling experimental reason for neutrinos to have a mass significant enough to compete with that of baryons (Mossbauer 1984). Therefore from that analysis on can predict that the Universe is open.

2 - The primordial abundance of helium puts constraints on the number of different neutrino families.

This can be seen on figure 5 but may be more conspicuously on figures 6a, 6b, 6c, from Olive et al (1981) where the number of allowed relativistic neutrinos N_ν appear to increase with Y at a rate of about 1 new neutrino family for any increase of 1% in Y.

This correlation between Y and N_ν comes from the fact that when N_ν increases the total density of the Universe increases too. The addition to this total density is

$$7/8\ g_f\left(\frac{T_f}{T_o}\right)^4 \rho_b$$ for any new fermion family f where

g_f is the statistical weight and T_f their temperature. There is than an increase by a factor of about 7/4 on the total density for each new neutrino family. As argued by Yang et al 1984 if $Y < 0.25$, $\eta_{10} < 7$ and if the neutron life time is about 10.6 ± 0.2 minutes then there should exist three neutrino flavours. This value of $N_\nu = 3$ makes the GUT lovers happy because there is then a strict correspondance between the 3 lepton and the 3 quark familes (Fayet 1984).

Therefore given the uncertainties on the primordial abundances of the light elements, the Standard Big Bang nucleosynthesis is a quite valid model which predicts a low baryon density such that $\Omega_B = 0.1\ ^{+0.11}_{-0.09}$ and three different types of neutrinos. As discussed by Delbourgo-Salvador et al (this volume), the uncertainties on the nuclear reaction rates do not affect significantly the present discussion. However one should be aware that this model works only under specific conditions :
a) the He abundance has to be such that $Y > 0.24$: in order to analyze the consequences of $Y < 0.24$ as suggested by Gautier and Owen (1983), in Audouze (1983) I used the theoretical curves ${}^7Li(\eta)$ (figure 7a) and He (η, N_ν, $\tau_{1/2}$) (figure 7b) provided by Schramm (1982). I deduced the η range from the Li observations of Spite and Spite 1982 (figure 7a) and reported it

on the He(η) (figure 7b). From this figure one can see that Y < 0.24 is barely consistent with N_ν = 3.

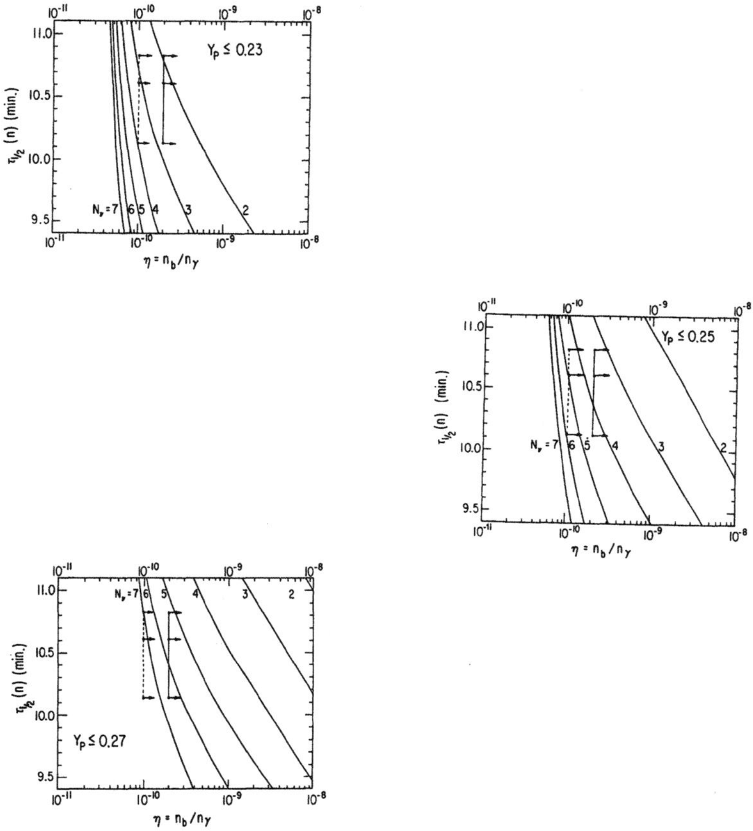

Figures 6a, 6b, 6c : Allowed number of different neutrino families for three upper limits of the primordial He abundances : Y $\leqslant$ 0.23 (fig. 6a) ; Y $\leqslant$ 0.25 (fig. 6b) and Y $\leqslant$ 0.27 (fig. 6c). The sketches are made after Olive et al (1981) in the plane ($\tau_{1/2}$, η) where $\tau_{1/2}$ is the neutron life time in minutes and η the baryon-photon density ratio. Solid lines correspond to $\eta \geqslant 2\ 10^{-10}$ (coming from the dynamics of binary systems or small groups of galaxies) while dashes lines correspond to $\eta \geqslant 10^{-10}$ which is the lowest upper limit coming from the mass-luminosity ratios determined in the Solar neighborhood. For Y $\leqslant$ 0.23 there is only room for 2 different types of neutrinos, for Y $\leqslant$ 0.25 (our prefered case) N_ν = 3 and for Y $\leqslant$ 0.27 N_ν = 5.

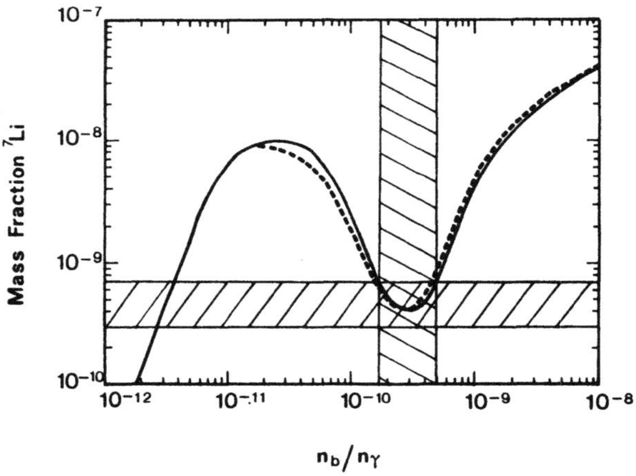

Figure 7a : Figure 7a shows the dependance of the calculated ^{7}Li abundance with the baryon-photon density ratio η on a plot drawn by Schramm (1982). One can deduce from it the allowed range for η from the Spite and Spite (1982) measurements concerning the halo stars (Audouze 1983).

b) The consideration of some contrasted models of chemical evolution of galaxies shows (Gry et al 1983) that if the models allows infall of external gas with primordial composition, the η ranges deduced from the abundances of He and D are strictly incompatible (figure 8). To make them compatible requires chemical evolution models where inflow of star processed gas takes place.

Although the Standard Big Bang nucleosynthesis is a quite attractive and simple model one should realize that it is not free of possible difficulties coming either from primordial abundances or from models of chemical evolution of galaxies. This is why it is legitimate to consider possible alternatives to this otherwise beautiful theory.

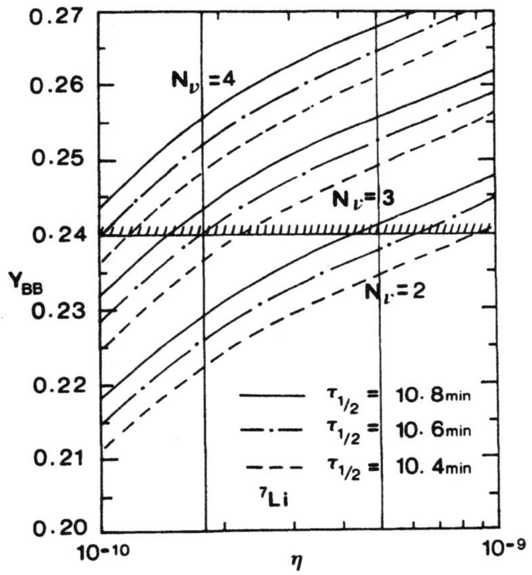

Figure 7b : on this figure 7b where the He abundance Y is plotted against η by Schramm (1982). One reports the η range deduced frome figure 7a for various values of N_ν and $\tau_{1/2}$ (respectively the number of neutrino families and the life time of the neutron). One can then notice that values of Y $\leqslant$ .24 are hardly compatible with N_ν = 3 (from Audouze 1983).

4 - Some competing models to synthetize the light elements

Many different works have attempted to propose models other than the Standard Big Bang nucleosynthesis (see Audouze 1984 for references). In this section I would like to provide a brief summary of two different proposals currently suggested by Audouze and Silk (1984). They concern 1) the possible partial photodisintegration of He to produce D by energetic photons coming from the decay of massive unstable neutrinos 2) the production of D during the pregalactic phase by spallation reactions.

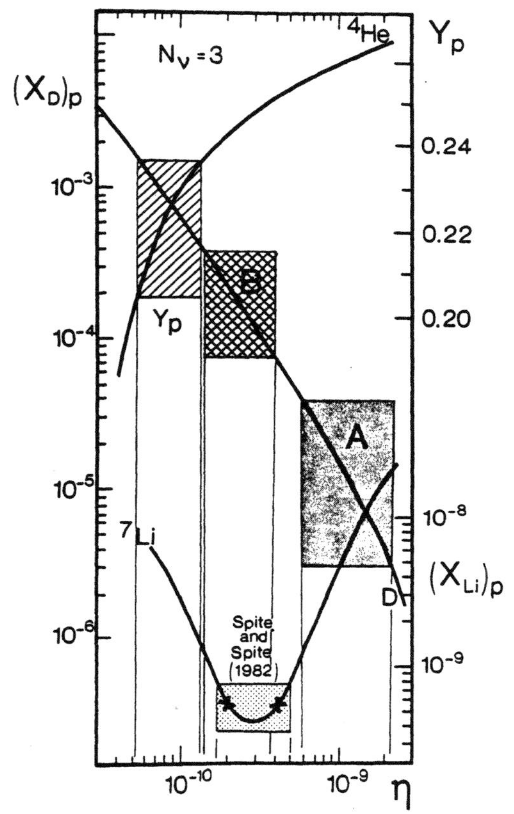

Figure 8 : Primordial abundances of D, ^{4}He and ^{7}Li as a function of η (the baryon over photon density ratio) and in the case of $N_\nu = 3$. The boxes come from the abundance ranges selected by Vidal-Madjar (1983). Box A (for D) corresponds to models of chemical evolution of galaxies with infall of external matter while Box B corresponds to models with inflow of star processed material (Gry et al 1983). Box A leads to η values deduced from D quite discrepant to those deduced from ^{4}He : in this case the Standard Big Bang nucleosynthesis is introuble. Models of chemical evolution in the inflow of star processed material might be the only one for which Standard Big Bang nucleosynthesis predicts consistent values for η together for D, ^{7}Li and ^{4}He.

4-1 Photodesintegration of He by energetic photons coming from massive neutrinos.

The existence of massive unstable neutrinos during the early phases of the Universe has already been proposed by several authors (see Cowsik 1981 for a review). As shown by Lindley (1979) these massive neutrinos in decaying release energetic photons which could photodisintegrate ^4He (if $50 < M_\nu < 250$ MeV) or D if ($10 < M_\nu < 50$ MeV). In an analysis inspired by that of Hut and White (1984) and illustrated in fig 9 we show that if the massive neutrinos have a life time τ such

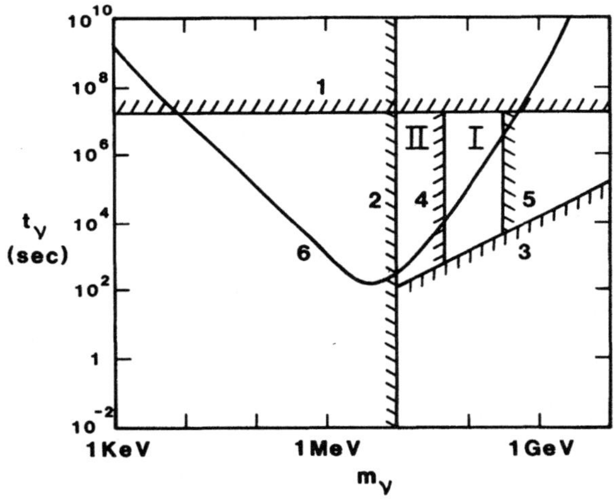

Figure 9 : Constraints on the mass M_ν and the life time $t_{1/2}$ of massive neutrinos which could have existed during the early phases of the Universe and decay into energetic photons which could either photodisintegrate He (region I) or D (region II). Line 1 corresponds to the thermalization of the decay photons in the background radiation ; line 2 is the limit of the mass of the unstable neutrinos coming from the supernova observations ; line 3 is the limit below wich the energetic photons are more likely to be transformed into e^+e^- pairs than to induce any photodesintegration reaction ; line 4 and 5 set the limits of the photon energy for which ^4He is more likely to be photodesintegrated than D. Finally line 6 corresponds to a photon density increased by a factor 3 with respect to the cosmic background radiation. (adapted from Hut and White 1983 and from Audouze and Silk 1984).

that $10^4 < \tau < 2 10^7$ sec and a mass M_ν such that $50 \leqslant M_\nu \leqslant 250$ MeV then the energetic photons coming from these massive neutrinos are able to photodisintegrate preferentially He than D (region 1 of fig 9). In this case the Big Bang may lead to a denser Universe where ^{4}He is formed during the Big Bang nucleosynthesis (but not D) while D is formed afterwards by these photodisintegration processes.

If the mass of neutrinos is such that $10 < E_\nu < 50$ MeV, with the same type of lifetime (region II of fig 9) then D is photodesintegrated by these energetic photons. One has then to consider models such as those of Rees (1983) where D can be formed in the matter surrounding black holes or by the model that we are considering now.

2- Pregalactic cosmic rays

The proposal according which D could be produced by spallation reactions during pregalactic phases of the Universe has already been presented by Epstein 1977 and Woltjer 1982. Our suggestion (already presented in Audouze and Silk 1983) is at variance of these early presentations in the sense that we have attempted to find ways to avoid the over production of ^{7}Li with respect to D by the He + He spallation reaction. Our hypothesis is as follows :

a) The Big Bang does not produce any light element : the pregalactic gas is then made of pure hydrogen.

b) From this pregalactic gas a first generation of massive (10^2-10^3 solar mass) stars is formed. As recalled during this symposium by Professor O. Godart, Mgr Lemaître was attracted by the idea of the existence of a first generation of massive stars born before the formation of the galaxies.

c) These massive stars release strong winds during their lifetime. The matter of these winds is enriched into He (and partially also into CNO). From these winds, the pregalactic cosmic rays are accelerated and lead to the spallation reaction He (pregalactic cosmic rays) + H (pregalactic gas → D + ... which can take place in principle in the absence of the He + He → Li reaction.

In Audouze and Silk (1984) the severe constraints acting against this proposal are discussed. While this process is energetically possible, it is only efficient for He pregalactic cosmic ray particles with energies > 300 MeV amu^{-1} which is a quite severe and may be lethal constraint regarding this proposal.

In order to keep the idea of a spallative origin for D, D.D. Clayton (private communication) had the exciting idea that pregalactic cosmic rays are accelerated before the formation of the first generation of stars and there might spall some of the He nuclei trapped in the atmospheres of these stars. The possible only difficulty with this entertaining idea is that the GCR proton flux might lead to an overproduction of gamma ray photons coming from the $p+p \rightarrow \pi_0 \rightarrow \gamma$ reactions.

5 - Conclusions

To end up this presentation made in honor of Mgr Lemaître let me borrow the following quotation coming from Georges Lemaître and reported by J. Barrow and J. Silk at the end of their book "The left hand of Creation" : Mgr Lemaître was asked like other prominent cosmologists what single question he would ask to an infallible oracle who could only answer by yes or no. Mgr Lemaître made this wise choice "I would ask the Oracle not to answer in order that a subsequent generation would not be deprived of the pleasure of searching for and finding the solution".

We are at about the same situation concerning the primordial nucleosynthesis : Standard Big Bang nucleosynthesis is certainly the most attractive possibility but nobody can swear that it will not encounter in a very near future some quite embarassing difficulties coming either from the primordial abundances of the light elements or their galactic evolution. We may not like much to-day to call on some magic massive neutrinos or some quite contrived pregalactic spallation processes. In cosmology the problems are sufficiently complex such that the next generations inspired by the example of scientists with a stature like that of Mgr George Lemaître may still have the pleasure of searching in problems as exciting as the primordial nucleosynthesis.

Acknowledgements

I am most grateful to the organizing committee and especially Professor A. Berger to have given me an opportunity to present my tribute to the memory of Mgr Georges Lemaître. This paper has been written during my visit at the Department of Astronomy of the University of California Los Angeles. I thank all the members of this department for their hospitality and especially Professor Michael Jura who kindly arranged this visit.

I am enjoying a very pleasant collaboration with Professor Joseph Silk whom I thank for some remarks on the draft of this manuscript. I am finally indebted to Mrs Brigitte Sockeel for her very skilled typing of this paper.

References

Audouze, J., 1983 in Primordial Helium eds PA Shever, D. Kunth and K. Kjar, ESO Garching p.3

Audouze, J., 1984 in Large Scale Structure of the Universe, Cosmology and fundamental Physics ed. L. Van Hove, CERN report to be published

Audouze, J., Boulade, O., Malinie, G., and Poilane, Y., 1983 Astron. Astroph. 127, 164

Audouze, J., and Silk, J., 1983 in Primordial Helium eds P.A. Shaver, D. Kunth and K. Kjar ESO Garching p.71

Audouze, J., and Silk, J., 1984 to be submitted to Astrophys. J.

Barrow, J.D. and Silk, J., 1983 The left hand of creation. Basic Books New York

Black, D.C., (1972) Geochim. Cosmochim. Acta 36, 347

Cowsik, R., 1981 in "Cosmology and particles" ed. J. Audouze, P. Crane, T. Gaisser, D. Hegyi and J. tran Thanh Van Editions Frontières p

Delbourgo-Salvador P., Vangioni-Flam, E., Malinie, G., and Audouze, J., (1984) this volume

Epstein, R.I. Astrophys.J. 212, 595

Fayet, P., 1984 to appear in Large Scale Structure of the Universe, Cosmology and Fondamental Physics ed L. Van Hove CERN Report

Gautier, D., and Owen, T., 1983 Nature 302, 215

Geiss, J., and Reeves, H., (1972) Astron. Astroph. 18, 126

Gry, C., Malinie, G., Audouze, J., and Vidal-Madjar, A., 1983 in evolution of galaxies and large structures in the Universe, eds J. Audouze and J. Tran Thanh Van p.279

Hut, P., and White S.D.M. 1984 to be published

Kunth, D., and Sargent, W.L.W. 1983 Astrophys.J. 273, 81

Laurent, C., (1983) in Primordial Helium eds P.A. Shaver, D. Kunth and K. Kjar, ESO-Garching p.335

Lindley, D., 1979 Month.Notices of the Roy. Astron. Soc. 188, 15p

Mossbauer R.L. 1984 in Large Scale Structure of the Universe, Cosmology and fundamental Physics ed L. Van Hove, CERN Report, to be published.

Olive, K.A., Schramm D.N., Steigman, G., Turner M.S. and Yang, J., (1981) Astrophys.J. 246, 557

Rees M. 1983 in Evolution of galaxies and large structures of the Universe eds J. Audouze and J. Tran Thanh Van Reidel-Dordrecht p.271

Schramm, D.N. 1982 Phil trans. Roy. Soc. London A 307, 36

Spite, F., and Spite , M., 1982 Astron. Astroph. 115, 357

Vidal-Madjar, A., and Gry, C., 1983 Astron. Astroph. (in press)

Vidal-Madjar, A., 1983 in Diffuse matter in Galaxies eds J. Audouze, J. Lequeux, M. Levy and A. Vidal-Madjar Reidel-Dordrecht p.57

Wagoner R.V. 1973 Astrophys.J. 179, 343

Wilson, T.L. Rood R.T. and Bania T.M. (1983) in Primordial Helium eds P. Shaver, D. Kunth and K. Kjar, ESO Garching p.

Woltjer, L., 1982 in Astrophysical cosmology eds H.A. Brück, G. Coyne and M.S. Longair, Pontificia Academiae Scientarum Scripta Varia p. 423

Yang, J., Turner, M.S., Steigman, G., Schramm D.N., and Olive K.A., 1984 Astrophys.J. (in press)

A GENERALIZATION OF THE LEMAITRE MODELS

Andrzej Krasiński

N. Copernicus Astronomical Center, Polish Academy of Sciences, Bartycka 18, 00716 Warszawa, Poland

Abstract A generalization of the Friedman-Lemaitre -Robertson-Walker (FLRW) models is obtained by weakening the assumptions under which they are derived from Einstein's relativity. It is assumed that each section t = const is homogeneous and isotropic while the space-time itself not necessarily has any symmetry. The resulting Stephani Universe has an undetermined function of time in place of the constant curvature index k. In this Universe, some spatial sections may be open while others will be closed. Its geometrical picture is presented and its physical properties are discussed.

1. GENERALIZATIONS ARE COMPATIBLE WITH OBSERVATIONS.

The Friedman-Lemaitre-Robertson-Walker (FLRW) solutions of Einstein's field equations (1 - 4) were derived under the very strong assumptions that the space-time is homogeneous and isotropic. These assumptions were not meant to reflect our knowledge about the Universe, but rather our ignorance: at that time (1930-ies) no structures larger than galaxies were known. The homogeneous and isotropic distribution of galaxies was thus a reasonable first hypothesis which at the same time made Einstein's equations tractable.

The models proved successful in describing several observable properties of the Universe, like Hubble's expansion law, the abundance of helium or the microwave background radiation. These successes are often understood as confirmations of the underlying assumptions.

A. Berger (ed.), The Big Bang and Georges Lemaitre, 63–73.

In fact, they only confirm that the Universe was much hotter and denser in the past than it is now and that it was very nearly isotropic at the time when the radiation last interacted with massive particles. Even within these classical models massive particles and radiation are considered as two independent components of matter which decoupled in the moment of last scattering and later evolved independently. Whatever happened to particles afterwards, did not affect the distribution of radiation. Therefore, the isotropy of radiation does not force upon us a model in which all matter is distributed so symmetrically.

Once isotropy is given up, the requirement of homogeneity is not compelling anymore. The Universe is assumed homogeneous because it would be unnatural if it were spherically symmetric only around us (5), but if it is not spherically symmetric at all, it might be inhomogeneous as well. This statement does not speak against the copernician philosophy. According to it, no place in the Universe should be preferred. This does not mean that all the places in the Universe should be exactly identical. The latter assumption fulfills the former, but is much stronger (see also Ref. 6).

A purely theoretical argument also shows that more general models of the Universe can be reconciled with the existing data (cf Fig. 1, in Ref. 7). Only the events lying on our past light cone are directly observed, and only directions to them can be measured with a satisfactory precision. All other data needed to calculate the spatial distribution of matter are inferred therefrom through a model-dependent procedure: 1. Through each event on the light cone we draw a world line representing the history of that portion of matter, e.g. a galaxy (the equations of those lines can only be calculated given a specific class of spacetimes); 2. Through the vertex of the light cone we draw a hypersurface S of events simultaneous with "now" (Even within a fixed model this depends on the reference system chosen. The reference system is usually attached to a physical structure in the spacetime, e.g. the congruence of matter world-lines); 3. The points of intersection of the world-lines with the hypersurface S represent the positions of the galaxies now (These positions depend on the slopes of the matter world-lines, i.e. on the velocity of expansion, given the model and given S. This velocity can be calculated from the observed redshift - provided we know precisely what part of the redshift is of cosmological origin). 4. Only now can we calculate the spatial distribution of matter. Thus a model is assumed before any observations are taken into account. It can be confirmed or refuted by these observations, but is in no way implied by them (see also Refs 6, 8, 9).

2. GENERALIZATIONS ARE IN FACT NECESSARY.

According to present data, galaxies are grouped into clusters and shells surrounding voids which contain no visible matter at all (10). Thus the Universe might possibly be homogeneous only on still larger scales (if at all). Such a large scale homogeneity coupled with small scale inhomogeneity is not properly described by a spacetime with a continuous transitive group of symmetry (curve a in Fig. 1) of which the FLRW spacetimes are examples. A more appropriate description would be a spacetime with a <u>discrete</u> group of symmetry in which matter density would be given by a function like curve b in Fig. 1 (see similar remarks by Ellis (6)). Such distribution of matter does not distinguish any single observer because, if the space is infinite, there exist infinitely many identical copies of any chosen finite portion of matter distributed regularly. Such a solution can only be found if the assumption of continuous homogeneity is relaxed altogether.

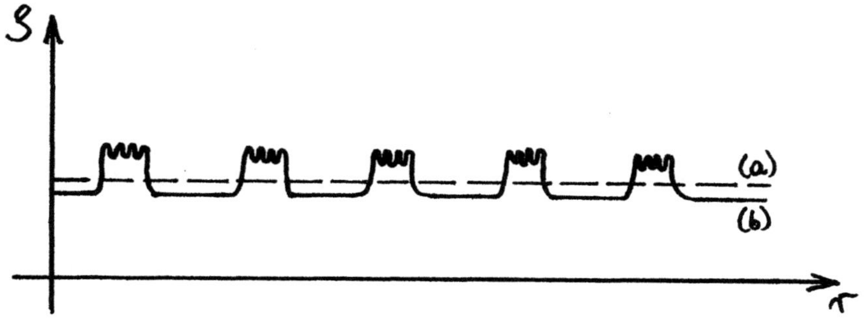

Fig 1. Matter density vs position in a 3-space that is homogeneous with respect to (a) a continuous group of symmetry, (b) a discrete group of symmetry.

Moreover, the FLRW models taken literally tell us that no galaxies may ever have formed out of a homogeneous and isotropic background. All theories of galaxy formation must consider perturbations of the FLRW models (see e.g. (11)). If we have to go beyond the FLRW models, it is equally reasonable to consider exact generalizations instead of approximate perturbations.

This author is in a definite minority, but not alone with his criticism of "standard cosmology". Similar concerns were expressed by Ellis (6, 8, 9, 12-14), MacCallum (15) and Mashhoon (16, see also this volume).

Since the FLRW models proved so successful, the more general new models should contain them as special cases, i.e. as first approximations. This paper will show how a certain generalization results if the assumptions underlying the FLRW models are slightly relaxed. This generalization does not go sufficiently far in order to be free from the above mentioned weaknesses. Its existence proves however that more general solutions can still be reasonably simple.

3. ASSUMPTIONS.

The considerations of the previous sections show that what is checked against astronomical observations is the 3-dimensional space t = now rather than the whole spacetime. It is then a natural question, to what extent the 3-geometries of the spaces t = const determine the 4-geometry of our spacetime. Let us assume, as is commonly done, that:

1. Each 3-space t = const is homogeneous and isotropic,
2. The spaces are orthogonal to the family of t-coordinate lines,
3. Matter moves along the t-lines,
4. The Einstein's field equations are fulfilled, the source being a perfect fluid,

but let us consider the possibility that:

5. The spacetime not necessarily has any symmetry.

4. THE SOLUTION.

The assumptions 1 to 5 produce the following solution of the Einstein's equations (17):

$$ds^2 = D^2 dt^2 - (R^2/V^2)(dx^2 + dy^2 + dz^2), \tag{4.1}$$

$$D = F \frac{R}{V} \frac{\partial}{\partial t}\left(\frac{V}{R}\right), \tag{4.2}$$

$$V = 1 + \frac{1}{4} k \{(x - x_0)^2 + (y - y_0)^2 + (z - z_0)^2\} \quad (4.3)$$

$$k = (C^2 - 1/F^2)R^2, \quad (4.4)$$

$$\kappa\varepsilon = 3C^2, \quad (4.5)$$

$$\kappa p = -3C^2 + 2C \frac{dC}{dt} \frac{V}{R} \Big/ \frac{\partial}{\partial t}\left(\frac{V}{R}\right), \quad (4.6)$$

$$\left.\begin{aligned} a^0 &= 0, \\ a^i &= (V^2/DR^2)D_i, \qquad i = 1, 2, 3. \end{aligned}\right\} \quad (4.7)$$

where F, R, C, x_0, y_0, z_0 are arbitrary functions of time, ε is the energy density, p is the pressure, and a is the acceleration field of the fluid flow.

This solution was first found by Stephani (18) in 1967, but was not investigated from the point of view of cosmology.

5. LOCAL PROPERTIES OF THE SOLUTION.

The solution has in general no symmetry at all. Its most striking property is the fact that k is a function of t, the sign of k being not determined. Since k is the curvature index of the 3-spaces t = const, one sees that in this spacetime some spacelike sections have positive curvature (and so should be closed) while some others have negative or zero curvature (and so should be open). Other differences with the FLRW solutions are the following:

1. Matter moves with acceleration, i.e. not on geodesic lines.

2. The equation of state is not of the form $\varepsilon = \varepsilon(p)$, but depends on the position in the space: $p = p(\varepsilon, x, y, z)$.

This last property means that a single thermodynamic function of state (e.g. pressure) does not suffice to describe matter in this model, at least one other function is necessary, e.g. temperature which would have different values in different places.

The Stephani Universe reduces to a FLRW model when any one of the following situations occurs:

1. The functions x_0, y_0, z_0 and k are constant.

2. The acceleration field vanishes (i.e. matter moves on geodesics).

3. The equation of state is of the form $\varepsilon = \varepsilon(p)$, i.e. it does not depend on position.

This solution is conformally flat, and moreover it is the most general conformally flat solution with a perfect fluid source and nonvanishing expansion (19).

6. GLOBAL PROPERTIES OF THE STEPHANI UNIVERSE.

Stephani has shown (18) that this solution can be embedded in a flat five-dimensional space. To construct the embedding explicitly would in general be too difficult because of the 6 arbitrary functions of time. It was more instructive to study a special case in which the embedding could be performed explicitly.

Such a special case results when C = const; the Stephani Universe reduces then to the deSitter solution. It was further assumed $x_0 = y_0 = z_0 = 0$, R = const, k = -t. In the case C = const these additional assumptions amount just to a choice of a simpler coordinate system (foliation).

The deSitter manifold is then a 4-dimensional one-sheet hyperboloid embedded in a 5-dimensional pseudoeuclidean space. The metric form of the 5-space is:

$$dS^2 = dZ^2 - dX^2 - dU^2 - dW^2 - dY^2 , \tag{6.1}$$

while the equation of the deSitter hyperboloid is:

$$Z^2 - X^2 - U^2 - W^2 - (Y - 1/C)^2 = -1/C^2 , \tag{6.2}$$

or, in parametric form:

$$Z = R (x^2 + y^2 + z^2) (C^2 R^2 + t)^{1/2} / 2V \tag{6.3}$$

$$(X, U, W) = (R/V) (x, y, z) \tag{6.4}$$

$$Y = CR^2 (x^2 + y^2 + z^2) / 2V \tag{6.5}$$

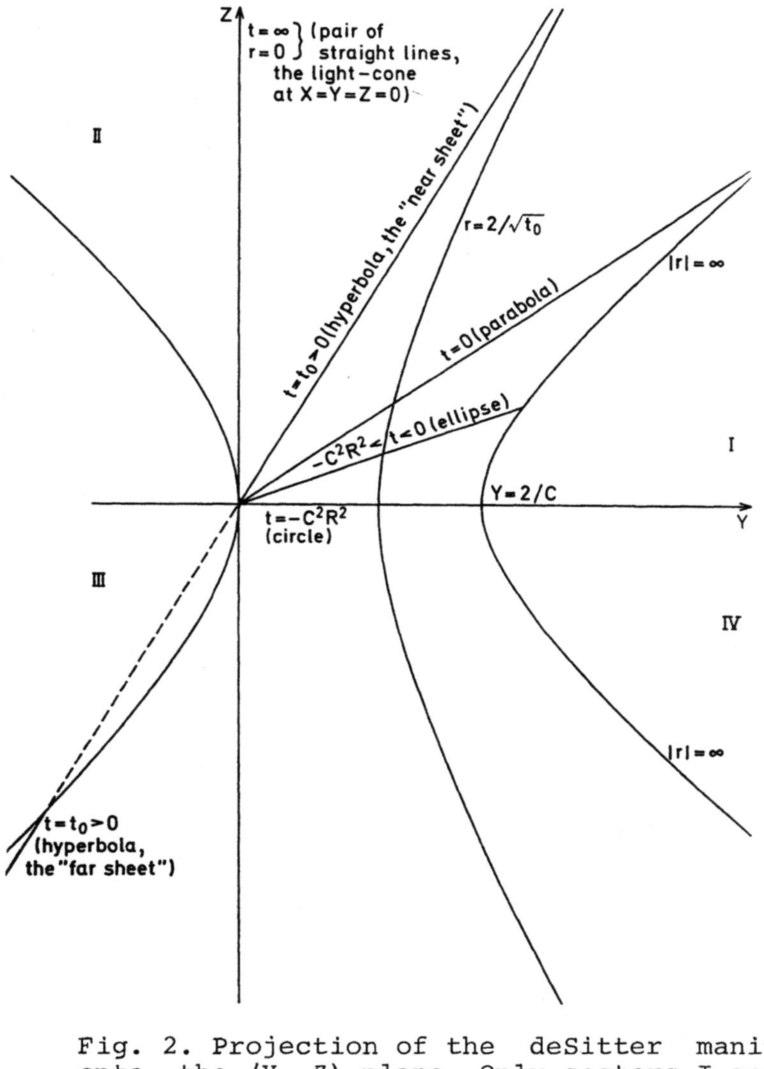

Fig. 2. Projection of the deSitter manifold onto the (Y, Z) plane. Only sectors I and IV are covered by the parametrization (6.3) - (6.5). See text for more remarks. (Adapted from Ref. 20 with the permission of the Plenum Publishing Corporation).

The projection of the hyperboloid (6.2) onto the (Y, Z) plane in the space (6.1) is shown in Fig. 2. On the figure one can see that a spacetime of a simple topology can result from the foliation introduced in sec. 3. The sections t = const of the spacetime are intersections of the hyperboloid (6.3) - (6.5) with the hyperplanes Z/Y = const. They all contain the (X, U, W) space (the X axis in Fig. 2) and their tilt to the (X, U, W, Y) hyperplane (the (X, Y) plane in Fig. 2) is determined by k(t) = -t. With $-C^2R^2 \leq t < 0$ we have $k(t) > 0$ and the tilt of the t = const hyperplanes is such that their intersections with the hyperboloid (6.2) are 3-ellipsoids (ellipses in Fig. 2) - closed spaces of positive curvature (in the special case $t = -C^2R^2$ it is a 3-sphere). With t = 0 we have k = 0 and the intersection is a 3-paraboloid (a parabola in Fig. 2) - the flat space. With t > 0 (k < 0) the intersections are two-sheet hyperboloids (hyperbolas in Fig. 2) - open spaces of negative curvature.

Fig. 2 faithfully represents not only the topology of the general Stephani solution, but also several details of its geometry - more than would be worth discussing in this place (see Ref. 20). Only the singularity at x = y = z = 0 seen on Fig. 2 looks differently in the general case. It is then a true curvature singularity, and it occurs at different values of (x, y, z) for every t. It is an additional singularity to the one predicted by the Hawking-Penrose theorems (21) which occurs also in the FLRW models. The additional singularity can be avoided when the functions k(t) and R(t) and their derivatives obey certain inequalities (20). If k(t) > 0 for all t, then the inequalities can be readily fulfilled. Otherwise, they imply that pressure must be negative somewhere. This, in turn, can only be avoided by matching the Stephani Universe to an empty space solution. In any case, however, the weak energy conditions of Hawking and Ellis (21), $\varepsilon \geq 0$ and $\varepsilon + p \geq 0$, can be fulfilled.

7. IN WHAT SENSE IS THIS UNIVERSE HOMOGENEOUS?

The pressure and acceleration scalar do depend here on spatial coordinates. On the other hand, we assumed in sec. 3 that all the 3-spaces t = const should be intrinsically homogeneous. Is this a contradiction?

No - because pressure and acceleration are not intrinsic properties of these 3-spaces. They are fields defined on the 4-dimensional spacetime (or on 4-dimensional subsets thereof). As such, they have well defined values over the spaces t = const. These values, however, can never be calculated if we are given only the geometry of a single 3-space t = const - they are determined by the whole 4-dimensional metric tensor through the Einstein´s field equations. The theory of relativity is telling us here, in its own language, the message known from statistical physics: it is impossible to determine pressure (in any kind of matter) by an instantaneous measurement. The measurement must always take a finite time (the pressure must be defined over a continuous family of t = const spaces), and only afterwards can we determine momentary values of pressure - as limits at $\Delta t \to 0$ of mean values over time-intervals Δt. This fits with the microscopic definition of pressure - as the mean momentum transferred by the gas particles to a unit surface in a unit of time.

Let us consider a more general spacetime in which the 3-spaces t = const are orthogonal to the t-lines, but have arbitrary intrinsic geometries:

$$ds^2 = D^2 dt^2 - h_{ij} dx^i dx^j , \tag{7.1}$$

where i, j = 1, 2, 3 and all the functions (D, h_{ij}) are arbitrary. Let us assume this metric fulfills the Einstein´s field equations with a perfect fluid source whose velocity field is tangent to the t-lines. The density of matter can then be calculated to be

$$\kappa\varepsilon = R(h)/2 + \{(h^{ij} h_{ij},_t)^2 + h^{ij},_t \, h_{ij},_t\}/8D^2 \tag{7.2}$$

where R(h) is the 3-dimensional scalar curvature of the metric h_{ij}. Eq. (7.2) shows that also matter density need not be spatially homogeneous when h_{ij} is. It happens to be so for the Stephani Universe by accident (and for the Bianchi type models by assumption).

8. IS THE STEPHANI MODEL COMPATIBLE WITH OBSERVATIONS?

Since the FLRW models are contained in this one as special cases, and are themselves believed to be good models of the observed Universe, the answer to the question asked above is immediate: yes, the functions

$k_0(t)$, $x_0(t)$, $y_0(t)$ and $z_0(t)$ can always be chosen to vary so slowly that no observation can distinguish them from constants to which they reduce in the FLRW limit. This statement raises a further question: what are the limits imposed by observations on the derivatives of these functions? This will be a subject of a separate study. An ultimate question is however: can the Stephani model describe anything that the FLRW models could not? The calculations in it are undoubtedly more involved, so does it pay off to use it?

To the author, it was interesting to learn that the classification of cosmological models into the open, the flat and the closed one is not required by Einstein's theory itself, but is an artifact of the very strong symmetry requirements imposed on the FLRW models a priori. The Stephani model had thus at least this conceptual advantage. Whether it has any others, remains to be seen. Further generalizations are needed in any case, since, with spatially homogeneous matter-density, the model cannot serve to describe the galaxy formation in a nonperturbative manner.

REFERENCES

(1) A. A. Friedman, Z. Physik 10, 377 (1922); 21, 326 (1924).
(2) G. Lemaitre, Ann. Soc. Sci Bruxelles 47A, 19 (1927); Mon. Not. Roy. Astr. Soc. 91, 483 (1931).
(3) H. P. Robertson, Proc. Nat. Acad. Sci 15, 822 (1929); Rev. Mod. Phys. 5, 62 (1933).
(4) A. G. Walker, Quart. J. Math. Oxford 6, 81 (1935).
(5) P. J. E. Peebles, Lecture given at this meeting.
(6) G. F. R. Ellis, Relativistic cosmology: its nature, aims and problems (invited talk at the GR10 meeting, Padua, July 1983). Preprint of the Max Planck Institute, Garching/Munich 1983.
(7) A. Krasiński, in: The birth of the Universe. Proceedings of the 17th Rencontre de Moriond. J. Audouze and J. Tran Tranh Van (eds.). Editions Frontieres, Gif sur Yvette 1982, p. 15.
(8) G. F. R. Ellis and J. Perry, Mon. Not. Roy. Astr. Soc. 187, 357 (1979).
(9) G. F. R. Ellis, Ann. N. Y. Acad. Sci. 336, 130 (1980).
(10) S. A. Gregory, L. A. Thompson, Astrophys. J. 222, 784 (1978); 243, 411 (1981); see also Sci. Am. 246, 88 (March 1982).

(11) B. J. T. Jones, in : The birth of the Universe, p. 215. See Ref. 7.
(12) G. F. R. Ellis, Quart. J. Roy. Astr. Soc. 16, 245 (1975).
(13) G. F. R. Ellis, R. Maartens, S. D. Nel, Mon. Not. Roy. Astr. Soc. 184, 439 (1978).
(14) G. F. R. Ellis, Gen. Rel. Grav. 11, 281 (1979).
(15) M. A. H. MacCallum, Relativistic cosmology for astrophysicists. Lectures from the 7th School on Cosmology and Gravitation, Erice 1981.
(16) B. Mashhoon, M. Hossein Partovi, Toward verification of large scale homogeneity in cosmology. Preprint SLAC 1982.
(17) A. Krasiński, Gen. Rel. Grav. 13, 1021 (1981).
(18) H. Stephani, Commun. Math. Phys. 4, 137 (1967).
(19) D. Kramer, H. Stephani, E. Herlt, M. MacCallum, Exact solutions of Einstein's field equations. Cambridge University Press 1980, p. 371 (theorem (32.15)).
(20) A. Krasiński, Gen. Rel. Grav. 15, 673 (1983).
(21) G. F. R. Ellis, S. W. Hawking, The large scale structure of the spacetime. Cambridge University Press 1974.

IS THE UNIVERSE HOMOGENEOUS ON A LARGE SCALE?

Bahram Mashhoon

Institut für Theoretische Physik, Universität zu Köln
D-5000 Cologne 41, Federal Republic of Germany

ABSTRACT. The present standard model of cosmology is based on the cosmological principle which has only limited observational support, especially in connection with the issue of large-scale homogeneity. The recent discovery of voids provides further impetus for the study of large-scale inhomogeneities. It is proposed to replace the hypothesis of spatial homogeneity by the assumption that the (matter and radiation) content of the universe is on the average uniform, i.e., the equation of state of the system is everywhere the same. It has been shown that if (a) the universe is spatially isotropic, (b) the content of the universe is approximated on the average by a perfect fluid obeying a physically reasonable equation of state $p = p(\mu)$, including $p \equiv 0$, and (c) the expansion or contraction of the universe is shear-free, then the only physically acceptable nonstatic cosmological solution of the Einstein-Maxwell equations is the Friedmann-Lemaître-Robertson-Walker (FLRW) universe. If (c) is relaxed, then Einstein's equations allow solutions which differ from the FLRW models by the existence of radial inhomogeneities due to shear. As a first step toward a general study of inhomogeneities, local models with radial inhomogeneities have been developed and the observational quantities for such models have been determined.

I. INTRODUCTION

To interpret cosmological observations, it is necessary to have knowledge of the cosmic gravitational field which, in turn, must be determined from the distribution and motion of matter over cosmological scales. The distribution of luminous matter in the universe appears quite nonuniform. The recent discovery of huge

A. Berger (ed.), The Big Bang and Georges Lemaitre, 75–81.

voids in the distribution of galaxies has led astronomers to contemplate that the galaxies may form cellular patterns rather than disjoint clumps (see Shandarin and Zeldovich, 1983). On the other hand, the present standard model of cosmology is based on the Friedmann-Lemaître - Robertson-Walker (FLRW) cosmological models which embody the hypothesis that the universe is homogeneous and isotropic on a large enough scale. The remarkable isotropy of the microwave background radiation, the isotropy of the X-ray background, and the isotropic distribution and recession of galaxies and radio sources provide impressive evidence in favor of the hypothesis of isotropy of the universe. The evidence in support of spatial homogeneity is not as strong, however. The problem of determination of cosmic distances is at the root of the difficulties that the verification of the hypothesis of spatial homogeneity has encountered. The observational limits on anisotropy place severe restrictions on any significant inhomogeneities that might exist, except, of course, for radial inhomogeneities (Barrow, Juszkiewicz and Sonoda, 1983). There is no reason to suppose that our cosmic neighborhood is in a central position with respect to the distribution of matter in the universe. On the other hand, significant local inhomogeneity might exist which could well influence cosmological observations. It should be emphasized that the assumption of homogeneity is strictly valid if the averaged-out content of the universe is scattered uniformly throughout space. But the averaged-out distribution may in fact be strongly inhomogeneous locally. This can happen, for instance, if the prevailing structure in the universe is of network type.

To investigate the possibility of existence of local large-scale inhomogeneities, it is proposed to replace the geometrical hypothesis of spatial homogeneity by the assumption of uniformity of the nature of matter that occupies space. For simplicity, the averaged-out content of the universe will be represented by a perfect fluid with proper energy density μ and pressure p. The new hypothesis of uniformity then implies that the matter satisfies the same equation of state $p = p(\mu)$, or $p \equiv 0$, everywhere.

The pioneering studies of the inhomogeneous cosmological models are due to McCrea and McVittie (1930), Lemaître (1931) and Tolman (1934). As a first step toward the study of general inhomogeneous spacetimes, it proves interesting to focus attention on locally isotropic models. To develop local models incorporating radial inhomogeneities and based on assumptions that are consistent with observations, nonstatic and isotropic solutions of the field equations will be considered for a perfect fluid satisfying an equation of state. The general solution of Einstein's equations satisfying these conditions is not known. The problem is considerably simplified, however, if it is assumed that the motion of matter has no shear, i.e., the rate of expansion (or contraction) is the same along the lateral and radial directions so that an infinitesimal sphere remains a sphere during the motion. On the other hand, shear would cause an infinitesimal sphere to become a spheroid as the universe expands (or contracts). It has been shown (Mashhoon and Partovi, 1980) that the only nonstatic shear-free solutions of the Einstein-Maxwell equations which are isotropic, satisfy an equation of state of the form $p = p(\mu)$, or $p \equiv 0$, and are physically reasonable (i.e., $\mu \geq 0$, $p \geq 0$, and $\mu \geq 3p$ or $\mu \geq p$) are the FLRW models. This result has been recently partially generalized (Collins and Wainwright, 1983; Mashhoon and Partovi, 1984): For neutral matter the assumption of isotropy may be replaced by the weaker hypothesis of irrotational motion and still the FLRW class remains the only physically reasonable solution. It follows from these results that in physically reasonable cases radial inhomogenities are associated with shearing motions. Shear inhomogeneities are therefore considered in the next section and their influence on cosmological observations are studied.

II. SHEAR INHOMOGENEITIES

Very little is known about exact isotropic solutions with shear; the Tolman solutions for dust provide the best known examples. If the metric of an isotropic spacetime is expressed as

$$-ds^2 = -a^2(t,r)dt^2 + b^2(t,r)dr^2 + R^2(t,r)d\Omega^2, \tag{1}$$

in comoving coordinates, then the shear is proportional to

$$\sigma(t,r) = \frac{1}{a}\left(\frac{1}{b}\frac{\partial b}{\partial t} - \frac{1}{R}\frac{\partial R}{\partial t}\right) . \tag{2}$$

In the case of dust, the shear must depend on the radial coordinate if inhomogeneities are to exist. For a general equation of state $p = p(\mu)$, the general solution of the field equations is not known. To simplify the problem, one may impose constraints on the functions a,b, and R and search for possible solutions. To this end, one may assume, e.g.,

$$\frac{1}{ab}\frac{\partial b}{\partial t} = h(t) \tag{3}$$

as an expression of a generalized Hubble law. With this assumption the comoving coordinate condition may be integrated once to yield

$$\frac{1}{a}\frac{\partial R}{\partial t} - h(t)R = s(t) \tag{4}$$

where s(t) is proportional to the shear. Among the Tolman solutions for dust, the condition (3) holds only for the homogeneous (FLRW) solutions. Various attempts to find simple shearing solutions satisfying (3) for the equation of state $\mu = 3p$ have been unsuccessful. Thus no isotropic solution of the field equations for radiation in equilibrium is known at present which contains shear inhomogeneities. Static solutions for radiation in equilibrium have been considered by Klein (1947). Moreover, the only inhomogeneous solution of the form $a = a(r)$ and $b = b(t)$ for an equation of state $\mu = (\gamma - 1)p$ with constant γ exists when $\gamma = 2$ and b is a constant. This solution has been studied by Wesson (1978).

In the absence of a general solution and in view of the fact only local inhomogeneities are of interest, expansions of the functions a,b, and R will be considered with respect to the radial coordinate r around the center of symmetry (r = 0). Einstein's equations for a perfect fluid with a reasonable equation of state imply (Partovi and Mashhoon, 1984)

$$a = 1 + \frac{1}{2}\alpha r^2 + \ldots \quad , \tag{5}$$

$$b = (1 + \frac{1}{2}\beta r^2 + \ldots)\ S(t) \quad , \tag{6}$$

and

$$R = r(1 + \frac{1}{2}\gamma r^2 + \ldots)\ S(t) \quad , \tag{7}$$

where α, β, and γ are functions of time only and the shear is given by

$$\sigma = \frac{1}{2}\frac{d}{dt}(\beta - \gamma)\ r^2 + \ldots \quad . \tag{8}$$

The comoving coordinate condition implies that

$$\frac{d}{dt}(\beta - 3\gamma) = -2H\alpha \quad , \tag{9}$$

where $\beta - 3\gamma$ is the spatial curvature and H is the Hubble parameter

$$H = \frac{1}{S}\frac{dS}{dt} \quad . \tag{10}$$

The expansion of the metric quantities has been considered only to second order in the radial coordinate since in our local analysis the expansion of the observational quantities in terms of the redshift parameter z is of interest only to second order ($z < 1$). In general, the leading terms in the expansion of the observational quantities in z are the same as in the FLRW models and the higher-order terms are modified by the presence of inhomogeneities. For instance the luminosity distance is given by

$$d_L = \frac{z}{H} + \frac{1}{2H}(1 - q - C)\, z^2 + \ldots \quad , \tag{11}$$

where

$$C = -\frac{\alpha}{(SH)^2} \tag{12}$$

is the inhomogeneity parameter and q,

$$qH^2 = -\frac{1}{S}\frac{d^2S}{dt^2} \quad , \tag{13}$$

is the usual deceleration parameter. Hence the physical parameter that would be determined from the luminosity distance-redshift relation, taking due account of the difficulties associated with source evolution, is $q + C$. On the other hand

$$q - C = \frac{1}{2}\left[\Omega(1 + 3\frac{p}{\mu})\right]_{r = 0} \quad , \tag{14}$$

where Ω is the density μ in units of the "closure" density $3H^2/(8\pi)$. It follows from (14) that

$$C < q \leq C + \Omega_o \tag{15}$$

if the equation of state satisfies the condition $\mu \geq 3p$. One can give a similar analysis for the $p \equiv 0$ case (Partovi and Mashhoon, 1984).

The absence of any firm observational upper limit on the shear inhomogeneity in our local cosmic neighborhood implies that significant local deviations from the FLRW models could exist. According to a singularity theorem (cf. Hawking and Ellis, 1973), the existence of a trapped surface in the universe leads via the gravitational field equations (and certain other reasonable conditions) to the prediction of a singularity in the spacetime. Trapped surfaces exist in FLRW models; in fact, the physical radius of the apparent horizon R_{AH} is greater than, equal to, or less than the Hubble radius H^{-1} for the spatially open, flat or closed model, respectively. If the universe is sufficiently homogeneous over length scales of the order of the Hubble radius, then a singularity must exist (e.g., an initial singularity). However, the possibility of existence of shear inhomogeneities can vitiate this argument.

III. NEWTONIAN COSMOLOGY

The meaning of shear-free motion, upon which the previous analysis is based, can be elucidated further by considering its Newtonian analog. The absence of isotropic shear, which leads to the uniqueness property of FLRW solutions, corresponds in the Newtonian theory to homologous motion. Various investigations of gravitational collapse in Newtonian theory have indicated that homologous motion develops in the late stages of collapse (cf. Cohn, 1980). A consistent treatment of the isotropic and homologous motion of a perfect fluid in Newtonian theory leads to the conclusion that the only cosmological (i.e., unbound) solution with an equation of state $p = p(\mu)$, or $p \equiv 0$, is the standard homogeneous solution of Newtonian cosmology. Thus the uniqueness property of FLRW solutions has an exact Newtonian analog. It is important to recognize, however, that the Newtonian result is not a weak-field limit of the relativistic theory. To clarify the relationship between these cosmological theories, it is useful to consider an immediate consequence of the principle of equivalence, namely that a weak-field approximation to any gravitational field may be obtained over a restricted spacetime domain except at a singularity of the gravitational field. This can be illustrated for the FLRW model,

$$-ds^2 = -dt^2 + S^2(t) f^{-2}(r)(dr^2 + r^2 d\Omega^2), \tag{16}$$

where the fundamental observers follow geodesics. Here

$$f(r) = 1 + \frac{1}{4}kr^2 \qquad , \tag{17}$$

and $k = +1$, 0, or -1 for the closed, flat or open models. To describe local observations, the observer can set up a local Fermi coordinate system (t,X,Y,Z) based upon the natural system of parallel-transported tetrads $\lambda^{\mu}{}_{(\alpha)}$. Thus in the (t,r,θ,ϕ) coordinate system, the only nonzero tetrad components of $\lambda^{\mu}{}_{(\alpha)}$ are those with $\mu = \alpha$. Near the observer, the spacetime deviates only slightly from the Minkowski spacetime and the metric is given to second order in spatial Fermi coordinates by

$$-ds^2 = -(1+qH^2\rho^2)dt^2 + d\rho^2 + \left[1-\frac{1}{3}(H^2+kS^{-2})\rho^2\right]\rho^2 d\Omega^{*2}, \tag{18}$$

where spherical coordinates (ρ,θ^*,ϕ^*) have been introduced in the Fermi frame,

$$\rho^2 = X^2 + Y^2 + Z^2 \qquad , \tag{19}$$

etc., and H and q have their usual meaning in terms of $S(t)$. Einstein's equations imply that

$$qH^2 = \frac{4\pi}{3}(\mu + 3p) \qquad , \tag{20}$$

and

$$H^2 + kS^{-2} = \frac{8\pi}{3}\mu \tag{21}$$

It should be noted that the metric in the Fermi frame does not depend upon r, θ, or ϕ as a consequence of the spatial homogeneity of the FLRW spacetime. Though the gravitational field is weak in the spacetime neighborhood of the observer, it is nevertheless fully relativistic. To obtain a Newtonian approximation, additional restrictions on the magnitude of velocities is necessary. Thus the Newtonian potential is given by

$$\Phi_N = -\frac{2\pi}{3}\mu\rho^2 \tag{22}$$

The weak-field solution is local, in contrast to cosmology which by its very nature must be global.

ACKNOWLEDGMENT

This work was supported by the Deutsche Forschungsgemeinschaft, Bonn.

REFERENCES

Barrow, J.D., Juszkiewicz, R., and Sonoda, D.H.: 1983, Nature 305, pp. 397-402.
Cohn, H.: 1980, Astrophys. J. 242, pp. 765-771.
Collins, C.B., and Wainwright, J.: 1983, Phys. Rev. D27, pp. 1209-1218.
Hawking, S.W. and Ellis, G.F.R.: 1973, The Large Scale Structure of Space-Time, Cambridge University Press, Cambridge, Chap. 10.
Klein, O.: 1947, Ark. Mat. Astr. Fys. A34, p. 1.
Lemaître, G.: 1931, Mon. Not. R. Astron. Soc. 91, p. 483, p. 490.
Mashhoon, B., and Partovi, M.H.: 1980, Ann. Phys. (N.Y.) 130, pp. 99-138.
Mashhoon, B., and Partovi, M.H.: 1984, to be published.
McCrea, W.H., and McVittie, G.C.: 1930, Mon. Not. R. Astron. Soc. 91, p. 128.
Partovi, M.H., and Mashhoon, B.: 1984, Astrophys. J., Jan. 1.
Shandarin, S.F., and Zeldovich, Ya.B.: 1983, Comments Astrophys. 10, pp. 33-45.
Tolman, R.C.: 1934, Proc. Nat. Acad. Sci. 20, pp. 169-176.
Wesson, P.S.: 1978, J. Math. Phys. 19, pp. 2283-2284.

VACUUM INHOMOGENEOUS COSMOLOGICAL MODELS

J.-L. HANQUIN

Boursier I.R.S.I.A., Institut d'Astrophysique,
Université de Liège, B-4200 Cointe-Ougrée,
Belgium

We present some results concerning the vacuum cosmological models which admit a 2-dimensional Abelian group of isometries : classifications of these space-times based on the topological nature of their space-like hypersurfaces and on their time evolution, analysis of the asymptotical behaviours at spatial infinity for hyperbolical models as well as in the neighbourhood of the singularity for the models possessing a time singularity during their evolution.

The spatially homogeneous and isotropic cosmological models, known as Friedmann-Lemaître-Robertson-Walker models,constitute a satisfactory description of the present stage of evolution of our universe. However in addition to the mathematical interest of more genera cosmological models, there are undeniable physical reasons to study such universes. They give us the possibility of examining if the properties of the simplest models are either a result of the symmetries or an intrinsic physical characteristic . They could also provide a better model for the first phases of our universe and in particular, information on the more general form of a space-time in the neighbourhood of the initial singularity[1].

The first step towards complexity is relaxing the assumption of isotropy. The spatial homogeneity is then expressed by imposing to the models to be invariant under the action of a 3-dimensional group of

A. Berger (ed.), The Big Bang and Georges Lemaitre, 83–92.

isometries, G_3, acting transitively on spacelike hypersurfaces. These are the Bianchi models classified in 9 types following the different possible G_3. These models were intensively studied during the last fifteen years[2].

More recently, inhomogeneous generalizations of certain models were also considered[3]. Because of the mathematical complexity of the generic models, the major attempts have been concentrated on a class of models with one-dimensional inhomogeneity. These models, characterized by an Abelian G_2 acting orthogonally transitively on 2-dimensional spacelike orbits, are described by a generalized Einstein-Rosen metric. In particular, if the two Killing vectors are hypersurface-orthogonal, this metric can be globally diagonalized and written in the following form

$$ds^2 = e^A(-dt^2+dz^2)+R(e^X dx^2+e^{-X}dy^2) \qquad (1)$$

where the (real) functions A,R and X depend on both variables t and z. The two commuting Killing vectors are then $(\partial/\partial x)$ and $(\partial/\partial y)$.

In this paper, we shall describe some results that we have recently obtained on the nature and the behaviour of vacuum cosmological models with the metric (1) (for more details, see [6] and [7]). First of all, one must concentrate on the global structure, that is the topological nature, of these space-times. For that purpose, the function R plays a primordial role. Indeed, the separated solutions[4] $R=f(t)\,g(z)$ of the field equation

$$\partial^2R/\partial t^2 = \partial^2R/\partial z^2 \qquad (2)$$

can only be of the following types[5] :

(i) $R=t$ or $t\,z$ corresponding to the plane topologies usually represented by a compactification of R^3, the three-torus topology, T^3.

(ii) $R=\sin t \sin z$ corresponding to the spherical topologies, S^3 (3-sphere) and S^1xS^2 (3-handle).

(iii) Both f and g are one of the following functions : sinh, exp or cosh. These nine possible cases correspond to the hyperbolical topologies, H^3 (3-hyperboloid), S^1xH^2 and a third kind of topology, denoted H_{sp}, in which both 2-dimensional surfaces (x,z) and (y,z) have the H_2-topology.

The plane and spherical cases are closed spaces in the sense that their spacelike hypersurfaces are compact. The plane model was intensively studied by Misner [8], Gowdy [9],[10] and Berger [11] and will not be considered here anymore. The spherical models were also studied in [9],[10] and [12] but with less details. In our knowledge, the hyperbolical models were not considered so far. However, they are interesting at several levels :

(i) They restore the parallelism between the spherical, plane and hyperbolical topologies already present in the simplest models.

(ii) As they form a large class, they offer more various behaviours not always present in other models.

(iii) They constitute a natural generalization of Bianchi models of types III,V and VI_h and they even contain spatially homothetic models of types $_fI$, $_fIII$, $_fV$ and $_fVI_h$ as particular cases.

(iv) Finally, being inhomogeneous open models, they pose the very complicated problem of their asymptotical behaviour with respect to the spacelike variable.

The classification of the spacelike hypersurfaces is then based on their topology but also, in cases H^3 and S^1xH^2, on the coordinate system used to describe this topology[6] (indicated in Table 1 by the first down index). Once these hypersurfaces are classified, it is necessary to classify their time evolution. This evolution essentially depends on the function f(t), a choice indicated by a second down index. The combined result of both these classifications as well as some other characteristics of the corresponding models, are given in Table 1.

We have calculated most of the exact solutions of the Einstein's field equations for these spacetimes[8] and then examined their regularity with respect to the spacelike variable z. This led us to compute, with the help of the algebraic program SHEEP, the components of the Riemann tensor in a Lorentz basis and the curvature invariant $R^{\alpha\beta\gamma\delta}R_{\alpha\beta\gamma\delta}$. The requirement that these expressions be bounded at spatial infinity imposes some regularity conditions on the solutions[9]. However, this is only a necessary

f(t)	g(z)	sinh(z)	exp(z)	cosh(z)	sin(z)
	I_z \ I_t	$[0, +\infty)$	R	R	$[0, \pi]$
sinh(t)	$(0, +\infty)$	mixed cs-r $(H^3_1)_s$ $(S^1 \times H^2_1)_1$	ti cs-r $(H_{sp})_1$	ti cs-r $(S^1 \times H^2_3)_1$	
exp(t)	R	sp r-r $(H^3_1)_2$ $(S^1 \times H^2_1)_2$	null r-r $(H_{sp})_2$	ti s-r $(S^1 \times H^2_3)_2$	
cosh(t)	R	sp r-r $(H^3_1)_3$ $(S^1 \times H^2_1)_3$	sp {r-r, s-s} $(H_{sp})_3$		
sin(t)	$(0, \pi)$				mi cs-cs S^3 $S^1 \times S^2$

Table 1 : Classification of the spherical and hyperbolical generalized Einstein-Rosen space-times with metric (1) based on the function R = f(t) g(z). The range of variation of the spacelike and timelike variables is indicated as I_z and I_t respectively. For each possible case, we specify the name of the model (precising its topology and the coordinate system describing it), on the upper left, the causal nature of $R_{,\mu}$, the gradient R, and finally, on the upper right, the nature of the time evolution of the model, cs, s and r denoting respectively a curvature singularity, a singularity (at least a whimper singularity) and regularity. Thus, a space-time characterized by cs-r, evolves from an initial curvature singularity to regularity.

requirement in order to ensure regularity since the corresponding space-time would be singular in another basis : a whimper singularity would then occur. This question of regularity or singularity of a general space-time is however a very complicated problem necessitating a global analysis of the space-time, which goes beyond our purpose (for a review, see [14]).

The behaviour of the curvature invariant and of the Riemann tensor components was then examined with respect to the timelike variable t; the results are also indicated in Table 1. For f=sinh(t), we have been able to show that a singularity occurs at t=0 by calculating the asymptotic solution in the neighbourhood of t=0[10]. If

$$X \sim S(z)\ \ln(t) + T(z) \tag{3}$$

where "$\sim$" means "asymptotic expansion to the first order in t", the metric functions can be written as follows

$$\begin{aligned} g_{xx} &\sim f_x(z)\,t^{1+S} \\ g_{yy} &\sim f_y(z)\,t^{1-S} \\ g_{zz} &\sim f_z(z)\,t^{(S^2-1)/2} \end{aligned} \tag{4}$$

where the functions f_i are known expressions of S and T. This expansion also allows us to characterize the behaviour of these spaces in the neighbourhood of their initial singularity (see Table 2). When one goes towards the singularity, two possibilities arise, either the model is contracting to zero in two directions and expanding to infinity in the third one ("cigarlike singularity", $S\neq\pm1$), or it is contracting to zero in one direction and reaches finite non-zero values in both others ("pancakelike singularity", $S=\pm1$). As z is running in I_z, S varies, so both types of singularities and directions of contraction and expansion can alternate. As for a fixed value of z, the singularity is of the Kasner type (from the name of the vacuum solution of Bianchi type I universe), we shall say this singularity is of the generalized Kasner type.

As particular cases of models H_{sp}, consider the spatially homothetic models of types ${}_f\mathrm{I}$, ${}_f\mathrm{III}$, ${}_f\mathrm{V}$ and ${}_f\mathrm{VI}_h$. These space-times are conformal to the standard metrics of Bianchi models of corresponding type in the synchronous basis (see [15] and [6]) and their general vacuum exact solution depends on two real parameters, λ and δ. The types ${}_fI$, ${}_fIII$ and ${}_fV$, particular cases of ${}_fVI_h$, correspond respectively to the following values of parameters, $\lambda=\delta=0$, $\lambda=\pm\delta$ and $\lambda=0$. When $\lambda^2+2\delta+1 \geqslant 0$, the space-time is of type $(H_{sp})_1$ and its metric can be expressed as follows [6]

$$(ds^2)=e^{(1-\delta)z}\{(\sinh t)^{\lambda^2+\delta}(\tanh (t/2))^{-\lambda|\lambda^2+2\delta+1|^{1/2}} (-dt^2+dz^2)+(\sinh t)^{1-\lambda}(\tanh (t/2))^{|\lambda^2+2\delta+1|^{1/2}} e^{(\delta-\lambda)z}dx^2+(\sinh t)^{1+\lambda}(\tanh (t/2))^{-|\lambda^2+2\delta+1|^{1/2}} e^{(\delta+\lambda)z}dy^2\}. \qquad (5)$$

The Ellis and MacCallum's solution for Bianchi types III,V and VI_h is obtained by posing $\delta=1$ [11]. In other cases ($\lambda^2+2\delta+1<0$), the space-time is of type $(H_{sp})_3$ and its metric is obtained from (5) by replacing ($\sinh t$) by ($\cosh t$) and ($\tanh (t/2)$) by ($\exp (\tan^{-1}(\sinh t))$). Finally, Figure 1 gives the evolution of the spatial volume of these models which can be of three different types in both cases,following the values of parameters λ and δ.

Notes

[1] Actually, the behaviour of the vacuum Bianchi IX model in the neighbourhood of its singularity is believed to be of the most general type [1], however, the debate on this subject is not yet closed [2].

[2] For more information on Bianchi models and other topics concerning exact solutions of Einstein's field equations, see [3].

[3] For a review on the inhomogeneous cosmological models, see [4] and [5].

[4] The general solutions of this wave equation is of course well known but suitable coordinate transformations on t and z can reduce non separated solutions to one of the possible separated forms.

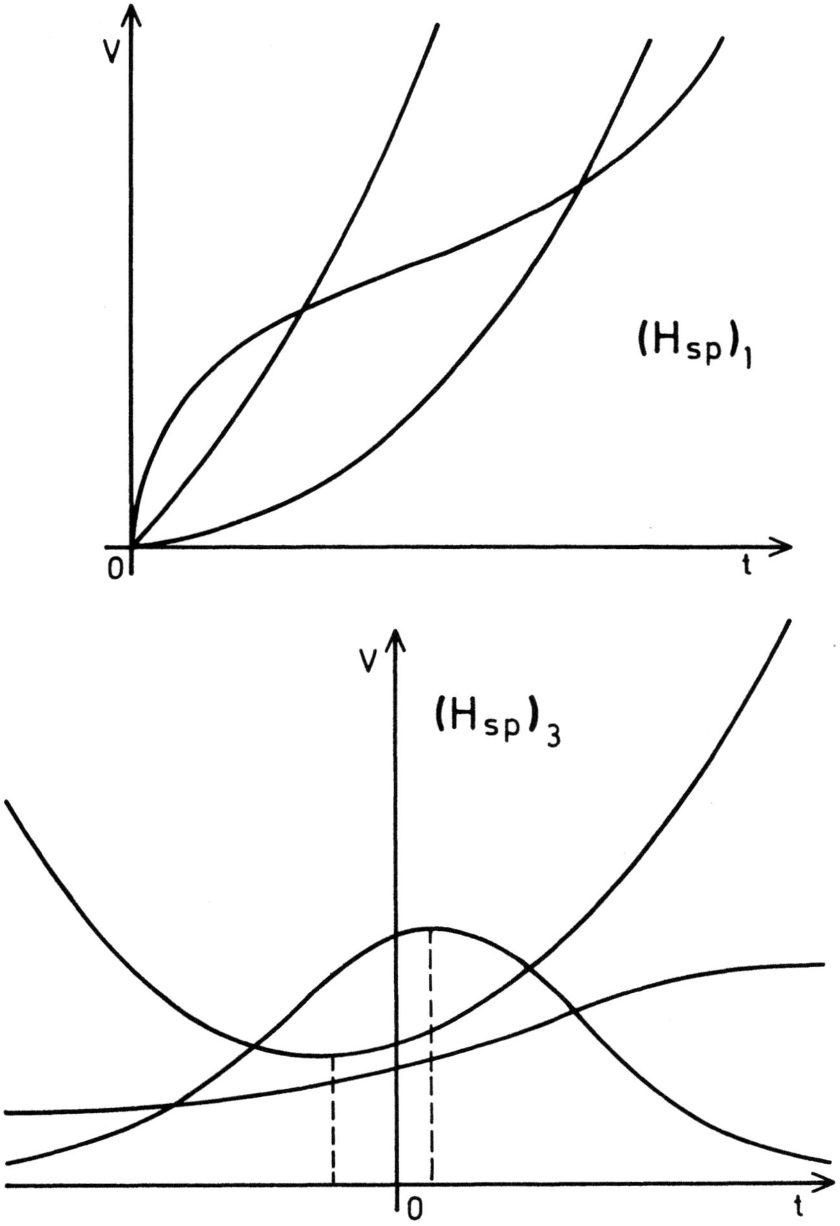

Figure 1 : Different possible evolutions of the spatial volume $V(t)=\alpha\int Re^{A/2}dz$ (α being an integration constant) of the spatially homothetic space-times.

S(z)		-1		1	
g_{xx}	∞	c	0	0	0
g_{yy}	0	0	0	c	∞
g_{zz}	0	c	∞	c	0

Table 2 : For f=sinh(t), the limit value of the metric functions for t going to 0 depends on the spacelike variable z through S (c means a non-zero constant).

5 These topological considerations are partly based on the following remark. If the topology of a 3-dimensional space is known (for instance, spaces of constant curvature, cylindrical spaces,...) and its 3-dimensional line element is written in the form $dl^2=\Sigma(\omega^i)^2$ (Latin indices run from 1 to 3, Greek ones from 0 to 3), a space whose line element is written $dl^2=\Sigma g_{ii}(\omega^i)^2$, will be considered as a deformation of the initial space and will have the same topology as that space.

6 Two coordinate systems describing the same topology are of course, equivalent for constant curvature spaces, but this is no more the case for more complex spaces. Note that the H^3_2 and $S^1xH^2_2$ spaces are particular cases of H_{sp}.

7 Thorne [13] interprets the causal nature of $R_{,\mu}$ as the direction that will be followed by hypothetical particles in a considered vacuum space-time. A timelike gradient will then be necessary to characterize a cosmological model while a spacelike gradient will rather correspond to a model where only gravitational waves would propagate. Certain models could present both types of behaviours, the null surface is then considered as a shock wave.

8 The $(S^1xH^2_3)_3$ space does not exist since it does not satisfy some integrability conditions of the field equations.

[9] In the (H_{sp_3}) space, these regularity conditions are never satisfied.

[10] In this case, the singularity is of the curvature type as all timelike geodesics (and even curves) going to this point are incomplete. A similar expansion is also possible at points t=0 and t= π for the spherical models [12].

[11] Note that the only possible solution for type I_f model is Minkowski space so there does not exist any homothetic generalization of Kasner's vacuum spatially homogeneous solution.

References

[1] Belinskii,V.A., Khalatnikov,I.M. and Lifshitz,E.M. : 1982, Adv. in Phys. 31, p.639.

[2] Barrow,J.D. and Tipler,F.J. : 1979, Phys.Reports 56, p. 372.

[3] Kramer,D., Stephani,H., MacCallum,M.A.H. and Herlt,E. : 1980, "Exact Solutions of Einstein's Field Equations", Cambridge University Press, Cambridge.

[4] MacCallum,M.A.H. : 1979, in "General Relativity : An Einstein Centenary Survey", ed. by S.W. Hawking and W. Israel, p.533, Cambridge Univ. Press, Cambridge.

[5] Carmeli,M., Charach,Ch. and Malin,S. : 1981, Phys. Reports 76, p.79.

[6] Hanquin,J.L. and Demaret,J. : 1984, to appear in Classical and Quantum Gravity.

[7] Hanquin,J.L. : 1984, in preparation.

[8] Misner,C.W. : 1973, Phys.Rev.D8, p.3271.

[9] Gowdy,R.H. : 1971, Phys.Rev.Lett. 27, p.826.

[10] Gowdy,R.H. : 1974, Ann.Phys., NY 83, p.202.

[11] Berger,B.K. : 1972, A cosmological model illustrating particle motion through quantum graviton production, PhD Thesis, University of Maryland (available from Univ.Microfilms,Ann Arbor, Michigan).

[12] Hanquin,J.L. and Demaret,J. : 1983, J.Phys.A 16, p. L5.

[13] Thorne,K.S. : 1965, Geometrodynamics of cylindrical systems, PhD Thesis, Princeton University (available from University Microfilms, Ann Arbor, Michigan).

[14] Ellis,G.F.R. and Schmidt,B.G. : 1977, Gen.Rel.Grav. 8, p.915.

[15] Eardley,D.M. : 1974, Commun.Math.Phys. 37, p.287.

[16] Ellis,G.F.R. and MacCallum,M.A.H. : 1969, Commun. Math.Phys. 12, p.108.

THE SIGNIFICANCE OF NEWTONIAN COSMOLOGY

D.Galletto and B.Barberis

Istituto di Fisica Matematica "J.-Louis Lagrange"
Via Carlo Alberto 10, 10123 Torino, ITALY.

ABSTRACT - Starting from the hypotheses that the physical space is Euclidean, that the Universe is infinite and homogeneous and that with regard to our galaxy its behaviour is isotropic, without resorting to Newton's law of gravitation we deduce Hubble's law, the law of motion of a typical galaxy, the equation of evolution of the Universe, that the force at a distance exerted between any two galaxies is expressed by Newton's law of gravitation, etc. Adding the hypothesis that the velocity of light is independent of its source, we obtain that the metric of space-time is necessarily given by the Einstein-de Sitter metric, that the tensorial form of the equations of Newtonian cosmology is given by Einstein's gravitational equations, etc.

1. - The results which are summarized here are based on the following hypotheses:

1. the physical space is the ordinary three-dimensional Euclidean space;
2. the Universe, at least on a large scale, is homogeneous and infinite;
3. with regard to our galaxy the behaviour of the Universe is isotropic, in the sense that with respect to the frame of reference with origin in the centre of mass of our galaxy and determined by three other distant galaxies the motion of a typical galaxy is radial.

Hypotheses 2 and 3 are suggested by astronomical observations.

To describe the Universe, the incoherent matter scheme is

A. Berger (ed.), The Big Bang and Georges Lemaitre, 93–102.

used and hence, taking into account hypothesis 2, the Universe is represented by means of an infinite homogeneous fluid without internal stresses (the cosmological fluid), which from now on will be indicated by $\mathcal{U}$.

Hypothesis 3 implies the existence of an element O of $\mathcal{U}$ and of a frame of reference $\mathcal{R}_0$ which has its origin in O and with respect to which the motion of any element P of $\mathcal{U}$ is radial.

Let $\mu(t)$ be the density of $\mathcal{U}$ and let us define

$$h(t) = -\frac{1}{3}\frac{\dot{\mu}}{\mu} \quad . \tag{1}$$

From the equation of continuity and from the principle of conservation of matter we obtain (cf. [6],2):

$$\frac{dOP}{dt} = h(t)\, OP \quad . \tag{2}$$

Let O' be any other element of $\mathcal{U}$ and let $\mathcal{R}_{0'}$ be the frame of reference which has its origin in O' and which is in translatory motion with respect to $\mathcal{R}_0$. With regard to $\mathcal{R}_{0'}$ as well law (2) is verified, in the sense that

$$\frac{dO'P}{dt} = h(t)\, O'P \quad ,$$

with the conclusion that all the frames of reference which have their origins in the elements of $\mathcal{U}$ and which are in translatory motion with respect to $\mathcal{R}_0$ are equivalent to one another. We shall call these frames **natural frames of reference**. The frame $\mathcal{R}_0$ is one of them and hence, at least from the kinematical point of view, it is completely indistinguishable from them. We can therefore say that:

I. The fluid $\mathcal{U}$ has the same kinematical behaviour with respect to any natural frame of reference. Whatever the natural frame $\mathcal{R}_0$ is, such behaviour with respect to $\mathcal{R}_0$ is described by the law expressed by (2).

Whatever the natural frame $\mathcal{R}_0$ is, from (2) follows

$$\frac{d^2OP}{dt^2} = (\dot{h} + h^2)\, OP \quad . \tag{3}$$

2. - Whatever the natural frame $\mathcal{R}_0$ is, from (3) it is possible to deduce (cf.[12] in which the deductions made in [6],3 and [11],3 and re-examined in [2] are fully revised and perfected) that the equation of motion of P with respect to $\mathcal{R}_0$, and therefore with respect to any natural frame of reference, is expressed by

$$\frac{d^2OP}{dt^2} = \kappa\mu\, OP \quad , \tag{4}$$

where κ is a constant. Defining

$$k = -\frac{3\kappa}{4\pi} ,$$

equation (4) becomes

$$\frac{d^2 OP}{dt^2} = -\frac{4}{3}\pi k \mu \, OP . \tag{5}$$

Equation (5) preserves its form unchanged in any natural frame of reference.

The considerations summarized above permit us to affirm that:

II. From the hypotheses 2 and 3 the explicit equation of motion of P with respect to any natural frame of reference follows, without resorting to Newton's law of gravitation.

III. With respect to any natural frame of reference the fluid $\mathcal{U}$ has the same dynamical behaviour, expressed by equation (5).

IV. All natural frames of reference are equivalent to one another, in the sense that the fluid $\mathcal{U}$ has, from both the kinematical and the dynamical points of view, the same behaviour with respect to them.

Furthermore, from the comparison of (5) with (3) the equation of evolution in Newtonian cosmology follows:

$$\dot{h} + h^2 = -\frac{4}{3}\pi k \mu , \tag{6}$$

from which follows

$$h = \sqrt{\frac{8}{3}\pi k \mu + a \mu^{2/3}} , \tag{7}$$

where a is the constant of integration, etc.

If we introduce the **deceleration parameter:**

$$q = -\frac{\dot{h} + h^2}{h^2} ,$$

from (6) follows

$$k = \frac{3}{4\pi}\frac{h^2 q}{\mu} ,$$

a relation which permits us to obtain the value of the constant k starting from the present values of h, μ, and q.

3. - Equation (5) is the same one that would be obtained by assuming the frame $\mathcal{R}_0$ to be inertial, the part of $\mathcal{U}$ external to

the material sphere $\mathcal{S}_{OP}$ with centre at O and radius $|OP|$ to give no contribution to the motion of P, and the forces at a distance (gravitational forces) exerted on P to be expressed by Newton's law of gravitation (where k is the gravitational constant). A priori, there is nothing, however, to authorise this procedure, which is the one that has been followed up to now in all treatises on cosmology made in Newtonian terms. On the contrary equation (5) has been obtained here without imposing a priori any limitation on the frame $\mathcal{R}_0$ and above all without resorting to Newton's law of gravitation and resorting instead only to the hypotheses made for $\mathcal{U}$.

However we can prove (cf. [12]) that:

V. From the hypotheses 2 and 3 Newton's law of gravitation necessarily follows, in the sense that these hypotheses imply that the action at a distance exerted between any two elements of $\mathcal{U}$ is necessarily expressed by Newton's law of gravitation.

An attempt to deduce Newton's law of gravitation from (1) has been made in [13], but this attempt is tautological (cf. [1]).

From (5) and V follows:

VI. Whatever the natural frame $\mathcal{R}_0$ is, as far as the motion of P with respect to it is concerned everything happens as if the frame $\mathcal{R}_0$ were inertial and as if the part of $\mathcal{U}$ external to the material sphere $\mathcal{S}_{OP}$ made no contribution to the motion of P.

This result implies (cf. [6],8; [11],6; [7],5) that:

VII. The fictitious force acting on P and the resultant of the gravitational forces exerted by $\mathcal{U}$ on P are inseparable. Whatever the natural frame $\mathcal{R}_0$ with respect to which the motion of P is considered is, their resultant is given by the resultant of the gravitational forces exerted on P by the material sphere $\mathcal{S}_{OP}$.

The results given in VI justify the procedure for the deduction of equation (5) which is followed in all treatises dealing with Newtonian cosmology: this procedure was followed in particular by Milne and McCrea, who in 1934, in [17] and [15], were the first to study cosmology in Newtonian terms. (Of course we do not mention Seeliger's attempt, which dates back to the end of the 19th century, because that attempt - since it was based on the belief that the Universe were static - contradicts astronomical observations as well as Newton's law of gravitation. For more details, see [9],1 and [3],2).

What has been established so far, and especially the results summarized in VI, prove that the criticism made by Layzer in [14] is incorrect: in that paper Layzer denies the possibility of formulating a Newtonian cosmological theory with an infinite homogeneous Universe. In particular what is proved to be incorrect is the general conviction that the results summarized in VI can be justified only by resorting to the general theory of relativity (cf., e.g., [14]; [19], p.475; [18], Chap.8, Sec.9; etc.).

In general what we have seen so far allows us to consider the so-called Newtonian paradox to be inconsistent: this paradox in essence states the impossibility of applying Newton's theory of gravitation to an infinite homogeneous fluid. For more details on this point, see [9] and above all [3].

4. If the instant t_0 is fixed once and for all, and if we define

$$R(t) = \exp \int_{t_0}^{t} h(t)\, dt \quad ,$$

from (2) follows

$$OP = R(t)\, OP_0 \quad , \tag{8}$$

where P_0 is the position assumed by P at the instant t_0. Owing to (8), from (1), (2) and (5) follow

$$\frac{\dot{R}}{R} + \frac{1}{3}\frac{\dot{\mu}}{\mu} = 0 \tag{9}$$

and

$$\frac{\ddot{R}}{R} = -\frac{4}{3}\pi k \mu \quad . \tag{10}$$

From (10), taking into account that

$$\frac{4}{3}\pi k \mu R^3 = \text{const.}$$

(as follows directly from (9)), we have

$$\frac{\dot{R}^2}{R^2} = \frac{8}{3}\pi k \mu + \frac{2\alpha}{R^2} \quad , \tag{11}$$

where α is the value of the energy constant which would belong to the element P (which we assume to have a unit mass) if it were at a distance R from the origin of the natural frame of reference with respect to which the motion is considered.

Equations (9) (the equation of continuity) and (10) (which in essence does not differ from the equation of evolution of $\mathcal{U}$, expressed by (6)), together with (11) (the energy integral, which in essence does not differ from (7)), are the equations of Newtonian cosmology. As has already been stressed, these equations have been deduced here without resorting to Newton's theory of gravitation, and making use only of the hypotheses made for the fluid $\mathcal{U}$.

5. - The results obtained in the preceding sections entail the following cosmological interpretation from the Newtonian point of view:

VIII. From the hypotheses 2 and 3 follow:

a) Hubble's law, expressed by (2), which is verified with respect to every natural frame of reference;

b) the law of motion of a typical galaxy, expressed by (5), and the law of evolution of the Universe, expressed by (6);

c) that the forces at a distance which determine the motion of any galaxy with respect to any natural frame of reference are necessarily expressed by Newton's law of gravitation;

d) that, with respect to any natural frame $\mathcal{R}_0$, the resultant of the forces acting on any galaxy P is equal to the resultant of the gravitational forces exerted on P by the part of the Universe contained in the sphere $\mathcal{S}_{OP}$ with centre at O and radius $|OP|$;

e) that all natural frames of reference are equivalent to one another, in the sense that the Universe has, from both the kinematical and the dynamical points of view, the same behaviour with respect to them.

This last result in essence expresses the so-called cosmological principle.

All the results hitherto obtained provide, among other things, sound mathematical support and proofs to the considerations made in [4].

6. - At this stage, if we take into account the property of the velocity of light revealed by the Michelson-Morley experiment, we have (cf. [10],3; [7],9):

IX. The local velocity of light is the same with respect to any natural frame of reference.

From this result and from the Galilean law of addition of velocities follows (cf. [10],4,5,7; [7],10):

X. The metric of the space-time manifold is necessarily expressed by the metric of the Einstein-de Sitter model of the Universe.

In other words, the metric of space-time is that particular case of the Friedmann-Robertson-Walker metric that Einstein and de Sitter suggested in 1932 for space-time (cf. [5]) in virtue of its simplicity, while working within the framework of general relativity.

At this point we can derive the usual formula for the red-shift (cf. [10],8,9; [7],11) and we have:

XI. In the present case the formula connecting the red-shift to the expansion of the Universe is a consequence of IX and of the Galilean law of addition of velocities.

Analogous considerations can be made about horizons (cf. [10],10), etc.

7. - If we impose the condition that the equations of Newtonian cosmology and the Einstein-de Sitter metric are compatible, in the sense that it would be possible to give an intrinsic form to these equations within the framework of this metric, it follows (cf. [7],12,13) that the energy constant α, which appears in (11), must be zero. In other words:

XII. The property of the velocity of light of being independent of the velocity of its source implies that the energy constant α is zero.

Therefore, while in the Newtonian framework we have no restriction on the energy constant α, the property of light of having the same local velocity in every natural frame of reference implies that this constant is zero.

Owing to (8), equality (2) implies

$$h = \frac{\dot{R}}{R}$$

and hence, with $\alpha = 0$, from (11) follows

$$h = \sqrt{\frac{8}{3}\pi k \mu} \quad , \tag{12}$$

an expression which, on the other hand, we could obtain directly from (7) because, from $\alpha = 0$ it follows that $a = 0$ and vice versa.

Therefore, as the metric of space-time is the Einstein-de Sitter metric if and only if the physical space is Euclidean and as this metric implies $\alpha = 0$, we find that equality (12) expresses the relation connecting h and μ which must be verified in order that the physical space is effectively Euclidean. In other words, if this equality is not verified (i.e. not confirmed by astronomical observations) the possibility that the physical space is Euclidean is excluded. For more details, see [7],16.

8. - Once it has been proved that $\alpha = 0$, we obtain (cf. [7],14,

15) that the intrinsic form imposed by the Einstein-de Sitter metric on the equations of Newtonian cosmology is precisely the one expressed by Einstein's gravitational equations. It can therefore be stated that:

XIII. The hypotheses 2 and 3, together with the property of light of having the same local velocity with respect to any natural frame of reference, necessarily lead to Einstein's equations of the general theory of relativity.

What has just been briefly discussed is therefore a deduction of Einstein's gravitational equations from astronomical observations. This deduction entails, among other things, that Einstein's gravitational constant χ is necessarily expressed by

$$\chi = \frac{8\pi k}{c^4} ,$$

where c is the local velocity of light, a result that in all treatises is obtained by resorting to approximation methods.

9. - With appropriate modifications the results summarized in the preceding sections may be extended to the case that the physical space, instead of being Euclidean, is a three-dimensional maximally symmetric space. The results given in I, II, III, IV, IX, XI and XIII are maintained, together with equations (9), (10) and (11). The energy constant α is no longer zero and determines the curvature of the physical space, whereas the metric of space-time is necessarily expressed by the Friedmann-Robertson-Walker metric. These results can be extended to the more general case of homogeneous and anisotropic universes (some of them have been dealt with in [8]).

All these results will be presented in detail in forthcoming papers.

All the considerations summarized here and the further ones which will be developed in forthcoming papers put Newtonian cosmology in a new light and, in clear contrast with [14] and [16], completely clarify its real meaning.

REFERENCES

[1] BARBERIS B., GALLETTO D., **Further Remarks on a Paper by P.T. Landsberg,** Atti Accad. Sci. Torino, 112 (1978), pp. 103-107.

[2] BARBERIS B., GALLETTO D., **On the Deduction of the Equation of Evolution in Newtonian Cosmology,** Atti Accad. Sci. Torino, 118 (1984) (in print).

[3] BARBERIS B., GALLETTO D., **On the Newtonian Paradox. I**, Atti Accad. Naz. Lincei Rend. (to appear).

[4] CALLAN C., DICKE R.H., PEEBLES P.J.E., **Cosmology and Newtonian Mechanics**, Amer. J. Phys., 33 (1965), pp. 105-108.

[5] EINSTEIN A., de SITTER W., **On the Relation between the Expansion and the Mean Density of the Universe**, Proc. Nat. Acad. Sci., 18 (1932), pp. 213-214.

[6] GALLETTO D., **Sui fondamenti della meccanica classica, della teoria newtoniana della gravitazione e della cosmologia newtoniana**, in: "Atti del 3° Convegno Nazionale di Relatività Generale e Fisica della Gravitazione, Torino, 18-21 settembre 1978", Accademia delle Scienze, Torino, 1981; pp. 111-157.

[7] GALLETTO D., **Des principes de la mécanique classique aux théories de la gravitation de Newton et Einstein**, in: "Atti del Convegno Internazionale 'Aspetti matematici della teoria della relatività', Roma, 5-6 giugno 1980", Accademia Nazionale dei Lincei, Roma, 1984; pp. 59-83.

[8] GALLETTO D., BARBERIS B., **Sui modelli newtoniani di universo omogenei e anisotropi**, in: "Atti del Convegno su 'Problemi attuali di fisica teorica', Torino, 12-13 dicembre 1980", Accademia delle Scienze, Torino, 1981; pp. 211-224.

[9] GALLETTO D., BARBERIS B., **Considerazioni sul paradosso newtoniano**, in: "Atti del Convegno celebrativo dell'80° anniversario della nascita di Renato Calapso, Messina-Taormina, 1-4 aprile 1981", Veschi, Roma, 1983; pp. 153-164.

[10] GALLETTO D., **Sulla metrica dello spazio-tempo. II**, Atti Accad. Sci. Torino, 115 (1981) (in print).

[11] GALLETTO D., **Premesse alla cosmologia razionale**, Rend. Sem. Mat. Univ. Padova, 68 (1982), pp. 295-324.

[12] GALLETTO D., BARBERIS B., **Newtonian Cosmology and Newton's Law of Gravitation** (to appear).

[13] LANDSBERG P.T., **A Deduction of the Inverse Square Law from Newtonian Cosmology**, Nature phys. Sci., 244 (1973), pp. 66-67.

[14] LAYZER D., **On the Significance of Newtonian Cosmology**, Astron. J., 59 (1954), pp. 268-270.

[15] McCREA W.H., MILNE E.A., **Newtonian Universes and the Curvature of Space**, Quart. J. Math. (Oxford Ser.), 5 (1934), pp. 73-80.

[16] McCREA W.H., **On the Significance of Newtonian Cosmology**, Astron. J., 60 (1955), pp. 271-274.

[17] MILNE E.A., **A Newtonian Expanding Universe**, Quart. J. Math. (Oxford Ser.), 5 (1934), pp. 64-72.

[18] NORTH J.D., **The Measure of the Universe**, Clarendon Press, Oxford, 1965.

[19] WEINBERG S., **Gravitation and Cosmology: Principles and Applications of the General Theory of Relativity**, Wiley, New York, 1972.

NUMERICAL SIMULATION OF EVOLUTION OF A MULTI-DIMENSIONAL HIGGS FIELD IN THE NEW INFLATIONARY SCENARIO

Katsuhiko Sato and Hideo Kodama

Department of Physics
Faculty of Science
The University of Tokyo
Tokyo 113, Japan

Numerical simulation of the evolution of the adjoint Higgs field in the new inflationary universe scenario based on the SU(5) GUT model with the Coleman-Weinberg potential is carried out by introducing a suitable viscosity term in the equation of motion. It is found that in order for a consistent scenario of the universe without too large density inhomogeneity to be constructed, the value of viscosity should lie in an appropriate range.

1. Introduction

Recently cosmological consequences of grand unified theories (GUTs) have been investigated actively. One of the most interesting and important results obtained in the investigation is that the universe might have undergone an exponential expansion caused by the vacuum energy of a metastable state when it made a first-order phase transition associated with the breakdown of a grand unification symmetry(Sato 1981, Guth 1981, Guth and Weinberg 1981,1983). The existence of this so-called inflationary stage gives a possible solution to the horizon, the flatness and the monopole problems because of the rapid increase in the cosmic scale factor and the particle horizon size.

It was shown, however, that the original scenario of the

A. Berger (ed.), The Big Bang and Georges Lemaitre, 103–112.

inflationary universe has serious difficulties: the phase transition never terminates and the universe does not get out of the exponential expansion stage due to the smallness of the nucleation rate of bubbles naturally predicted in the conventional GUTs(Guth and Weinberg 1983): furthermore even if the phase transition terminates, formation of large bubbles results in a universe with too large density fluctuations which conflict with the homogeneity of the 3K microwave background radiation(Sasaki et al. 1982, Kodama et al. 1982).

In order to avoid these difficulties, Linde(1982) and Albrecht and Steinhardt(1982) proposed a new version of inflationary universe scenario, which is based on the Coleman-Weinberg mechanism of symmetry breaking. In this new scenario the Higgs field spends a lot of time near the metastable symmetric point, where the potential energy is still very large, because of the flatness of the potential. As a result each coherent region in which the Higgs field has a nearly uniform nonvanishing expectation value expands exponentially for a sufficiently long time while the Higgs filed rolls down to the absolute minimum point of the potential.

In most of the investigation until now it has been assumed that the vacuum expectation value of the Higgs field evolves directly to the SU(3)xSU(2)xU(1) direction from the start of rolling down. Recently, however, some people have pointed out (Moss 1983, Breit et al. 1983, Kodaira and Okada 1983) that the Higgs field goes toward SU(4)xU(1) state, which may give rise to serious difficulties in the new inflationary scenario. The purpose of the present work is to investigate this point in more detail by examining the evolution of a Higgs field Φ belonging to the adjoint representation of SU(5) in the full 24-dimensional space by numerical simulation and to elucidate whether the new inflationary scenario is consistent with cosmological observations or not.

2. Formulation

In practical computations it is not necessary to calculate the evolution of the full 24 components directly. It is quite natural to assume that the time derivative of the Higgs field vanishes(Abbott et al. 1983), i.e., $\dot{\Phi}=0$, just when the Higgs field acquires non-vanishing classical expectation values in the first stage of the phase transition. Then the Higgs field represented by an arbitrary 5x5 hermitian traceless matrix can be diagonalized in each coherent region by a global gauge transformation, keeping $\dot{\Phi}=0$. The equation of motion of the Higgs field guarantees that Φ remains diagonal in the course of its evolution if Φ is diagonal and $\dot{\Phi}=0$ at the start. Thus in the calculation of evolution we can restrict the form of the Higgs field without loss of generality as

$$\Phi = \mathrm{diag}[\ \phi_1, \phi_2, \phi_3, \phi_4, \phi_5] \qquad (1)$$

with the constraint $\mathrm{Tr}\ \Phi = \sum_{i=1}^{5} \phi_i = 0$.

In this representation the Coleman-Weinberg potential with the one-loop correction by the gauge boson (the Higgs boson contribution to the one-loop correction is not included) is given by(Abbott et al. 1983)

$$V(\Phi) = \frac{3g^4}{256\pi^2}\Big[\ C\Big\{\sum_{i=1}^{5} \phi_i^4 - \frac{7}{30}\Big(\sum_{i=1}^{5} \phi_i^2\Big)^2\Big\} + \sum_{i,j=1}^{5} (\phi_i - \phi_j)^4 \Big\{ \ln\Big(\frac{\phi_i - \phi_j}{\mu}\Big)^2 - \frac{1}{2} \Big\}\Big], \qquad (2)$$

where C is an arbitrary parameter of this potential and μ is a renormalization parameter related to the vacuum expectation value of the Higgs field at the SU(3)xSU(2)xU(1) minimum point $\Phi = \sigma \times \mathrm{diag}\ [1,1,-3/2,-3/2]$ ($\sigma \approx 4.5\times10^{14}$GeV) as $\mu = 5\sigma/2$. In the present work we neglect the effect of temperature on the potential because the inflation begins after the cosmic temperature becomes less than the GUT temperature ($\approx 10^{14}$GeV) and the essential fate of the Higgs field is determined before the universe is heated up again to the GUT temperature.

Because of the traceless condition $\sum_{i=1}^{5} \phi_i = 0$, five components of the Higgs field ϕ_i, i=1,2,... 5, are not independent. This makes the numerical computation complicated if we calculate the time evolution of these components directly. In order for the convenience of numerical computation, we introduce the following four components fields which are completely independent each other,

$$\psi_i = \phi_i + \sum_{j=1}^{4} \phi_j/(1+\sqrt{5}\ ),\ (i=1,2,3,4) \qquad (3)$$

In Fig. 1, contours of the potential on a plane ($x = \sqrt{3}\,\psi_1 = \sqrt{3}\,\psi_2 = \sqrt{3}\,\psi_3$, $y = \psi_4$) are displayed for the case of the potential parameter C=1. In this plane there are two SU(3)xSU(2)xU(1) minima and four SU(4)xU(1) minima. As is easily understood from the potential Eq.(2), local minima in the SU(4)xU(1) direction can exist for C<15, and these minima become global minima for C<-15ln(1.5)(Breit et al. 1983)

The equation of motion for the fields ψ_i are given by

$$\ddot{\psi}_i + 3\dot{R}/R\,\dot{\psi}_i + \partial V/\partial \psi_i + C_{vis}|\psi_i|\dot{\psi}_i = 0 \quad , \qquad (4)$$

where a viscosity term $C_{vis}|\psi_i|\dot{\psi}_i$ is introduced in order to convert the energy of the Higgs field into thermal energy. The scale factor of the universe R is calculated by the expansion equation of the universe

$$(\dot{R}/R)^2 = 8\pi(\rho_r + \rho_\phi)/3 \quad ^{1)}, \qquad (5)$$

where we have assumed that the universe is spatially flat, which is adequate in the early universe even if it is not flat exactly. The change of radiation energy density ρ_r and the energy density of the Higgs field ρ_ϕ are described, respectively, by

$$d(\rho_r R^4)/dt = \sum_{i=1}^{4} C_{vis}|\psi_i|\dot{\psi}_i^2 R^4 \quad , \qquad (6)$$

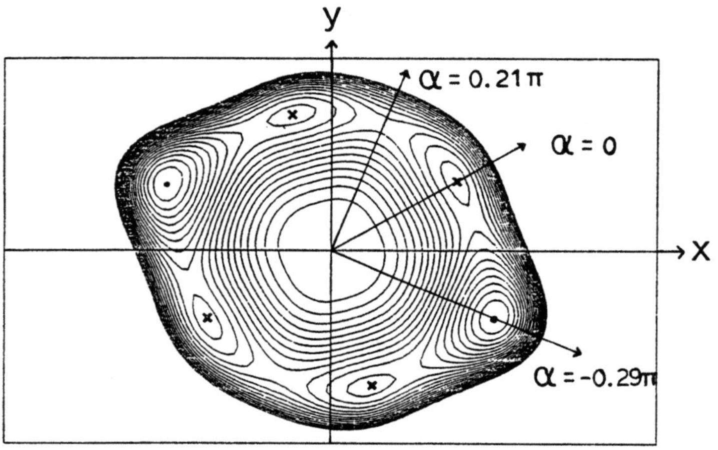

Figure 1. Contour map of the Coleman-Weinberg potential for the case C=1 is displayed on the plane ($x=\sqrt{3}\,\psi_1=\sqrt{3}\,\psi_2=\sqrt{3}\,\psi_3, y=\psi_4$). On this plane, there exist four SU(4)xU(1) minima (+) and two SU(3)xSU(2)xU(1) minima (•). In this plane, $\alpha=0$ means SU(4)xU(1) direction, $\alpha=-0.29\pi$ SU(3)xSU(2)xU(1) direction and $\alpha=0.21\pi$ the direction vertical to the SU(3)xSU(2)xU(1) direction. Numerical computation is carried out in the range $-0.29\pi < \alpha < 0.21\pi$.

and

$$\rho_\phi = \frac{1}{2}\sum_{i=1}^{4} \dot{\psi}_i^2 + V \quad . \tag{7}$$

The initial value of the Higgs field has four degrees of freedom ; one is the norm of Higgs field $\|\psi\| = (\sum_{i=1}^{5}\phi_i^2)^{1/2} = (\sum_{j=1}^{4}\psi_j^2)^{1/2}$ and the others are the direction of the vector (ψ_1, ψ_2, ψ_3, ψ_4) in the four dimensional ψ space. In the present investigation we take $\varepsilon = \|\psi\|_0 = 8\times10^{-6}\sigma$ as the initial value of $\|\psi\|$, which is about 0.2H, where $H=(8\pi GV(0)/3)^{1/2}$. In order to parametrise the initial direction of ψ we utilize three angles α , θ and ψ. α represents the deviation angle from the SU(4)xU(1) direction $\psi_1=\psi_2=\psi_3=\psi_4(>0)$ on the plane $\psi_1=\psi_2=\psi_3$ as shown in Fig.1, and θ and ϕ represent the deviation angles off this plane. Though we do not limit the range of α essentially, we restrict θ and ϕ in the very narrow range $|\theta| < 10^{-4}$ and $|\phi| < 10^{-4}$ for the convenience of the analysis of the numerical computation as a first step.

3. Result of Numerical Simulation

In Fig.2-a$\sim$2-c, some results of numerical computation for the case of the potential parameter C=1 are shown. As demonstrated in these figures, the Higgs field goes to the SU(4)xU(1) direction $\psi_1=\psi_2=\psi_3=\psi_4>0$ at first independent of the value of viscosity parameter C_{vis} and the initial angle α , provided that $-0.29\pi < \alpha < 0.21\pi$. This is obviously a natural consequence of the fact that the potential has the steepest gradient along this direction where the norm $\|\psi\|$ is small(see Fig.1)(Breit <u>et al.</u> 1983). In Fig.3, the dependence of the degree of inflation on the initial angle α is shown. Here we define the degree D as the ratio of the cosmic scale factors, $D=R_2/R_2$, where R_1 is the value when the inflation begins, i.e., when the vacuum energy density becomes greater than that of the radiation, and R_2 is the value when the Higgs field arrives near the SU(4)xU(1) minimum. Note that inflation begins again when the Higgs field settles down at an SU(4)xU(1) state because of the remaining vacuum energy density. Of course this inflation is not taken into account in this definition. As shown in Fig.3, the degree of inflation is very sensitive to the initial angle α , but almost independent of the potential parameter C and the viscosity parameter C_{vis} provided that $C_{vis}<0.1$. The evolution after the arrival to this minimum, of course, depends on the value of C_{vis}.

Generally speaking, the Higgs field settles down in a very direct way to the SU(4)xU(1) minimum after short time oscillation independent of the angle α provided that $C_{vis}>1$ as is illustrated in Fig.2-a. For the smaller values of the viscosity

Figure 2-a.

$C_{vis} = 0.1$

$\alpha = 0.1\pi$

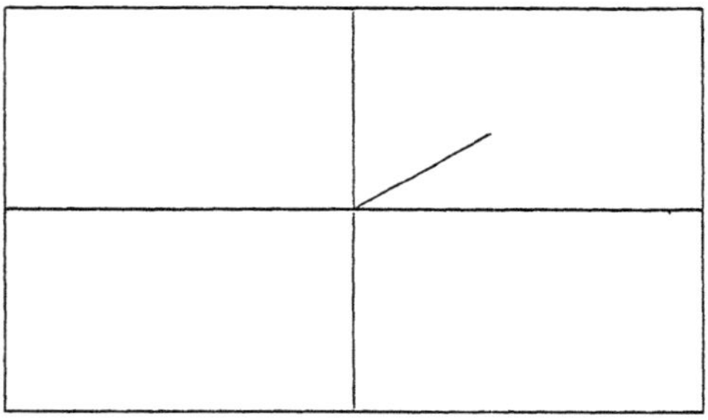

Figure 2-b.

$C_{vis} = 0.01$

$\alpha = -0.2\pi$

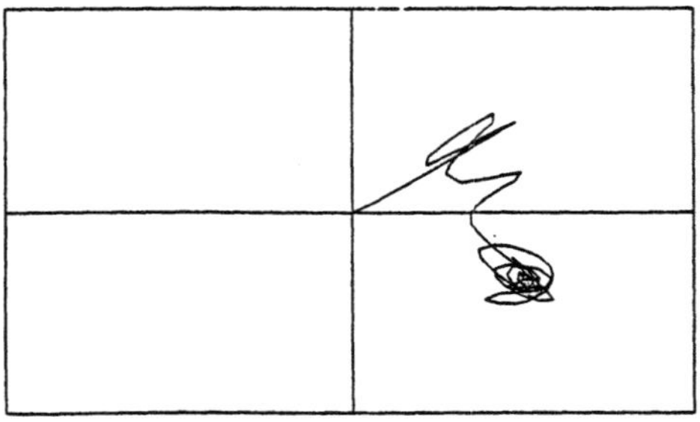

Figure 2-c.

$C_{vis} = 0.001$

$\alpha = 0.2\pi$

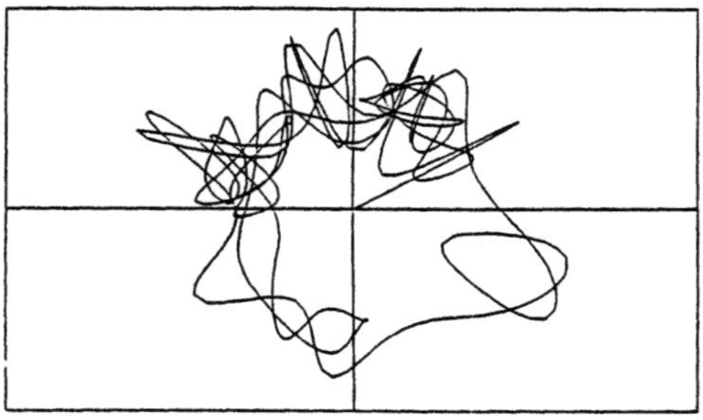

Fig.2-a ∿ Fig.2-c   Time evolutions of the Higgs field projected on the same plane as shown in Fig.1 for three characteristic pairs of values of C_{vis} and α. C=1 for all the three cases.

parameter C_{vis}, however, it can depart from this local minimum and further evolve to an SU(3)xSU(2)xU(1) minimum state . As shown in Fig.2-b ($C_{vis}=10^{-2}$ and $\alpha=0.2\pi$), the Higgs field evolves to the nearest SU(3)xSU(2)xU(1) state via the SU(4)xU(1) state and settles down to this state after oscillation around it. Note that, however, details of the evolution of the Higgs field are different for the different initial angles even if the value of the viscosity parameter C_{vis} is the same. For example, if we take $\alpha=-0.1\pi$, the Higgs field settles down to the SU(4)xU(1) state after large amplitude oscilations. When we take the smaller values for C_{vis}, the Higgs field begins to circulate in this (x,y)-plane and wanders around a lot of SU(3)xSU(2)xU(1) states and SU(4)xU(1) states. After a few time circulations, the Higgs field goes out from this plane and begins to wander around more numbers of SU(3)xSU(2)xU(1) and SU(4)xU(1) states in the four dimensional space of the Higgs field. This result suggests that the eventual state of the Higgs field is changed greatly by the very small deviation of the initial direction of the Higgs field.

In Fig.4 final states of the Higgs field are summarized in the plane of the initial angle α and the viscosity parameter C_{vis}. As has been discussed, final states of the Higgs field

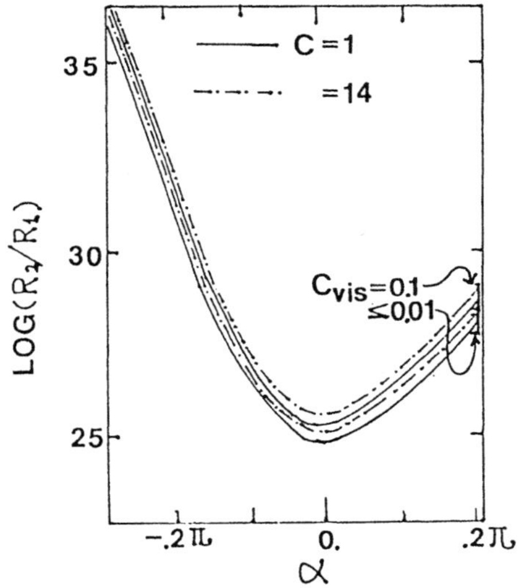

Figure 3. The dependence of the degree of inflation $D=R_2/R_1$ on the initial angle α, where R_1 and R_2 represent the values of the cosmic scale factor R when the inflation begins and when the Higgs field arrives near the SU(4)xU(1) minimum, respectively. The initial modulus of the Higgs field is assumed to be $(\psi_1^2 + \psi_2^2 + \psi_3^2 + \psi_1^2)^{1/2} = 8x10^{-6}\sigma$.

depend on the value of the viscosity parameter C_{vis} strongly. The final state is an SU(4)xU(1) minimum if C_{vis} is greater than a critical value. Although the critical value depends on the initial angle α as shown in Fig.4, we may conclude that the final state is SU(4)xU(1) if $C_{vis}>10^{-2}$. This state is, however, unstable if we take into account the quantum tunnelling effect. Although the Higgs field can reach a nearby SU(3)xSU(2)xU(1) state by this tunnelling, the same difficulties as appeared in the original inflation scenario(Sato 1981, Guth 1981, Guth and Weinberg 1981,1982, Sasaki et al. 1982, Kodama et al. 1982) arise because this phase transition is of first order, i.e., large scale inhomogeneity is created by bubbles formed by this phase transition as pointed out by Breit, Gupta and Zacks(1983).

When we take the value of the viscosity parameter in the range $2\times10^{-3}<C_{vis}<10^{-2}$, the Higgs field can settle down to the nearest SU(3)xSU(2)xU(1) state steadily without traveling to other SU(3)xSU(2)xU(1) or SU(4)xU(1) states. In this case, the inflation scenario works well with no trouble as has been investigated by many people.[2)]

On the other hand, if we take values $C_{vis}<2\times10^{-3}$, the Higgs field travels around a lot of minimum states as illustrated in Fig.2-c. Eventual states to which the Higgs field settles down are changed by fluctuations of initial values of the Higgs field. This result strongly suggests that a coherent region, which is

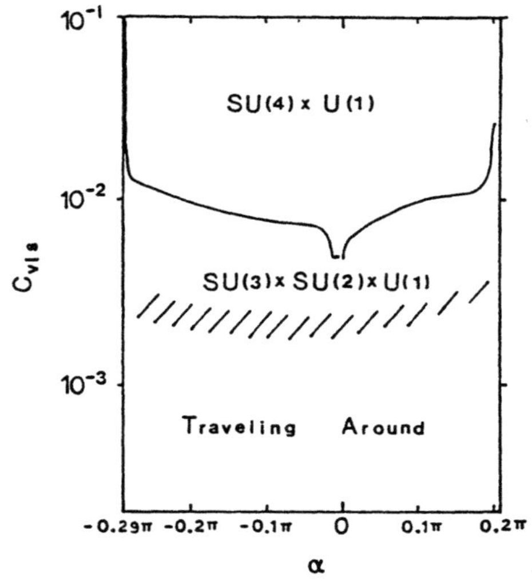

Figure 4. Summary of final states of the Higgs field in the time evolution calculations.

formed by nucleation of bubbles or spinodal decomposition in the early stage $\|\psi\| < H$, will be fragmented into many SU(3)xSU(2) xU(1) and SU(4)xU(1) states by the fluctuations of the Higgs field associated with the initial state. Even if the classical fluctuations of the Higgs field are extremely small, the quantum ones might fragment the original coherent region into small pieces of different state(Breit et al. 1983). Thus this result suggests that large scale inhomogeneities also appear for the too small values of C_{vis}, which conflict with present observation.

We have carried out numerical computations for different values of the potential parameter C (Eq(2)) in the range $0<C<14$. The result displayed in Fig.4 (the case C=1), does not change qualitatively for the other values of C in this range except that the critical value of the viscosity parameter C_{vis}, which decides whether the final state is SU(4)xU(1) or not, depends more sensitively on the initial angle α. We have also carried out the simulation by using the viscosity of the form $C_{vis}\|\psi\|\dot{\psi}_i$ instead of $C_{vis}|\dot{\psi}_i|\dot{\psi}_i$. The results were essentially the same both qualitatively and quantitatively except for the small change in the critical value of C_{vis}.

4. Conclusion

In the present work, we have found that in order for an inflationary scenario of the universe consistent with observation to be constructed, the value of the viscosity must be in an adequate range, otherwise large scale inhomogeneities which conflict with observations arise. Recently Abbott et al.(1983) and Hosoya and Sakagami(1983) estimated the strength of viscosity. At present, it is hard to judge whether the value of viscosity lies in the range adequate for the new inflationary scenario or not, because the result of Abbot et al. is very qualitative and the viscosity obtained by Hosoya and Sakagami is a thermal viscosity. In order to make clear whether a consistent scenario can be constructed or not, more precise evaluation of viscosity is necessary.

The authors thank D.W. Sciama, C.J. Pethick, I. Wasserman, and A. Hosoya for valuable discussion. This work is supported in part by the Grant in Aid for Science Research Fund of the Ministry of Education, Science and Culture No.56340021 and also by Asahi Scholastic Promotion Fund. Numerical computations were carried out with FACOM M-190 of LICEPP.

Note

1) The absolute units $c=\hbar=G=1$ are used throughout this paper.
2) We have checked that at least in this range of C_{vis} the

universe is heated up again to a temperature which is high enough for the baryosynthesis to occur.

References

Abbott, L.F., Farhi, E., and Wise, M.B. 1983, Phys. Lett. **117B**, 29.

Albrecht, A., and Steinhardt, P.J. 1982, Phys. Rev. Lett. **48**, 1220.

Breit, J.D., Gupta, S. and Zacks, A. 1983, Phys. Rev. Lett. **51**, 1007.

Guth, A.H. 1981, Phys. Rev. **D23**, 347.

Guth, A.H. and Weinberg, E.J. 1981, Phys. Rev. **D23**, 876.

Guth, A.H. and Weinberg, E.J. 1983, Nucl. Phys. **B212** (1983), 321.

Hosoya, A. and Sakagami, M. 1983, preprint OU-HET-57 .

Kodaira, J. and Okada, J. 1983, Kobe Univ., preprint 83-08.

Kodama, H., Sasaki, M. and Sato, K. 1982, Prog. Theor. Phys. **68**, 1979.

Linde, A.D. 1982, Phys. Lett. **108B**, 389.

Moss, I.G. 1983, Phys. Lett. **128B**, 385.

Sasaki, M., Kodama, H. and Sato, K. 1982, Prog. Theor. Phys. **68**, 1561.

Sato, K. 1981, Phys. Lett. **99B**, 66 ; Month. Notices Roy. Astron. Soc. **195**, 467.

PRIMORDIAL NUCLEOSYNTHESIS AND NUCLEAR REACTION RATES UNCERTAINTIES

P. Delbourgo-Salvador, E. Vangioni-Flam, G. Malinie and J. Audouze

Institut d'Astrophysique du CNRS
98 bis Bd Arago
Paris-France

I INTRODUCTION

Recent measurements concerning abundances of elements of cosmological interest (D, ^{4}He, ^{7}Li) have been performed respectively by Vidal-Madjar et al [1], Kunth and Sargent [2], Spite and Spite [3].

Vidal-Madjar and Gry [4] pointed out some incompatibilities between the measured values and theoretical predictions of the standard Big-Bang nucleosynthesis.

On the other hand, the values of the nuclear reaction rates involved in nucleosynthesis processes have been updated very recently by Harris et al [5].

In order to examine if the reaction rates uncertainties could account for the discrepancy between theoretical predictions and observations, we performed independent calculations of the abundances of these elements in the frame of the hot Standard Big-Bang (J. Audouze, this volume). But here we took also into account the uncertainties on the nuclear reaction rates involved in these computations.

Section 2 describes the computational techniques and the reaction rates used in the work, in Section 3 a comparison is made between the theoretical results and the observations in Section 4, the prediction regarding the maximun number of neutrino families (J. Audouze, this volume) is presented while Section 5 contains our conclusions.

A. Berger (ed.), The Big Bang and Georges Lemaitre, 113–122.

II THE NUMERICAL CODE

In the frame of the hot standard Big-Bang model, we have to follow the evolution of the abundances in a network of 50 nuclear reactions involving 16 nuclei (n, p, D, T, ^{3}He, ^{4}He, ^{6}Li, ^{7}Li, ^{8}Li, ^{7}Be, ^{9}Be, ^{8}B, ^{9}B, ^{10}B, ^{11}B, ^{11}C).

The abundance N_i of an element i is given by :

$$\frac{dN_i}{dt} = -\sum_j N_i N_j \langle\sigma v\rangle_{ij} + \sum_{kl} N_k N_l \langle\sigma v\rangle_{kl}$$

where $\langle\sigma v\rangle$ is the nuclear reaction rate by pair of elements i and j. The first term of the r.h.s. represents the rate of destruction of i by the all the reactions $i+j \rightarrow k+l$ and the second the formation of i by all the reactions $k+l \rightarrow i+j$. This system of 16 non linear differential equations must be solved by an implicit scheme because of the very strong temperature dependence of the reaction rates. We used the classical computational method described by Arnett and Truran [8]. The timestep is variable and adjusted in order to have a maximum relative variation of abundances of 5.10^{-3}. The amount of computing time required for a typical run is about 60s CPU on a Cyber 750 computer.

The physical conditions (temperature and density profiles) during the nucleosynthetic phase occuring just after the Big-Bang are described by Weinberg [9]. The network and references for the reaction rates are given in table I.

The parameters governing the final abundances are N_ν the number of neutrinos families, ρ_0 the present density of the Universe and $\tau_{1/2}$ the lifetime of neutron.

Fig.I displays the values of primordial abundances as a function of ρ_0 for N_ν = 2, 3 and 4.

The results are in good agreement with those obtained by Wagoner [10] and Beaudet and Yahil [11].

Table I

Reactions	Rates
p(n,γ)D	Fowler et al 1967
D(n,γ)T	" "
^{3}He(n,γ)^{4}He	" "
^{6}Li(n,γ)^{7}Li	" "

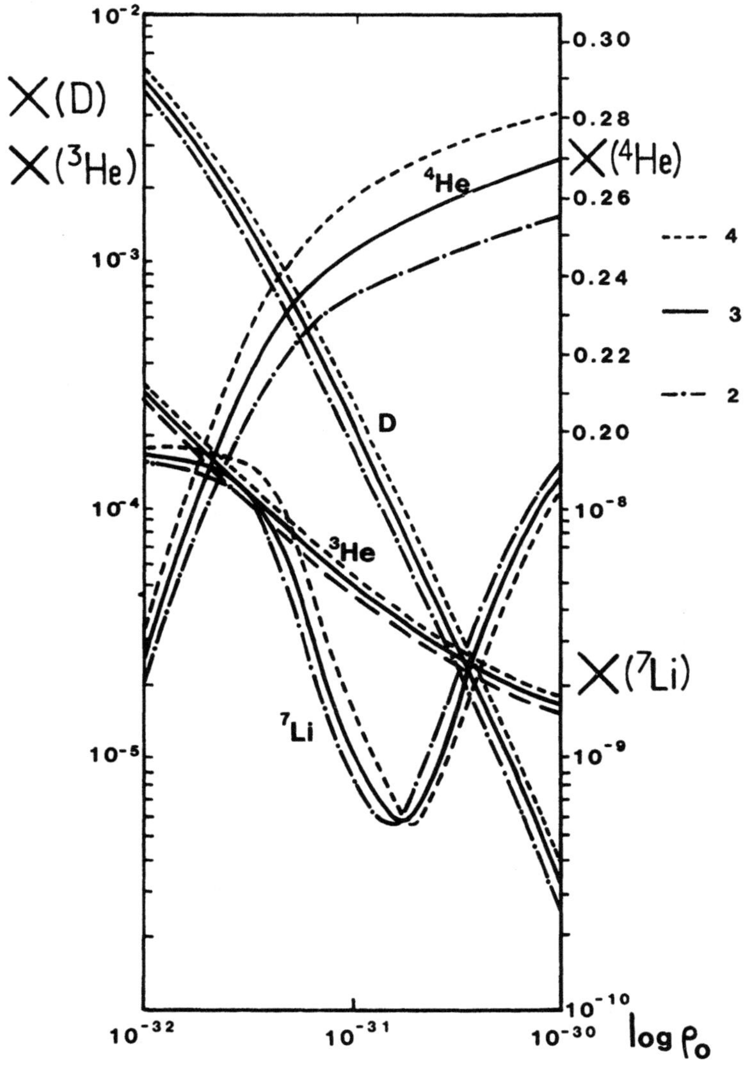

Figure I
Primordial abundances by masses of D, ^3He, ^4He, ^7Li as functions of ρ_0, present density of the Universe, and for 2, 3, 4, families of neutrinos.

$\tau_{1/2}$ = 10,61 min

$^{10}B(n,\gamma)^{11}B$	Fowler et al 1967
$T(p,n)^{3}He$	Fowler et al 1973
$^{7}Li(p,n)^{7}Be$	" "
$^{11}B(p,n)^{11}C$	" "
$^{9}Be(p,n)^{9}B$	" "
$D(p,\gamma)^{3}He$	Harris et al 1983
$T(p,\gamma)^{4}He$	Fowler et al 1973
$^{6}Li(p,\gamma)^{7}Be$	Harris et al 1983
$^{7}Be(p,\gamma)^{8}B$	" "
$^{9}Be(p,\gamma)^{10}B$	Fowler et al 1975
$^{10}B(p,\gamma)^{11}C$	" "
$T(\alpha,n)^{6}Li$	" "
$^{8}Li(\alpha,n)^{11}B$	Wagoner 1969
$^{4}He(\alpha,n)^{7}Be$	" "
$^{7}Li(\alpha,n)^{10}B$	Fowler et al 1975
$D(\alpha,\gamma)^{6}Li$	Harris et al 1983
$T(\alpha,\gamma)^{7}Li$	Fowler et al 1975
$^{6}Li(\alpha,\gamma)^{10}B$	Harris et al 1983
$^{7}Li(\alpha,\gamma)^{11}B$	Fowler et al 1975
$^{7}Be(\alpha,\gamma)^{11}C$	Harris et al 1983
$D(D,n)^{3}He$	Fowler et al 1975
$T(D,n)^{4}He$	" "
$D(D,n)T$	" "
$^{3}He(D,p)^{4}He$	" "
$^{3}He(^{3}He,2p)^{4}He$	" "
$^{4}He(\alpha n,\gamma)^{9}Be$	" "
$^{7}Be(D,p)2\,^{4}He$	" "
$^{11}B(p,\alpha)2\,^{4}He$	Harris et al 1983
$^{7}Li(T,2n)2\,^{4}He$	Fowler et al 1975
$^{7}Li(^{3}He,np)2\,^{4}He$	" "
$^{7}Be(^{3}He,2p)2\,^{4}He$	" "
$D(p,n)2p$	" "
$T(T,2n)^{4}He$	" "
$^{3}He(T,D)^{4}He$	" "
$^{3}He(T,np)^{4}He$	" "
$^{4}He(np,\gamma)^{6}Li$	" "
$^{7}Be(T,np)2\,^{4}He$	" "
$^{7}Li(D,n)2\,^{4}He$	" "
$^{8}B(n,2\,^{4}He)p$	Wagoner 1969
$^{11}C(n,2He)^{4}He$	" "
$^{8}Li(p,2\,^{4}He)D$	" "
$^{9}Be(p,2\,^{4}He)D$	" "

III COMPARISON WITH OBSERVATIONS

A compilation of observations provides the following range (Table II) of primordial abundances. (J. Audouze, this volume)

Table II

^4He	$0.22 \lesssim X \lesssim 0.25$
D	$X(D) = (2 \pm 2)\ 10^{-5}$
^3He	$X(^3He) = (2 \pm 2)\ 10^{-5}$
^7Li	$X(^7Li) = 5\ 10^{-10}$

The corresponding uncertainty boxes are shown on figure II where the abundances have been computed with respect to ρ_o for $N_\nu = 3$ and $\tau_{1/2} = 10.6$ min.

One can find ρ_o values consistent with all primordial abundances of the light elements only if X(D) primordial > 10^{-4}. This requires that models of chemical evolution of galaxies as specific as those of Gry et al [13] where $X(D)_{prim}/X(D)_{interstellar} \gtrsim 20$ do apply. This requests inflow of already processed material inside the galactic zones where chemical evolution is analysed.

If this is not the case, there is then a discrepancy between the ρ values deduced from D and ^4He abundances respectively as noticed e.g. by Vidal-madjar and Gry [4]. In order to explain this possible discrepancy one could assume that the standard Big-Bang model does not apply (i) Audouze and Silk [12] are currently analyzing possible photodesintegration processes able to produce D from ^4He or the effect of pregalactic cosmic rays in the case of a cold Big-Bang model (ii) the effect of anisotropy of the Universe has been considered by Barrow [15], Juskiewicz et al [16] and also by Gorski and Delbourgo-Salvador [17].

In any case (discrepancy between the ρ values or not) it is worth to examine the effect of nuclear rate uncertainties.

IV INFLUENCE OF THE REACTION RATES AND OF THE NUMBER OF NEUTRINO FAMILIES

- Influence of the new reaction rates

In table III, the modifications of the computed abundances of the light elements when the most recent nuclear rates are used instead of the older ones, are displayed. The calculations have been performed only for the two most important reactions : $D(p,\gamma)^3He$ and $^3He(^4He,\gamma)^7Be$.

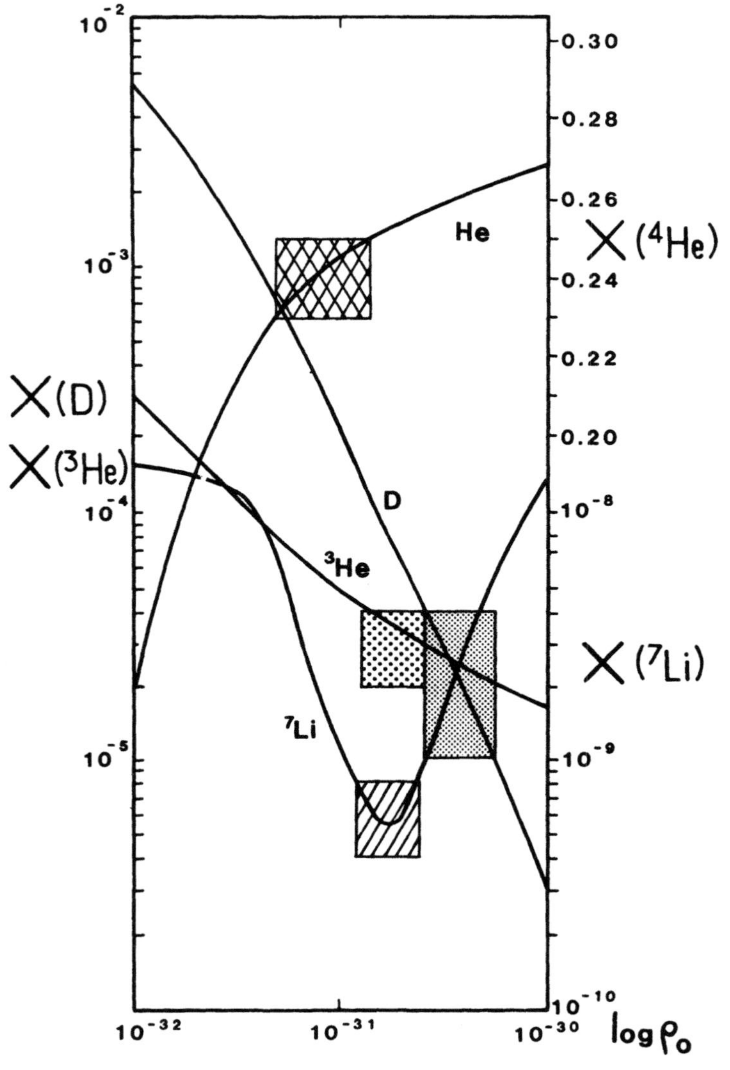

Figure II
Primordial abundances of D, ^3He, ^4He, ^7Li as functions of ρ_o for 3 families of neutrinos.
The boxes show the ranges deduced from observations.
$\tau_{1/2}$ = 10,61 min

Table III

- $D(p,\gamma)^3He$

ρ_0	^4He	D	^3He	^7Li
10^{-32}	0%	0%	0%	0%
10^{-31}	0%	0,6%	1,3%	0%
10^{-30}	0%	18%	8,6%	9%

- $^3He(^4He,\gamma)^7Be$

ρ_0	^4He	D	^3He	^7Li
10^{-32}	0%	0%	0%	0%
10^{-31}	0%	0%	0%	0,4%
10^{-30}	0%	0%	0%	1%

- Influence of the uncertainties on the reactions rates

The deviations of the primordial abundances of the light elements corresponding to 10% variations of the nuclear rates are shown on Table IV.

Table IV

- $p(n,\gamma)D$ ±10%

ρ_0	^4He	D	^3He	^7Li
10^{-32}	5%	1,5%	1%	8%
10^{-31}	0,5%	7%	3%	8%
10^{-30}	0%	1%	0,6%	4%

- $D(D,n)^3He$ ±10%

ρ_0	^4He	D	^3He	^7Li
10^{-32}	0%	3,5%	2,5%	3,5%
10^{-31}	0%	4%	2,6%	12%
10^{-30}	0%	10%	1,7%	4,1%

- $T(^4He,\gamma)^7Li$ ±10%

ρ_0	^4He	D	^3He	^7Li
10^{-32}	0%	0%	0%	10%
10^{-31}	0%	0%	0%	9%
10^{-30}	0%	0%	0%	0%

- $T(D,n)^4He$ ±10%

ρ_0	^4He	D	^3He	^7Li
10^{-32}	0,7%	0%	1%	9%
10^{-31}	0%	0%	0,2%	10%
10^{-30}	0%	0,5%	0,5%	0,7%

From tables III and IV one can see that the variation of abundances are ⩽ 1% for ^4He, ⩽ 20% for D, ⩽ 10% for ^3He and ⩽ 20% for ^7Li.

Fig III displays the variation of abundance of ^4He in fonction of ρ_0 for 3 values of $\tau_{1/2}$. The primordial value of ^4He depends more strongly on the parameter $\tau_{1/2}$ than on the others reactions rates. The primordial abundance of ^4He can be obtained directly from the ratio n/p between the neutron and the proton density at the begining of nucleosynthesis (Yang et al [14]) :

$$X_{^4He} = 2 \frac{n/p}{1 + n/p}$$

n/p depends on $\tau_{1/2}$

On the same figure, the primordial abundances of D, ^3He and ^7Li have been represented with the maximum variation i.e. ± 20% due to the nuclear reaction rate uncertainties.

The uncertainties ranges of observations are drawn on this figure and one can see that the discrepancy between ^4He and D still exists (if the inflow chemical evolution model of Gry et al [13] does not apply.

- Influence of the number of neutrino families

The possible discrepancy between the ρ values deduced from D or ^4he increases if $N_\nu = 4$ and decreases if $N_\nu = 2$ (which could be the actual case if the tau neutrino is not relativistic during the early phases of the Universe.

V CONCLUSION

There are no discrepancy between the values for the present density ρ_0 of the Universe deduced from ^4He and D only if the chemical evolution models allowing inflow of processed material in the considered zone (Gry et al [13]) are those which actually apply.

In the case of discrepant ρ_0 values, the consideration of uncertainties on the nuclear reaction rates cannot solve this difficulty.

We are indebted to Robert Mochkovitch for fruitful discussions.

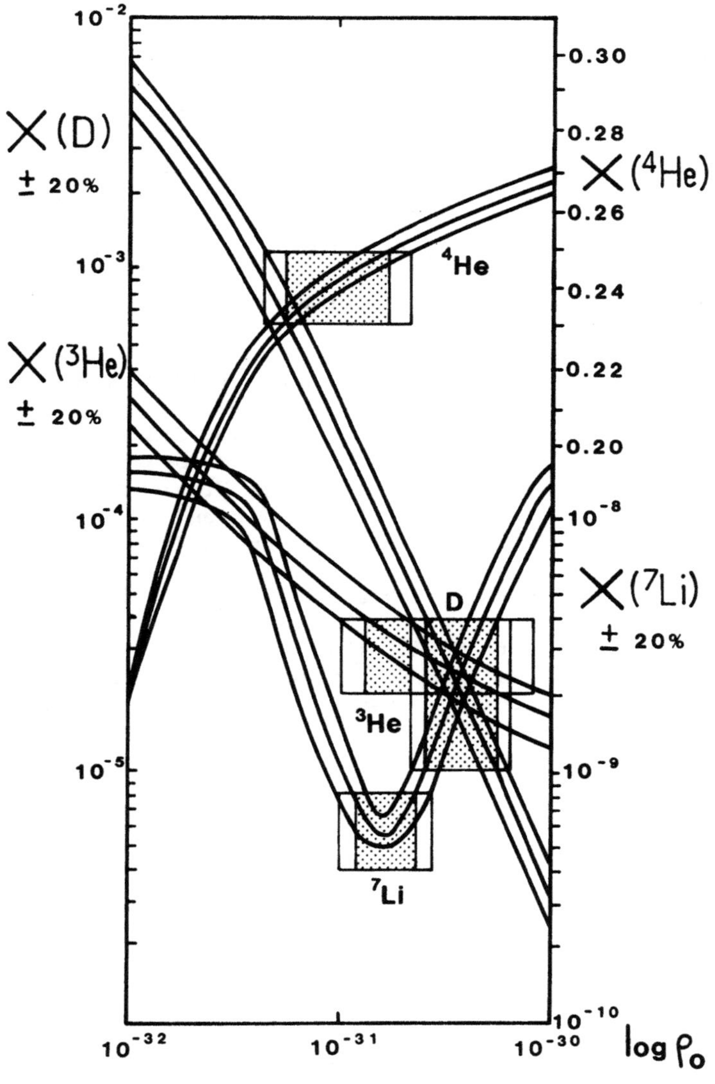

Figure III

Variation of primordial abundance of ^4He as a function of ρ_0 for three values of $\tau_{1/2}$: 10.81 min ; 10.61 min ; 10.41 min (the abundance increase when $\tau_{1/2}$ increase). On the same figure : Primordial abundances of D, ^3He, ^7Li with a ± 20% variation. The boxes show the consequences on the range of ρ_0 given by the observationnal datas. The number of families of neutrinos is three.

REFERENCES

(1) Vidal-Madjar A., Gry C., Bruston P., Ferlet R., York D.G., Astron. & Astroph., 120, 58 1983

(2) Kunth D., Sargent W., Submitted to Astrophys. J.

(3) Spite F., Spite M., Astron. & Astroph. 115, 357 1982

(4) Vital-Madjar A., Gry C., Submitted to Astron. & Astroph.

(5) Harris M., Fowler W., Caughlan C., Zimmerman B., Ann. Rev. of Astron. & Astroph 1983

(6) Fowler W., Caughlan G., Zimmerman B., Ann. Rev. of Astron. & Astroph. 1975

(7) Fowler W., Caughlan G., Zimmerman B., Ann. Rev. of Astron. & Astroph. 1967

(8) Arnett W., Truran J., Astrophys. J. 157, 339 1969

(9) Weinberg S., Gravitation and Cosmology 1972 Wiley and Sons

(10) Wagoner R., Astroph. J. 218, 253, 1977

(11) Beaudet G. & Yahil A., Astroph.J. 218, 253 1977

(12) Audouze J. & Silk J., 1984, submitted to Astroph. J.

(13) Gry C., Malinie G., Audouze J., and Vidal-Madjar A., "Formation and evolution of galaxies and large structure of the Universe" ed. J. Audouze and J. Tran Thanh Van 1983, Reidel Dordrecht

(14) Yang J., Steigman G., Schramm D., Rood R., 1975, Astroph.J. 227, 697

(15) Barrow J.D., 1976, MNRAS, 175, 359

(16) Juskiewicz R., Bajtlik S., Gorski K., 1983, MNRAS, 204, 63P

(17) Gorski K., Delbourgo-Salvador P., 1984, in preparation

TIME and SINGULARITY*

Jacques Demaret
Institut d'Astrophysique, Université de Liège,
B-4200 Cointe-Ougrée, Belgium

Mark J. Gotay
Department of Mathematics and Statistics, University of Calgary, Calgary, Alberta, Canada T2N 1N4

Abstract: We show that the occurrence of quantum gravitational collapse and, more generally, the validity of Wheeler's "rule of unanimity" are inextricably linked to the classical choice of time. The crucial distinction is between "fast" and "slow" times, that is, between times which give rise to complete or incomplete classical evolution respectively. We conjecture that unitary slow-time quantum dynamics is always non-singular, while unitary fast-time quantum dynamics inevitably leads to collapse. These findings are illustrated by an analysis of the dust-filled Friedmann-Lemaître-Robertson-Walker universes.

* Dedicated to the memory of a colleague and friend, Arsène Boury (J.D.)

A. Berger (ed.), The Big Bang and Georges Lemaitre, 123–131.

It was a most ancient ... tradition amongst the Pagans ... that the cosmogonia ... took its first beginning from a chaos.

Cudworth (1678)

What is the genesis of the Universe? This fundamental question has never ceased to arouse man's curiosity and fascinate theologians and scientists alike. That there are two distinct issues here, the *beginning* of the Universe as opposed to its *creation*, was already pointed out by St. Thomas Aquinas 700 years ago. The creation of the Cosmos is an a priori philosophical concept, and its elucidation belongs to the realm of metaphysics. In contrast, the beginning of the Universe is an empirical concept and thus amenable to scientific analysis.[1]

With the prodigious development of relativistic cosmology since 1915, the traditional Western belief in the permanence of the heavens has gradually yielded to the notion that the Universe had an absolute beginning. Indeed, the first successful relativistic model of our expanding Universe, due to Friedmann, possessed an infinite density, infinite curvature cataclysm a finite proper time in the past. Insofar as one can reject the possibility of a closed cyclic universe with an infinitude of past cycles -- the "phoenix" universe of Monseigneur Lemaître -- such an *initial singularity* must represent the beginning of the Universe.[2]

Although the first detailed analysis of this phenomenon of catastrophic spacetime collapse was given by Monseigneur Lemaître, it was not realized until the late 1960s to what extent singularities form an essential element of modern cosmology [4]. In view of the celebrated theorems of Hawking, Penrose and Geroch, it is now clear that singularities must occur in all "physically reasonable" spacetimes. Moreover, the existence and isotropy of the cosmic microwave background strongly imply the presence of a singularity in the past of our Universe.

Still, the case for the initial singularity is not ironclad, since the singularity theorems are classical constructs and as such do not take into account quantum phenomena which are expected to be important during the exotic early stages of the Universe. As one extrapolates further into the past, it is therefore conceivable that quantum effects could modify -- or perhaps prevent altogether -- gravitationally induced spacetime collapse. On the other hand, Wheeler [5] has recently proposed a "rule of unanimity" which, if valid, would shatter this hope: "Given that all solutions of the equations of motion run into a singularity (or are free of singularity) except a set of measure zero. Then all solutions of the corresponding quantum-mechanical problem are singular (or free of singularity)."

In the absence of a complete consistent quantum theory of the gravitational field and its interactions, research on quantum singularity avoidance has proceeded along two lines: a search for semi-classical mechanisms for suppressing the formation of singularities and model calculations within the framework of quantum cosmology. Despite considerable effort, however, a satisfactory resolution of the quantum collapse problem remains a chimera. On the semiclassical level, in which the matter is quantized but the gravitational field is treated classically, attempts to eliminate the classical singularities by inducing violations of the positive energy conditions in the singularity theorems remain inconclusive [6]. Quantum cosmological studies, which include the quantum effects of both matter and gravity in the analysis (albeit at the expense of "freezing out" all but a finite number of degrees of freedom), are similarly beset with a host of technical and conceptual difficulties [7]. The foremost among these stems from the freedom in the classical choice of time: different such choices often lead to wildly divergent quantum behaviors [7-15].

The work which we now briefly summarize [8] is devoted to a study of the relationship between quantum gravitational collapse and the choice of time. We find that whether *quantum* collapse occurs is effectively predetermined, on the *classical* level, by this very choice. The crucial distinction is between *fast* and *slow* times, that is, between times which give rise to complete or incomplete classical evolution respectively.[3] More precisely, we conjecture that *unitary slow-time quantum dynamics is always non-singular, while unitary fast-time quantum dynamics inevitably leads to collapse.* These results indicate that the quantum collapse question is really quite intricate and also help to reconcile the heretofore bewildering array of "answers" to this question.

We substantiate these contentions with an analysis of the classically collapsing Friedmann-Lemaître-Robertson-Walker (FLRW) universes in two time gauges, one fast and the other slow (cf. [8]).[4] These homogeneous and isotropic cosmologies are described by the metrics

$$ds^2 = -N(t)^2 dt^2 + e^{2\mu(t)} d\Sigma^2 \ ,$$

where $d\Sigma^2$ is the line element for a 3-manifold of constant curvature $k = +1$, 0 or -1. The matter content is taken to be dust with density ρ and 4-velocity $u = -d\varphi$, φ being the only nonvanishing Seliger-Whitham velocity potential. The super-hamiltonian constraint, characteristic of the general relativistic Hamiltonian formalism, is

$$p_\varphi - \frac{1}{24} e^{-3\mu} p_\mu{}^2 - 6ke^{\mu} = 0 \ , \tag{1}$$

where

$$p_\mu = -\frac{12}{N}e^{3\mu}\dot{\mu} \quad \text{and} \quad p_\varphi = \rho u^0 N e^{3\mu}$$

are the momenta canonically conjugate to μ and φ, respectively. Since these models are in parametrized form, they admit an Arnowitt-Deser-Misner reduction. This consists of two steps: choosing a time t and then solving the constraint (1) in the form $p_t - H = 0$, thereby determining the effective Hamiltonian H.

We first choose the time from among the matter variables: $t = -\varphi$. This is essentially cosmic time, and hence slow. After performing the canonical transformation

$$x = \frac{4}{3}\sqrt{6}e^{3\mu/2}, \quad p_x = \frac{\sqrt{6}}{12}e^{-3\mu/2}p_\mu,$$

reduction yields the phase space $(0,\infty) \times \mathbb{R}$ and the Hamiltonian

$$H(x,p_x) = p_x{}^2 + Kx^{2/3}, \tag{2}$$

where $K = \frac{3}{2}\,{}^3\sqrt{6}k$. The dynamics is thus equivalent to that of a particle on the half-line $(0,\infty)$ moving in a potential $V(x) = Kx^{2/3}$.

Upon quantizing we find that the quantum Hilbert space is $L^2(0,\infty)$ and that the operator corresponding to the Hamiltonian (2) has an infinite number of self-adjoint extensions

$$\hat{H}_\alpha = -\hbar^2\frac{d^2}{dx^2} + Kx^{2/3}$$

determined by the boundary conditions

$$\psi'(0) = \alpha\psi(0), \tag{3}$$

where the parameter $\alpha \in (-\infty,\infty]$. Since the operators $\hat{H}_\alpha$ are rather complicated, we illustrate the qualitative features of the dynamics via the motion of a $k = 0$ wave packet.

Fix $\beta = b + iB$, $b > 0$, and consider the initial state

$$\psi(x,0) = \left(\frac{8b}{\pi}\right)^{\frac{1}{4}}e^{-\beta x^2}. \tag{4}$$

Taking $\alpha = 0$, this evolves according to

$$\psi(x,t) = \left(\frac{8b}{\pi}\right)^{\frac{1}{4}}[1 + 4i\hbar\beta t]^{-\frac{1}{2}}e^{-\beta x^2/[1+4i\hbar\beta t]}. \tag{5}$$

To check for collapse, we study the expectation value

$$\langle\psi(t)|\hat{x}|\psi(t)\rangle = (2\pi b)^{-\frac{1}{2}}[1 - 8B\hbar t + 16(b^2+B^2)\hbar^2t^2]^{\frac{1}{2}}$$

of the quantum "radius" operator, recalling that classically $x \propto e^{3\mu/2}$ measures the expansion of our model.

If $B > 0$, this wave packet represents a universe which is initially contracting. But as t approaches the "turn-around time"

$$T = B/4(b^2+B^2)\hbar \ ,$$

quantum effects are decisive: the universe decelerates, "bounces," and expands thereafter! In contrast, the classical model corresponding to the initial state (4) contracts uniformly and collapses after a time $T_c = 1/4B\hbar > T$. This behavior is displayed below.

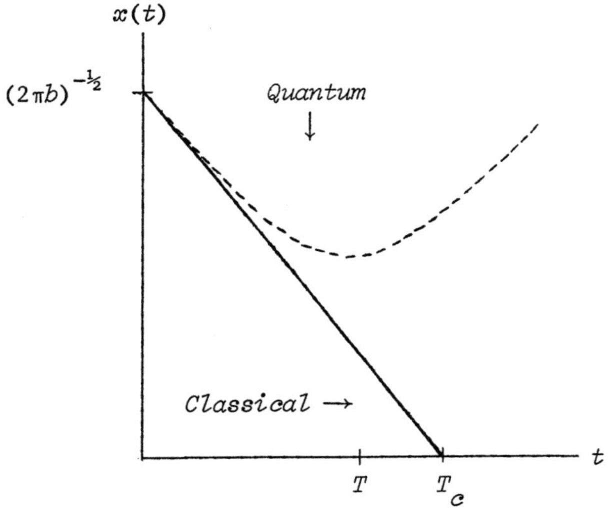

Classical/quantum correspondence for the wave packet (5)

In fact, this instance of quantum singularity avoidance is not exceptional: since $\hat{x}$ is a positive operator and as each $\hat{H}_\alpha$ is self-adjoint, $\langle\psi(t)|\hat{x}|\psi(t)\rangle$ can *never* vanish in finite time for *any* evolving state $\psi(t)$. Consequently, *no nontrivial state can evolve into a singularity* so that, within this dynamical framework, *quantum gravitational collapse is strictly forbidden.*

An unexpected corollary is that *this phenomenon of quantum singularity avoidance is independent of the choice of boundary condition* (3). This is contrary to widespread belief, which holds that an evolving state $\psi(x,t)$ is non-singular if and only if $\psi(0,t) = 0$ for all t [9]. In particular, note that the wave packet (5) is certainly non-singular, even though $\psi(0,t) \neq 0$ always.

Another important consequence of our analysis is the breakdown of Wheeler's rule of unanimity. To regain it there are only two options: modify either the classical or the quantum formalism. But which one, and how? The key observation is that, ultimately, the cause of the disparity between the classical and the quantum predictions is that *the quantum evolution persists eternally, whereas the classical evolution does not*. Since self-adjointness *guarantees* that the quantum dynamics is defined for all time,[5] it is apparent that we should "complete" the classical dynamics.[6]

As this "paradox" of incomplete classical versus complete quantum evolution arises whenever one makes a *slow* choice of time, one might expect results more in agreement with the unanimity principle when one quantizes in a *fast*-time gauge. Then both the corresponding classical and quantum dynamics are complete, although the physical implications of this completion are rather surprising. Classically, of course, the system is still singular. Quantum mechanically, however, completeness in fast time has a quite different meaning than it does in slow time. Eternal slow-time quantum evolution implies that collapse is impossible. But quantum completeness in fast time, being physically equivalent to incompleteness in slow time, can only signal the presence of a singularity. In other words, it is plausible that fast-time quantum dynamics incorporates collapse in much the same way that slow-time dynamics prohibits it.

We verify this assertion for the $k = -1,0$ dust-filled FLRW universes in the intrinsic-time gauge $t = \mu$. Since $t = -\infty$ corresponds to the initial singularity, $t = \mu$ is a fast clock. The reduced phase space is $\mathbb{R} \times (0,\infty)$ and, from (1), the effective Hamiltonian is

$$H(\varphi,p_\varphi,t) = 2\sqrt{6}e^{3t/2}[p_\varphi - 6ke^t]^{\frac{1}{2}} . \tag{6}$$

Quantizing in the momentum representation, the time-dependent quantum Hamiltonian $\hat{H}(t)$ is represented by multiplication by $H(t)$ on the Hilbert space $L^2(0,\infty)$.

It is straightforward to check that the resulting quantum dynamics is unitary, so that these models must evolve to the $t = -\infty$ limit. Furthermore, since classically $H(t) \to 0$ as $t \to -\infty$ and as $\hat{H}(t)$ is a positive operator, the expectation value $\langle\psi(t)|\hat{H}(t)|\psi(t)\rangle$ is a good indicator of quantum collapse. Then (6) and the dominated convergence theorem yield

$$\lim_{t \to -\infty} \langle \psi(t) | \hat{H}(t) | \psi(t) \rangle = 0 \ ,$$

so that these models asymptotically collapse. But asymptotic collapse in this fast intrinsic time means that *these quantum models become singular in finite proper time* so that, within this dynamical setting, *quantum gravitational collapse* is *inevitable*.

As this analysis demonstrates, the validity of Wheeler's rule of unanimity depends critically upon the choice of time. This classically innocuous choice is the decisive factor governing the occurrence of quantum gravitational collapse. Although our conclusions are motivated in the context of the FLRW models, a moment's reflection shows that they will apply, *mutatis mutandis*, to any spatially homogeneous cosmology.

Our claim that quantum collapse is strictly forbidden within the slow-time dynamical framework is supported by the work of DeWitt [9], Lund [10] and Lapchinskii and Rubakov [11] on the FLRW universes and Demaret's analyses [12] of several Bianchi models. On the other hand, our contention that the fast-time version of quantum cosmology does not significantly alter the classical behavior near the singularity is consistent with the findings of Misner and Ryan [13], Gotay and Isenberg [14] and Brill [15]. Thus, our conjectures are confirmed for a wide range of both cosmological models and (intrinsic-, extrinsic- and matter-) time gauges.

Of course, it remains to determine which of these classical/quantum formalisms is "correct". Philosophical considerations [2-4] aside, the answer must likely await the development of a complete quantum theory of gravity. Until then, one can only wonder, like philosophers of all ages, whether indeed "... the world has a beginning in time."

Notes

[1]This is so despite Scriven's claim [1] that the origin of the Universe "... is not within the power of science to determine, nor will it ever be." North refutes this assertion in Chap. 18 of [2].

[2]Whether it can be demonstrated, on a purely philosophical basis, that the Universe has either a finite or an infinite past remains open to question [3].

[3]We call a time variable t a *fast time* if the singularities always occur at either $t = -\infty$ or $t = +\infty$. If this is not the case, then t is said to be a *slow time*.

[4]We chose units so that $c = 1$ and $16\pi G = 1$.

[5]It is possible [8] to relax the requirement that the quantum Hamiltonian be self-adjoint by letting α in (3) be complex. The operators $\hat{H}_\alpha$ with Im $\alpha < 0$ will then generate *contraction semigroups* rather than unitary groups. In this case, the quantum models may *asymptotically collapse* in the sense that $\langle\psi(t)|\hat{x}|\psi(t)\rangle \to 0$ as $|t| \to \infty$, although it still cannot be ensured that an initially contracting state will collapse in *finite* time.

[6]This is in keeping with Lund's suggestion [10] that one should always quantize on a geodesically complete minisuperspace. Here, however, the completion consists of modifying the choice of time rather than the minisuperspace itself.

References

[1] Scriven, L. (1954), Brit. J. Phil. Sci., *5*, p. 181.

[2] North, J.D. (1965), *The Measure of the Universe* (Clarendon Press, Oxford).

[3] Whitrow, G.J. (1978), Brit. J. Phil. Soc., *29*, p. 39; Popper, K. (1978), Ibid., *29*, p. 47; Bell, J. (1979), Ibid., *30*, p. 161; Craig, W.L. (1979), Ibid., *30*, p. 165; Weingard, R. (1979), Ibid., *30*, p. 170.

[4] Misner, C.W. (1969), Phys. Rev., *186*, p. 1328.

[5] Wheeler, J.A. (1977), Gen. Rel. Grav., *8*, p. 713.

[6] Barrow, J.D. and Matzner, R.A. (1980), Phys. Rev. D *21*, p. 336; Birrell, N.D. and Davies, P.C.W. (1982), *Quantum Fields in Curved Space*, Sec. 7.4 (Cambridge U. Press, Cambridge).

[7] MacCallum, M.A.H. (1975), in *Quantum Gravity*, eds. Isham, C.J., Penrose, R., and Sciama, D. (Clarendon Press, Oxford).

[8] Gotay, M.J. and Demaret, J. (1983), Phys. Rev., D*28*, p. 2402.

[9] DeWitt, B.S. (1967), Phys. Rev., *160*, p. 1113.

[10] Lund, F. (1973), Phys. Rev., D*8*, p. 3253.

[11] Lapchinskii, V.G. and Rubakov, V.A. (1977), Teor. i. Mat.

Fiz., *33*, p. 364.

[12] Demaret, J. (1979), Gen. Rel. Grav., *11*, p. 453; (1980), Phys. Lett., *95*B, p. 413; (1980), Bull. Soc. Math. Belgique, *32*, p. 99.

[13] Misner, C.W. (1969), Phys. Rev., *186*, p. 1319; Ryan, M. (1972), *Hamiltonian Cosmology*, Lecture Notes in Physics No. 13 (Springer, New York).

[14] Gotay, M.J. and Isenberg, J.A. (1980), Phys. Rev. D*22*, p. 235; (1980), In Lecture Notes in Physics No. 135, p. 418 (Springer, New York).

[15] Brill, D.R. (1975), In *Quantum Theory and the Structures of Space and Time*, eds. Castell, L., Drieschner, M., and von Weizsächer, C.F. (Carl Hansen Verlag, München).

QUASAR ENERGY FROM FROZEN FUSION VIA MASSIVE NEUTRINOS ?

J. Steyaert

Institut de Physique, Université Catholique de Louvain, B-1348 Louvain-la-Neuve, Belgium

"... negative hydrogen ions. This is an instance of an important astrophysical problem demanding for its solution an atomic phenomenon which plays no important part in terrestrial physics and the understanding of which depends upon a considerable refinement of atomic theory".

W.Y. McCrea, Physics of the Sun and Stars, p. 70, 1950, Hutchinson, London

The above sentence, discovered by chance when finishing this article, is quoted to illustrate the importance of individual processes for the understanding of macroscopic ones.

Speculation is dangerous and speculative articles, like this one, should be targeted. "Proving they are wrong could be more enlightning than agreeing on their correctness". If this sentence is maybe from Feynman, it is a good place to recall Lemaître's opinion about the double frustration with mistakes : firstly when you discover it, secondly when you wonder why you didn't discover it earlier.

Intuition leading to objectivity is a difficult task but intuition has a role to play in the scientific walk.

Contradictory experiments in atomic physics about a simple charge exchange reaction

$$H^+ + H^- \rightarrow H + H \qquad [1]$$

A. Berger (ed.), The Big Bang and Georges Lemaitre, 133–138.

where the starting point of the present speculation. It has been called the H-H problem by the physicists (1) who first performed a detailed calculation before the complete experiment was executed and found to agree reasonably well (2).

Contrary to the detection of both H atoms in coincidence, previous experiments (3) detecting only one H, showed dramatic peaks in the cross section between 20 eV and 500 eV center-of-mass energy of both H.

At the Baddeck Conference (Nova Scotia, Canada), the discrepancy became public (4) and it was proposed that it could be due to the reaction

$$H^+ + H^- \rightarrow H + H^+ + e^- \qquad [2]$$

Unfortunately, this reaction amounts to a few percent of [1] shows no peaks as was recently reported (5).

At about the same time, a measurement of electron-neutrino mass was announced (6) giving $14\ eV < m_\nu < 44\ eV$. This experiment is on the way to be confirmed; it gives a most probable value $m_\nu = 34$ eV which coincides rather curiously with the low energy peak of [1], of which the other peaks are at about 54 eV and 150 eV.

This coincidence led us to propose a variant of the p-e-p reaction

$$H^+ + H^- \rightarrow D + \nu \qquad [3]$$

as a complement to reaction [1].

We have no idea about the details of reaction [3] which should be considered only for its energetics. Its probable cross section of 10^{-13} cm^2 is about 30 order of magnitude different from the 10^{-44} cm^2 cross section expected if only weak interaction is present. However, it should be pointed out that present neutrino physics is derived from relativistic neutrinos and no-one knows the dynamics of massive neutrinos with a kinetic energy around or below $m_\nu c^2$.

The peaks observed in the cross section of reaction [1] are maybe an experimental artefact as further investigations by Peart and Dolder have not been able to reproduce them, in spite of a considerable and careful experimental work (7). Nevertheless, peaks have been observed at a given time by at least two independent groups and their origin remains unexplained. The data should be reexamined and, if possible, found again with the help of investigating the role of detectors, ion sources, residual gas, the age of the hydrogen bottle (8 or 12 years old), origin of the hydrogen, presence of electric or magnetic fields in reaction area, etc.

Up to now, only atomic physicists are concerned with an embarrasing experimental conflict and above ideas could be kept as a mere clipboard curiosity.

Nevertheless there is a small chance that the above process is of interest to astrophysicists. The black hole hypothesis for quasar formation should still be considered as a working one. Just from the energetics the reaction

$$H + H \rightarrow D + \nu \qquad [4]$$

$$E_{\nu} = 1.4 \text{ MeV} \qquad E_D = 553 \text{ eV}, \; v_D \sim 230 \text{ km s}^{-1}$$

and less probably [3], are possible contenders. More over, it could well be that only radio-quiet quasars are concerned. The radio-louds ones, where electrons are accelerated (or decelerated), are maybe depending of the still more exotic reactions [5] cited a little further.

If the processes are really present we will call them "frozen fusion" as the concept "cold fusion" is already used in heavy ion nuclear physics.

Neutrinos are playing an increasing role in astrophysics : source of the missing mass in Galaxies (7), diffuse gas embedded in clusters of Galaxies (8).

If reaction [4] is really producing neutrinos it is maybe of importance to the physics of proto-galaxies, of which the quasars seem to be part of.

Supposing the quasar to be a hydrogen plasma with an atomic temperature T of a few ten to a few hundred eV, the charged components could undergo reaction [3] and the neutral ones reaction [4]. The reaction rate being $r = N^2 \langle \sigma v \rangle$, we get for $v \sim 10^7$ cm s^{-1}, $\sigma \sim 10^{-14}$ cm^2 and $N \sim 10^4$ cm^{-3} (as suggested by the strength of O[II] with respect to O[III] lines (9)) a power of $\sim 10^{41}$ W is obtained in a sphere of 1 pc radius, supposing most of E_D converted in radiation. This is consistent with the power emitted by a typical quasar, about 10^{39} W, or 10^{46} erg/s (10). For such a N the reaction rate would be 10 cm^{-3} s^{-1}. In about 1000 s the neutral gas (or the neutral component of the plasma) would be depleted in H and enriched in D. Frozen fusion could proceed further with D and so on, but there is no experimental indication for the following reactions, selected among others :

$$H + D \rightarrow T + \nu, \qquad [5a]$$

$$D + D \rightarrow {}^{4}He^{+} + e^{-} \qquad [5b]$$

$$^{4}He + {}^{4}He + {}^{4}He \rightarrow {}^{12}C^{+} + e^{-} \qquad [5c]$$

$$^{12}C + {}^{12}C \rightarrow {}^{24}Mg^{+} + e^{-} \qquad [5d]$$

$$D + {}^{4}He \rightarrow {}^{6}Li^{+} + e^{-} \qquad [5e]$$

$$^{6}Li + {}^{6}Li \rightarrow {}^{12}C^{+} + e^{-} \qquad [5f]$$

If they exist, they could complete with big-bang reactions and modify the proportions of primeval elements.

The rapid depletion of neutral hydrogen in a fraction of an hour, of a mass of about $10^{3} M_{\odot}$ will probably generate turbulence, a deuteron stream, a neutrino wind. If pieces of a 1 kpc radius gas sphere are brought successively into the core the process could nevertheless last 10^{9} times longer or about 10^{5} years for a total mass of $10^{12} M_{\odot}$. The neutral matter of the future galaxy would then be processed quite rapidly. Big-bang generation of D and ^{4}He would then be questionable, at least in the amounts we presently infer from observation. Evidence of the red-shifted deuterium equivalent to 21 cm line i.e. $\Delta\nu(D) = 327.38402$ MHz (91.6 cm) would give some clue about D generation and distribution (11).

The frozen fusion reaction could not take place in the big-bang phase before recombination, an era dominated by $e^{-}H^{+}$ plasma. Cool plasma, or atomic gas, is necessary to have enough H for reaction [4].

The kinetic energy necessary to start the frozen fusion could come from the gravitational potential. An hydrogen atom falling from the periphery at 1 kpc toward the cloud's center of total mass $10^{12} M_{\odot}$ would gain an energy of ~ 40 keV if no collisions occur. Taking them into account as well as radiation pressure would give a more realistic value.

It has been shown recently (12) that no quasars seem to exist with a red shift $z > 3.7$ and, more precisely, the abrupt limit where quasars disappear (or appear) is related to their intrinsic magnitude. We present here a conjectured time evolution of QSO core, following H temperature :

- $T_{H} > 500$ eV : pre-quasar state with "normal" atomic physics.
- 20 eV $< T_{H} < 500$ eV : quasar state dominated by reaction [3] or [4], possibly ending in BL Lacertae objects (13), Seyfert galaxy (14), N-galaxy or the center of our galaxy where a hot

galactic wind has been suggested (15) and short galactic arm found expelled at about 100 km s^{-1} with a mass loss expected of $10M_{\odot}$ per year (16).

- $T_H < 20$ eV : post-quasar state with a return to "normal" atomic physics as seen outside of galaxy cores.

Ingredients are present for a tentative explanation of main quasar properties : high power output, compact shape, non-thermal radiation, emission lines, evolution. Before detailed calculations could be undertaken, atomic physicists will have to prove that frozen fusion really exist, measure it cross section, explore its existence for elements like He, C, O, Ne ...

The particle physicist will be interested in the presence of frozen fusion reactions. They will enable him to study neutrinos at low energy or low momentum and determine if neutrinos are bradyons, tachyons of even light tachyonic monopoles.

The engineer will be happy to burn hydrogen with hydrogen producing energy, with no induced radioactivity, just by shooting a gas with a keV neutral injector.

REFERENCES

(1) Sidis, V., Kubach, C., Fussen, D., 1981. Phys. Rev. Letters 47, pp. 1280.
(2) Szucs, S., Karemera, M., Terao, M., Brouillard, F., 1984. Experimental study of the mutual neutralization of H^+ and H^- between 5 and 2000 eV, J. Phys. B 17, in press.
(3) Peart, B. et al., 1976. J. Phys. B9, pp. L369. Moseley, J.T. et al., 1970. Phys. Rev. Letters 24, pp. 435.
(4) Dolder, K.T., 1982. Physics of Ion-Ion and Electron-Ion Collisions, pp. 373-414, (Baddeck 1981) NATO ASI B83, Brouillard, F., Mc Gowan, J.W. (Eds), Plenum.
(5) Karemera, M., 1983. Thesis, University of Louvain (unpublished).
(6) Lyubimov, V.A. et al., 1980. Phys. Lett. 94B, pp. 266.
(7) Dolder, K.T., 1984. Private communication to F. Brouillard.
(8) Marx, G., 1982. The First Million Years. proc. of 8th Intern. Workshop on Weak Interactions and Neutrinos, p. 188, Javea, A. Marales (ed.), World Scientific.
(9) Stittmatter, P.A., Williams, R.E., 1976. Ann. Rev. of Astr. Astroph. 14, pp. 307-338.
(10) Schmidt, M., 1978. Physica Scripta 17, pp. 135-136.
(11) Kunsch, P., 1975. Phys. Rev. 100, pp. 1188.
(12) Schmidt, M. The Space Distribution of Quasars in the Universe. These Proceedings.

(13) Stein, W.A., O'Dell, S.L., Strittmayer, P.A., 1977. Ann. Rev. of Astr. Astroph. 14, pp. 173-195.
(14) Weedman, D.W., 1977. Ann. Rev. of Astr. Astroph. 15, pp. 69-95.
Anderson, K.S., 1981. Astroph. J. 246, pp. 13-19.
(15) Burton, W.B., 1976. Ann. Rev. of Astr. Astroph. 14, pp. 275-303, 303.
(16) Oort, J.H., 1977. Ann. Rev. of Astr. Astroph. 15, pp. 295-363, 321.

ESTIMATION OF GALACTIC MASSES USING THE ZERO ENERGY-MOMENTUM COSMOLOGICAL PRINCIPLE

D.K. Callebaut, N. Van den Bergh and P. Wils

Phys. Dept., U.I.A., Universiteit Antwerpen
B-2610 Wilrijk-Antwerpen. Belgium.

Summary

The cosmological zero energy-momentum principle is briefly explained. It is then applied in a quasi-Newtonian model to derive Jeans' instability criterion, once without and once with electrical charges. The calculation of the masses of galaxies from it yields a reasonable agreement if the estimated values for the temperature in galaxies are used. However, since the zero energy principle rather implies that we use the CBBR temperature we obtain about 5×10^8 solar masses. An expanding Newtonianlike universe combined with the zero energy principle gives presently no definite conclusion about the stability; but if unstable the relevant wavelengths and masses seem qualitatively acceptable.

1. INTRODUCTION

The zero energy-momentum cosmological principle introduced by one of us (1) allows to explain several incidences. However, if really a valid principle it should explain many other cosmic phenomena. It should be used e.g. to give a more acceptable derivation of Jeans' instability criterion and as a consequence to yield reasonable estimates of the galactic masses.

The plan of the paper is as follows. First the idea of the zero energy-momentum cosmological principle is explained (section 2). Then an improved analysis of Jeans' instability criterion is briefly derived, although still in a static and quasi-Newtonian approximation (section 3). It yields galactic masses that are too small. In section 4 an expanding Newtonianlike model using the zero energy principle is developed. The last

A. Berger (ed.), The Big Bang and Georges Lemaitre, 139–148.

section contains some conclusions, comments and suggestions for further extensions.

2. THE ZERO ENERGY-MOMENTUM COSMOLOGICAL PRINCIPLE

It was advanced by one of us (1) that the total energy and total momentum in the universe is zero. Several arguments in favor of this principle can be given. First of all it seems very objectionable to accept that the total mass and energy of the universe should have been contained in an extremely small volume (in the limit a point of zero dimensions). Such a singularity can be a useful approximation in the later stages of the universe; undoubtedly the idea of an expanding universe has been a most fruitful one. It was first considered from the viewpoint of mathematical physics by Friedmann, advanced by Hubble to explain the red shift of the galaxies and studied from the viewpoint of theoretical physics by Lemaître (i.e. looking for the real universe and not for several mathematical possibilities allowed by the general theory of relativity). However a singularity is unsatisfactory as an ultimate scientific explanation of whatever physical feature; and certainly this initial singularity is unsatisfactory as the very beginning of an otherwise physically well behaving universe. Moreover quantum fluctuations may be called for help for the initiation, but fluctuations of such an enormous magnitude are rather unbelievable, even if the initial conditions and physics are different from the present ones. Moreover fluctuations resulting in a net amount of energy from nothing remain much less acceptable than fluctuations which do not involve a net creation of energy but only the creation of positive and negative energy balancing each other perfectly to zero at all times.

The assumption of a zero total momentum seems very plausible in view of spatial symmetry; however from the relativistic viewpoint a zero total momentum is also an indication that the total energy may be zero.

There is a very strong Newtonian argument in favor of this cosmological principle. Indeed the gravitational energy of a mass m at the center of a gravitational sphere of radius R, of homogeneous density ρ and of mass M is:

$$m\varphi = -\frac{3}{2}\frac{GMm}{R} = -2\pi G\rho R^2 m \tag{1}$$

where G is the gravitational constant.
Clearly this (negative) energy is tremendous per unit mass if φ corresponds to the potential of the universe. On the other hand the rest energy mc^2, is also an enormous energy per unit mass but positive. According to the zero energy principle their sum should be zero. This entails immediately several consequences. First it explains where the tremendous rest energy comes from and also where the field energy went to. At the same time it

fixes the arbitrary constant in the potential; this constant in itself has usually no importance but for the total potential energy of the cosmos it does matter.

Moreover it shows that the inertial mass (m_i), which occurs in the rest energy, has to be equal to the gravitational mass, which occurs in the gravitational energy; otherwise there would be a surplus of energy. Moreover, in agreement with Mach's principle, besides the cosmological interpretation of the rest energy, it gives also a cosmological interpretation of the speed of light in vacuum:

$$c^2 = -\varphi = 2\pi G\rho R^2 \tag{2}$$

Using conservative values for the density and the radius of the universe ($\rho = 3\times10^{-28}$ kg/m^3 and $R = 2\times10^{26}$ m) one obtains $c = 7.10^7$ m/s, which is about a fourth of the real value. Moreover, the value for ρ may probably be increased by a factor 3 and maybe even by a factor 10 in view of the newly discovered extensions of the galaxies according to their rotational velocity curves. The value for R is also an underlimit. So the coincidence is rather good and probably will still improve. On the other hand one has to realize that this is only a Newtonian approximation and in addition one should take into account other generators of potential energy besides the gravitation.

The full expression of the cosmological zero energy-momentum principle requires that the right hand side ($-\kappa T_{\mu\nu}$) in Einstein's field equation is zero in the universe.

$$R_{\mu}^{\nu} - \frac{1}{2} R\, g_{\mu}^{\nu} + \Lambda g_{\mu}^{\nu} = 0 \tag{3}$$

This is clearly revisiting the De Sitter's universe, however with a plausible explanation for putting the right hand side equal to zero. This cosmology is fully in the line of Einstein's theory of gravitation, except for the additional assumption of the cosmological zero energy-momentum principle. It may be remarked that these appear to be inconsistent since the Einstein field equation is precisely the expression of the equality between field tensor and matter tensor. One of us has given the explanation of this paradox, but this would lead too far in the present context.

3. IMPROVED JEANS' STABILITY CRITERION.

Jeans studied the stability of an infinite, Newtonian gravitating medium of homogeneous density and uniform pressure (2). He obtained the following critical wavelength:

$$\lambda_J = \sqrt{\frac{\pi}{G\rho}\frac{dp}{d\rho}} \tag{4}$$

in which $dp/d\rho$ corresponds to the square of the sound velocity (about the thermal velocity of the molecules). According to Jeans a perturbation with wavelength λ larger than λ_J is (linearly) unstable and would lead to condensations (galaxies, stars, ...), while a perturbation with wavelength smaller than λ_J would be stable and would behave as a sound wave, modified more or less due to the gravitation in relation to the ratio λ/λ_J. However the equilibrium which Jeans had considered can not exist. To explain this we consider the Poisson equation and the equilibrium condition in Newtonian mechanics:

$$\Delta\varphi = 4\pi G\rho \tag{5}$$

$$\nabla p = -\rho\nabla\varphi \tag{6}$$

If p is uniform then $\nabla\varphi$ is zero everywhere (unless ρ is zero). This requires $\Delta\varphi$ = o and thus requires anyway ρ = o.

In fact the difficulty was basically the same as the one which Einstein met in 1917 in order to make a cosmological model. Einstein introduced the cosmological constant to cope with the problem. (This corresponds in the Newtonian approximation to adding a term $\Lambda\varphi$ in (5); the vanishing of grad φ then yields $\Lambda\varphi = 4\pi G\rho$.) Later the expansion of the universe fulfilled the same purpose (This corresponds to adding a term $\rho dv/dt$ in (6), then grad φ does not vanish because grad p vanishes). A third alternative is to use the cosmological zero ~~energy~~-momentum principle. This solves the difficulty and it will be used, in the following to give a better basis to the Jeans' stability criterion. However, it should be clear that we do not at all claim that the expansion of the universe or the cosmological constant are superfluous. In fact in the next section we will deal with an expanding universe in which the zero energy principle is also valid.

The relevant equations are

$$\partial_t\bar{v} + \bar{v}.\nabla\bar{v} = -\frac{1}{\rho}\nabla p - \nabla\varphi \tag{7}$$

$$\partial_t\rho + \text{div}\,\rho\bar{v} = o \tag{8}$$

$$p = K\rho^{\Gamma} \tag{9}$$

$$\Delta\varphi = 4\pi G(\rho-\rho_f) \tag{10}$$

where the last one is the form of the Poisson equation in which the mass density ρ_f corresponding to the field is added. The equilibrium quantities are characterized by the index o. One has $\bar{v}_o$ = o, $\rho_o = \rho_{f_o}$ (= ρ_f) and $p_o = K\rho_o^{\Gamma}$ are independent of space and time and φ_o is independent of space.

The perturbed quantities are characterized by the index 1.

ρ_f is considered as not changing markedly during the perturbation. After linearization and elimination (3) one obtains the differential equation

$$\partial^2_{tt}\,\rho_1 - \frac{\Gamma p_o}{\rho_o}\,\Delta\rho_1 - 4\pi G\rho_o\rho_1 = o \qquad (11)$$

Using a Fourieranalysis

$$\rho_1 \sim e^{i(\omega t+\bar{k}.\bar{r})} \qquad (12)$$

yields the dispersion relation

$$\omega^2 = \frac{\Gamma p_o}{\rho_o}\;k^2 - 4\pi G\rho_o \qquad (13)$$

which is the same as obtained by Jeans. This yields, by putting $\omega = o$, the critical λ_J of eq. (1). However a somewhat more correct analysis yields in his theory a factor 8 instead of 4 in the last term of eq. (13).

The same analysis can be performed taking into account that the material consists of electrons and ions (once ionized). This leads to a more involved dispersion relation, combining plasma oscillations and gravitational oscillations or instabilities (4). Instabilities now arise for wavelengths larger than the critical wavelength

$$\lambda_{J_p} = \frac{1}{\rho_o}\left(\frac{2\pi}{G}\;(\Gamma_i p_{io} + \Gamma_e p_{eo})\right)^{1/2} \qquad (14)$$

which is larger than λ_J by about $\sqrt{2}$. (Indices i and e are for ions and electrons respectively)
The critical Jeans'mass is

$$M_J = \rho_o\;\lambda_J^3 = \frac{1}{\rho_o{}^2}\left(\frac{\pi\Gamma p_o}{G}\right)^{3/2} = \frac{1}{\sqrt{\rho_o}}\left(\frac{\pi\Gamma k_B T}{G\;m}\right)^{3/2} \qquad (15)$$

with m say the mass of a proton. However one has to take at least a wavelength corresponding to $2\lambda_J$ in view of the corresponding growth rates. This makes M_J larger by about an order of magnitude. One obtains a reasonable mass for a galaxy using a temperature of 100K as is actually observed in our galaxy. However in this model with the zero energy principle it seems indicated to use the present 3K of the CBBR. This yields only an average galactic mass of about 2×10^8 solar masses. Using the plasma value brings this to about 5×10^8 solar masses, which is still two to three orders of magnitude too small.

4. EXPANDING NEWTONIABLIKE MODEL WITH ZERO ENERGY PRINCIPLE

The relevant basic equations are eqs. (7), (9), (10) supplemented by the following:

$$\partial_t \rho + \operatorname{div} \rho\bar{v} = \partial_t \rho_f + \operatorname{div} \rho_f \bar{v}_o \tag{16}$$

$$\rho_f = \frac{C^2}{R(t)^2} \tag{17}$$

R(t) is the radius of the universe in this Newtonianlike model. Since in the Newtonian approximation $\rho_f \sim \varphi R^{-2}$ (eq. (2) and since $-\varphi$ should correspond to c^2 the choice of eq. (17) is plausible. In fact the possibility of a variable G is not considered here. The equation of continuity seems logical, with $\bar{v}_o$ (the velocity of the unperturbed state) instead of $\bar{v}$ in the right hand side. This corresponds to the fact that we take ρ_f as not influenced by the perturbation, which simplifies the analysis. Also K and Γ are taken as constants which do not change neither with place nor during the evolution.

Assume $\bar{v}_o = \bar{r}f(t)$. We want to have ρ_o, p_o and φ_o independent of space. That also φ_o is independent of space means that in this model the so-called cosmological principle (that all places in the universe should be alike) is even better satisfied than in customary Newtonian models, where φ_o cannot be independent of $\bar{r}$, unless a cosmological constant is used, because of the Poisson equation.

From eq. (7) it follows then that $\dot{f} + f^2 = 0$ and thus

$$f = \frac{a}{t} \tag{18}$$

where a is a dimensionless constant. In an expanding universe one has $a > 0$. Thus:

$$\bar{v}_o = \frac{a\bar{r}}{t} \tag{19}$$

and

$$V \equiv v_{oR} = \frac{aR}{t} \tag{20}$$

It follows then from eqs. (17), (9) and (16):

$$\rho_o = \rho_{f_o} = \left(\frac{Ca}{Vt}\right)^2 \tag{21}$$

$$p_o = K\rho_o^\Gamma = K\left(\frac{Ca}{Vt}\right)^{2\Gamma} \tag{22}$$

$$\dot{\rho}_o + 3\rho_o \frac{a}{t} = \left(\frac{Ca}{V}\right)^2 \frac{(3a-2)}{t^3} \tag{23}$$

Note that the source term vanishes if $a = 2/3$. In spite of the source term each density decreases like t^{-2}, each starting at infinity (But there this Newtonianlike approximation is no more applicable). The total mass of each kind increases like t from zero on.

For the perturbed equations one obtains:

$$\partial_t \bar{v}_1 + \bar{v}_1 . \nabla \bar{v}_o + \bar{v}_o . \nabla \bar{v}_1 = - \frac{1}{\rho_o} \nabla p_1 - \nabla \varphi_1 \tag{24}$$

$$\partial_t \rho_1 + \text{div}\, \rho_1 \bar{v}_o + \text{div}\, \rho_o \bar{v}_1 = 0 \tag{25}$$

$$p_1 = \frac{\Gamma p_o}{\rho_o} \rho_1 \tag{26}$$

$$\Delta \varphi_1 = 4\pi G \rho_1 \tag{27}$$

Although our basic equations differ markedly from the Newtonian ones, the perturbed ones are formally the same. Cf. Weinberg (5), who refers to Lifshitz and Bonnor. Of course the unperturbed solution is different, so that the perturbed one is also different, although similar.

Put

$$\rho_1(\bar{r},t) = \rho_1(t) \exp\left(\frac{ia\, \bar{r}.\bar{k}}{Vt}\right) \tag{28}$$

and similarly for p_1, $\bar{v}_1$ and φ_1. $\bar{k}$ stands for a dimensionless wavevector. Taking $a = 1$ and eliminating p_1 and φ_1 yields

$$\dot{\bar{v}}_1 + \frac{v_{1r}}{t} \bar{e}_r = \frac{i}{Vt} \left(- \frac{\Gamma p_o}{\rho_o^2} + \frac{4\pi G V^2 t^2}{k^2}\right) \bar{k} \rho_1 \tag{29}$$

$$\dot{\rho}_1 + \frac{3\rho_1}{t} = - \frac{i\rho_o}{Vt} \bar{k}.\bar{v}_1 \tag{30}$$

Putting

$$\bar{v}_1 = v_{1r} \bar{e}_r + v_{1k} \frac{\bar{k}}{k} \tag{31}$$

yields $$\dot{v}_{1r} + v_{1r}/t = 0 \tag{32}$$

$$\dot{v}_{1k} = \frac{i}{Vt}\left(-\frac{\Gamma p_o}{\rho_o^2} + 4\pi G V^2 t^2\right) k\rho_1 \tag{33}$$

$$\dot{\rho}_1 + \frac{3\rho_1}{t} = \frac{-i\rho_o}{Vt}\left(v_{1k}k + v_{1r}\bar{e}_r\cdot\bar{k}\right) \tag{34}$$

From (32) it follows that v_{1r} dies out. We drop then the corresponding term in (34). Eliminating $\dot{v}_{1k}$ yields after some algebra:

$$\ddot{\rho}_1 + \frac{6\dot{\rho}_1}{t} + \left[\left(6 - \frac{4\pi G C^2}{V^2}\right)\frac{1}{t^2} + \frac{K\Gamma}{C^2}\left(\frac{C}{Vt}\right)^{2\Gamma} k^2\right]\rho_1 = 0 \tag{35}$$

For the particular case $\Gamma = 1$ (isothermal) one has solutions of the form $\rho_1 = At^{\alpha}$ with A an arbitrary constant and

$$\alpha_{\pm} = \frac{1}{2}\left(-5 \pm \sqrt{25 - \frac{4}{V^2}\left(6V^2 - 4\pi G C^2 + Kk^2\right)}\right) \tag{36}$$

This yields instability if

$$6V^2 - 4\pi G C^2 + Kk^2 < 0 \tag{37}$$

It seems likely to take $V = c$; also $C^2 = c^2/2\pi G$ seems a reasonable approximation. Inequality (37) then reads

$$4c^2 + Kk^2 < 0 \tag{38}$$

which is never satisfied.

When $\Gamma \neq 1$ one can express the solution of (35) by means of Besselfunctions:

$$\rho_1 = At^{-5/2}\, Z_p\left(\frac{\sqrt{K\Gamma}}{(\Gamma-1)}\,\frac{C^{\Gamma-1}}{V^{\Gamma}}\, kt^{\Gamma-1}\right) \tag{39}$$

$$\text{with } p^2 = \left(\frac{16\pi G C^2}{V^2} - 24\right)/25(\Gamma-1)^2 \tag{40}$$

It is clear that for gravitational instability leading to galaxies and the like we are interested mainly in extremely small k. In the limit of k zero all solutions lead to the same result as $\Gamma = 1$, independently of Γ, as can be seen from eq. (35) itself.

Hence this model doesn't yield a clear instability as long as we cannot insert more appropriate values for V and for C. This is not wholly unexpected since all analyses related to the Jeans' one are always a bit marginal. Lifshitz (5) found instability but with a growth like a power of t and not an exponential growth. In fact in his analysis this doesn't allow a statistical fluctuation to develop sufficiently. In our analysis this is a minor inconvenience since we start with a small total mass.

However it has become clear what to expect in a refined (i.e. general relativistic) analysis. Indeed k will have to be of order of $cK^{-1/2}$ or c/v_s with v_s the sound velocity

$$v_s^2 = \frac{\Gamma p_o}{\rho_o} = \frac{\Gamma k_B T}{m} \qquad (41)$$

The corresponding wavelength is

$$\lambda = \frac{2\pi R}{k} = 2\pi R \frac{v_s}{c} \qquad (42)$$

With $R = 2.10^{10}$ ly and $v_s = 150$ m/s (corresponding to 3K and atomic hydrogen), one obtains:

$$\lambda \approx 6 \times 10^4 \text{ly} \qquad (43)$$

This is qualitatively of the right order of magnitude, but again too small.

It may be remarked that in this expanding model with the zero energy principle, we may expect that the radiation density $1/3\, aT^4$ is roughly proportional to the mass density. Hence $T \sim t^{-1/2}$. It follows that λ would be in this approximation proportional to $t^{3/4}$.

5. CONCLUSION

The Jeans' criterion may be correctly derived in a static model with the zero energy principle. It would yield a reasonable average mass for the galaxies if one might use the observed temperatures of the galaxies (say 100 K). However the model rather requires the use of the CBBR (3 K) and yields masses which are too small by 2 or 3 orders of magnitude.

The expanding model with the zero energy principle leads to a very interesting analysis, similar to those of Lifshitz and Bonnor. But with the approximations at hand we cannot clearly decide about the stability. If unstable, however, the critical wavelength will be about the one given in eq. (42), yielding a plausible result but also too small. However, several interesting features of the model became clear in the analysis. Probably an analysis in the framework of general relativity will be able to give a more decisive result.

REFERENCES

1. Callebaut, D.K., 1982, Proc. Sec. M. Grossmann Meeting on Gen. Rel., Ruffini, R. (ed.), North Holland Publ. Co, pp. 871.
2. Jeans, J., 1902, Phil. Trans. Roy. Soc., 199A, pp. 49.
3. Callebaut, D.K. and Wils, P., 1981, Proc. Workshop III on (nonlinear) Stability, Callebaut, D. (ed.), U.I.A. press, pp. 70-78.
4. Callebaut, D.K. and Wils, P., 1981, Proc. Workshop III on (nonlinear) Stability, Callebaut, D. (ed.), U.I.A. press, pp. 79-90.
5. Weinberg, S., 1972, Gravitation and Cosmology, Wiley, N.Y., pp. 571-580.

Part II
Celestial Mechanics

DYNAMICS OF ORBITING DUST UNDER RADIATION PRESSURE

Andre Deprit

National Bureau of Standards,
Washington D.C. 20234 U.S.A.

For a three-dimensional Keplerian system in the presence of a homogeneous field possibly in uniform rotation, action and angle variables are introduced by canonical transformation in the averaged Hamiltonian truncated at the first order. After substitution, the first order averaged system proves to be integrable. More precisely, it is shown how the orbit space decomposes into a pair of spheres in a three-dimensional space, on which the representative curves are the small circles induced by a finite rotation about a fixed axis. From this intuitive geometric picture follow simple formulas for solving the initial value problem.

DEDICATION

The lectures on integral invariants which Elie Cartan gave at the Sorbonne in the schoolyear 1920-1921 fascinated Msgr Lemaitre, for he sensed that Cartan's geometric approach opens new vistas on classical problems. Cartan's techniques, he was convinced, could be made into a tool for producing and analyzing models of dynamic systems. Of all the applications dealt with in Cartan's textbook, the problem of three bodies caught Lemaitre's imagination. It is typical of the gap left open in Cartan's Reduction Theory: on the one hand, a differentiability criterion expressed in the intrinsic language of Exterior Calculus implies that the original phase space may be reduced; on the other hand, no indication is given on how the actions of the reducing Lie group could lead to an atlas of symplectic maps on the reduced manifold. Thus it is that whenever Cartan's theory of integral invariants announces a

A. Berger (ed.), The Big Bang and Georges Lemaitre, 151–180.

possible reduction, the physicist receives the news as a challenge to create a good set of coordinates for converting the mathematical abstraction into a workable dynamic system. Msgr Lemaitre thrived on this kind of challenge because he knew he was very good at meeting them. In the problem of three bodies, he invented precisely the kind of coordinates that are missing in Cartan's exposition. Not only do they make the reduced manifold amenable to numerical integration, but they have the incomparable advantage of transforming the binary collisions from moving singularities into fixed critical points to be regularized straightaway, all three of them at once, merely by a conformal mapping.

In the reduced problem, Msgr Lemaitre studied closely two situtations: the collinear case where the masses stay forever aligned, and the isoceles configuration which Chazy had been analyzing by Tauberian arguments. The reason is that both systems have only two degrees of freedom, and Msgr Lemaitre hoped they could be handled in much the same way as he had treated the Stoermer problem with remarkable success twenty years earlier. Might it not be, after all, that manifolds of asymptotic orbits emanating from unstable periodic orbits stake out accessible and forbidden regions in the phase space? On this hunch, a campaign began in Louvain for locating periodic orbits with real characteristic exponents, or perhaps it was rather a succession of brilliant raids bringing back outstanding periodic orbits but, alas, never one with four real characteristic exponents(*). So little do we understand of the mechanisms that produce unstable periodic orbits in a dynamic system. At any rate, Lemaitre proved right in thinking that Cartan's Geometric Dynamics could be made computable. There had never been a doubt in his mind that the physicist's imagination is the spark that throws off mathematical abstractions which then flare into brilliant mechanical models. Technology and mathematics, the enormous increase of mathematical sophistication and the ever galloping progress in electronic computers are now fostering the Nouvel Age in mechanics and mathematical astronomy of which Lemaitre, assuredly, was a precursor.

We chose the topic of our communication, not only because it bears on a current problem in planetology, but mainly because it offers a simple example of the gap Msgr Lemaitre bridged in Cartan's Reduction Theory when it is applied to the Problem of Three Bodies.

(*)C. Cauwe, 'Recherche d'orbites périodiques particulières dans le mouvement rectiligne de trois masses égales,' Mémoire présenté pour obtenir le grade de licencié en sciences, Université catholique de Louvain, 1951.

J. Meeus, 'Exposants caractéristiques des configurations équilatérales de Lagrange d'excentricité non nulle dans le cas de masses égales,' Mémoire présenté pour obtenir le grade de licencié en sciences, Université catholique de Louvain, 1952.

A. Bartholomé, 'Recherches sur le problème des trois corps dans le cas de masses égales,' Mémoire présenté pour obtenir le grade de licencié en sciences, Université catholique de Louvain, 1952.

A. Deprit, 'Configurations isocèles du problème plan des trois corps. Equation d'une orbite périodique d'éjection et ses exposants caractéristiques,' Mémoire présenté pour obtenir le grade de licencié en sciences, Université catholique de Louvain, 1953.

G. Steenbeckeliers, 'Configurations isocèles du problème plan des trois corps. Orbites symétriques d'éjection,' Mémoire présenté pour obtenir le grade de licencié en sciences, Université catholique de Louvain, 1957.

E. De Vylder, 'Les variables régularisantes du Chanoine Lemaitre et la théorie de la lune de de Pontécoulant,' Mémoire présenté pour obtenir le grade de licencié en sciences, Université catholique de Louvain, 1958.

1. Introduction

The dynamics of the two-dimensional Stark effect is characterized by the Hamiltonian

$$H = \frac{1}{2}(X^2 + Y^2) - \frac{\mu}{r} + \varepsilon x \tag{1}$$

describing a Keplerian system with constant parameter μ to which is added a small perturbation of constant magnitude and of fixed direction in the particle's orbital plane, the small parameter ε having the physical dimensions of an acceleration. Averaged over the mean anomaly and then truncated to the first order in ε, Hamiltonian (1) gives rise to an amazingly elementary system. Its orbit space is a two-dimensional sphere, and the phase motions consist of uniform rotations around a fixed axis [9]. It will now be shown how this simple picture carries on for the class of Hamiltonians

$$H = \frac{1}{2}(X^2 + Y^2 + Z^2) - \frac{\mu}{r} + \varepsilon (x \cos mt + y \sin mt) \tag{2}$$

representing three-dimensional Keplerian systems undergoing a small perturbation of constant magnitude but, in this case, rotating at a constant angular velocity m in a fixed plane that need not be the particle's orbital plane. The model is an abstraction contrived to account for the effect of radiation pressure -- in the precise sense of Burns [3, p. 8, second column] -- on dust particles orbiting about an idealized planet revolving as it were around the sun on a circle in a fixed plane (cfr e.g. [1, pp. 87 - 88]). In that context, the origin of coordinates is set at the center of mass in the pair sun-planet; the coordinate plane (x, y) is identified with the plane of the planet's orbit, and m stands for the mean motion of the planet around the sun. The plane of the planet is so oriented that, without loss of generality, m may be assumed to be $\geqslant 0$. The particular case m = 0 was examined sixty years ago in classical quantum mechanics where (2) was meant to model the effect of a homogeneous electric field on a charged particle in a Coulomb field (see e.g. [2, pp. 262 - 269]).

By virtue of Cauchy's uniqueness theorem, a particle starting in the planet's orbit with an initial velocity contained in that plane will never leave the plane. As a matter of fact, the class of co-planar orbits is the set of solutions for the canonical equations derived from the reduced version

$$H = \frac{1}{2}(X^2 + Y^2) - \frac{\mu}{r} + \varepsilon\,(x \cos mt + y \sin mt) \tag{3}$$

of Hamiltonian (2). It describes a dynamical system with only one degree of freedom which admits an integral. Averaging it over the mean anomaly and retaining only the first order in the approximation, the obtains a model of the long term evolution that is <u>integrable</u> [13]. For m = 0, Mignard's solution reproduces the conclusions reached by a different method for the two-dimensional Stark effect in Hamiltonian (1).

As one should expect from a perturbed Keplerian problem in three dimensions [14, 18] the orbit space after reduction by averaging is composed of two spheres in a three-dimensional real space. The outstanding feature of the title problem, gathered by Mignard [13] from from repeated numerical integrations, is that on each sphere the motions consist of rotations about a fixed axis at a constant angular velocity, <u>the same on both spheres</u>. The main motivation of the present communication is to prove Mignard's conjecture. In an approach that is somewhat unusual in celestial mechanics, although it is routine in quantum mechanics, the Cartesian components of the angular momentum and of the Runge-Lenz vector serve as the coordinates for the reduced system. From the point of view of Geometric Dynamics, the result is an elementary application of the Reduction Theorem going back to Cartan [4] but nowadays named after Kirillov, Souriau and Kostant. Actually it will be demonstrated that the reduction can be carried out by constructing a pair of integrals generating a set of action - and angle - variables. A more descriptive solution is obtained by rearranging the phase space so that the rotations on one of the spheres account for the motions of the particle's ascending node in the plane of the planet's orbit while those on the second sphere justify the angular displacements of the pericenter. In the extreme case when $\varepsilon = 0$, movements of both node and pericenter are evidently all circulations; but as the frequency inherent to the radiation pressure increases with respect to the planet's mean motion, there arise librations both in node and in pericenter. Eventually, at the other extreme when $m = 0$, node and pericenter both librate for whatever set of initial conditions is selected. The transition from general circulation (pure Keplerian problem in a rotating frame) to general libration (pure Stark effect in a fixed frame) is smooth and exempt from bifurcations.

With a view towards assisting astronomers in the task of assessing the impact of radiation pressure on orbiting dust, the initial value problem has been worked out in detail to present a complete solution in a form as simple as possible.

Is there more than a fortuitous coincidence in the fact that the two-dimensional variant (3) after averaging over the mean anomaly corresponds strictly to the Hamiltonian by which Pauwels [16, eq. 34] models the secular orbit/orbit resonance between the satellites Rhea and Titan of Uranus? The similitude seems to suggest that (2) is representative of a class of Keplerian systems in which the perturbation introduces a 1-1 semi-simple resonance between the mean motion of the node and of the perigee.

2. The Synodic Frame

The manner in which Hamiltonian (2) depends explicitly on time suggests adopting a frame of reference that rotates at the constant angular velocity m so as to maintain the moving axis of abscissae aligned in the direction of the radiation pressure [5]. To this effect, a time dependent canonical transformation $(X, Y, x, y, t) \to (U, V, u, v)$ is introduced by the implicit equations

$$X = \frac{\partial S}{\partial x}, \qquad Y = \frac{\partial S}{\partial y}, \qquad u = \frac{\partial S}{\partial U}, \qquad v = \frac{\partial S}{\partial V}$$

from the generating function

$$S \equiv S(U, V, x, y, t) = U\,(x \cos mt + y \sin mt) + V\,(-\,x \sin mt + y \cos mt).$$

Attention ought to be paid to the meaning of the new variables. Let $(\underset{\sim}{e}_1, \underset{\sim}{e}_2, \underset{\sim}{e}_3)$ be an orthonormal basis with $\underset{\sim}{e}_1$ and

$\underset{\sim}{e}_2$ fixed in the planet's orbit, and $\underset{\sim}{e}_3$ normal to it. Then $\underset{\sim}{x} = x\,\underset{\sim}{e}_1 + y\,\underset{\sim}{e}_2 + z\,\underset{\sim}{e}_3$ is the particle's position vector; furthermore, since

$$\dot{x} = \frac{\partial H}{\partial X} = X, \qquad \dot{y} = \frac{\partial H}{\partial Y} = Y, \qquad \dot{z} = \frac{\partial H}{\partial Z} = Z,$$

the vector $\underset{\sim}{X} = X\,\underset{\sim}{e}_1 + Y\,\underset{\sim}{e}_2 + Z\,\underset{\sim}{e}_3$ is the particle's velocity relative to the inertial frame $(\underset{\sim}{e}_1, \underset{\sim}{e}_2, \underset{\sim}{e}_3)$. The vectors $(\underset{\sim}{f}_1, \underset{\sim}{f}_2, \underset{\sim}{f}_3)$ such that

$$\begin{aligned} \underset{\sim}{f}_1 &= \underset{\sim}{e}_1 \cos mt + \underset{\sim}{e}_2 \sin mt, \\ \underset{\sim}{f}_2 &= -\underset{\sim}{e}_1 \sin mt + \underset{\sim}{e}_2 \cos mt, \\ \underset{\sim}{f}_3 &= \underset{\sim}{e}_3. \end{aligned}$$

constitute an orthonormal basis of the rotating or synodic frame of reference. On the one hand, by definition of the canonical mapping,

$$\begin{aligned} u &= \frac{\partial S}{\partial U} = x \cos mt + y \sin mt, \\ v &= \frac{\partial S}{\partial V} = -x \sin mt + y \cos mt, \end{aligned}$$

and these relations imply evidently that the position vector is the sum $\underset{\sim}{x} = u\,\underset{\sim}{f}_1 + v\,\underset{\sim}{f}_2 + z\,\underset{\sim}{f}_3$. On the other hand,

$$X = \frac{\partial S}{\partial x} = U \cos mt - V \sin mt, \qquad Y = \frac{\partial S}{\partial y} = U \sin mt + V \cos mt,$$

which means that the velocity vector is also the sum $\underset{\sim}{X} = U\,\underset{\sim}{f}_1 + V\,\underset{\sim}{f}_2 + Z\,\underset{\sim}{f}_3$. In words, the moments (U, V, Z) canonically conjugate to the synodic coordinates (u, v, z) are the synodic components of the velocity for the particle's motion relative to the inertial frame. The point insisted upon is that these moments are not the synodic components of the velocity relative to the synodic frame itself.

The differential identity

$$X\,dx + Y\,dy = U\,du + V\,dv - \frac{\partial S}{\partial t}\,dt + d(S + Uu + Vv)$$

shows that $\partial S/\partial t = -m\,(uV - vU)$ is a remainder of the transformation. Thus, in the synodic variables (U, V, Z, u, v, z), the canonical equations are to be derived from the Hamiltonian

$$K = H + \frac{\partial S}{\partial t} = \frac{1}{2}(U^2 + V^2 + Z^2) - \frac{\mu}{r} - m\,(uV - vU) + \varepsilon\, u \qquad (4)$$

which, as it was intended, is now independent of the time. One may notice that the Hamiltonian admits a few discrete symmetries: the symplectic mapping $(U, V, Z, u, v, z) \to (U, V, -Z, u, v, -z)$ which is the reflection in the plane of the planet, as well as the contact transformations $(U, V, Z, u, v, z, t) \to (-U, V, Z, u, -v, -z, -t)$ and $(U, V, Z, u, v, z, t) \to (-U, V, -Z, u, -v, z, -t)$ in the phase space-time. However useful they may be in solving the problem by numerical integration, they are of no assistance in the geometric analysis.

There are two particular situations in which the problem put by Hamiltonian (4) is elementary: when $\varepsilon = 0$, the problem is <u>separable</u> in the cylindrical coordinates (r, ϕ, z) such that $x = r \cos \phi$ and $y = r \sin \phi$, and when $m = 0$, it is "separable" in the parabolic coordinates $(u, v\ \phi)$ such that

$$x = (u - v)/2, \qquad y = \sqrt{uv}\cos\phi, \qquad \text{and} \qquad z = \sqrt{uv}\sin\phi$$

[10, 17]. Otherwise, the problem appears intractable in its full generality, and far - reaching restrictions must be imposed to make it fit for analytical treatment.

a) It will be assumed that not only the acceleration ε but also the angular velocity m is a small parameter, and that ε and m are <u>both</u> of the first order. Such an asymptotic scaling of the parameters justifies decomposing Hamiltonian (4) into the sum $K = K_0 + K_1$; the first term

$$K_0 \equiv K_0(U, V, Z, u, v, z) = \frac{1}{2}(U^2 + V^2 + Z^2) - \frac{\mu}{r}$$

represents an ordinary Keplerian system in the synodic frame, the second term

$$K_1 \equiv K_1(U, V, u, v; m, \varepsilon) = -\,m\,(uV - vU) + \varepsilon\, u$$

stands for a perturbation of the first order.

b) Only those orbits along which K_0 remains < 0 at all times will be considered. Accordingly, with the synodic Cartesian coordinates (U, V, Z, u, v, z) will be associated a set of Delaunay elements (L, G, N, ℓ, g, ν) in the usual manner. Nonetheless, to dissipate ambiguities one ought to review attentively how the Delaunay elements are defined geometrically. Variables G, N, ν are attached to the vector

$$\underset{\sim}{G} = (vZ-zV)\, \underset{\sim}{f}_1 + (zU-uZ)\, \underset{\sim}{f}_2 + (uV-vU)\, \underset{\sim}{f}_3$$

which, on account of the remarks made made earlier, is precisely the angular momentum (per unit of mass) $\underset{\sim}{x} \times \underset{\sim}{X}$ relative to the inertial frame. Then the moment G is the norm of the vector $\underset{\sim}{G}$, and $N = \underset{\sim}{G}.\underset{\sim}{f}_3$ is the projection of $\underset{\sim}{G}$ on the normal to the planet's orbit. Let $\underset{\sim}{n}$ be a unit vector such that $\underset{\sim}{G} = G\,\underset{\sim}{n}$; the angle I such that $\underset{\sim}{f}_3.\underset{\sim}{n} = \cos I$ with $0 \leqslant I \leqslant \pi/2$ is the inclination of the particle's orbital plane over the planet's orbit. Also let $\underset{\sim}{\ell}$ be a unit vector such that $\underset{\sim}{f}_3 \times \underset{\sim}{n} = \underset{\sim}{\ell} \sin I$. As a direction in the planet's orbital plane, it may be decomposed into the sum $\underset{\sim}{\ell} = \underset{\sim}{f}_1 \cos \nu + \underset{\sim}{f}_2 \sin \nu$, a relation which defines the longitude ν of the ascending node for the particle's orbital plane reckoned from the synodic axis $\underset{\sim}{f}_1$. Decomposing the angular momentum in the synodic frame, one obtains readily that

$$G_1 = \underset{\sim}{G}.\underset{\sim}{f}_1 = G\,n_1, \qquad n_1 = \underset{\sim}{n}.\underset{\sim}{f}_1 = \sin \nu \sin I,$$

$$G_2 = \underset{\sim}{G}.\underset{\sim}{f}_2 = G\,n_2, \qquad n_2 = \underset{\sim}{n}.\underset{\sim}{f}_2 = -\cos \nu \sin I,$$

$$G_3 = \underset{\sim}{G}.\underset{\sim}{f}_3 = G\,n_3, \qquad n_3 = \underset{\sim}{n}.\underset{\sim}{f}_3 = \cos I.$$

Astronomers usually associate the angle g with the Laplace vector function

$$\underset{\sim}{A}^* = \underset{\sim}{X} \times \underset{\sim}{G} - \frac{\mu}{r}\,\underset{\sim}{x}$$

which has the same physical dimensions as the Keplerian parameter μ, namely length^3/time^2; but quantum physicists favor the Runge-Lenz vector

$$\underset{\sim}{A} = \frac{L}{\mu}\left(\underset{\sim}{X} \times \underset{\sim}{G} - \frac{\mu}{r}\,\underset{\sim}{x}\right)$$

not only for the reason that $\underset{\sim}{A}$ and $\underset{\sim}{G}$ are similar in dimensions, but mainly because the Poisson brackets involving $\underset{\sim}{A}$ and $\underset{\sim}{G}$ are simpler than those in $\underset{\sim}{A}^*$ and $\underset{\sim}{G}$. In the above formula, L is the action defined by the relation $K_0 = -(\mu^2/2L^2)$, which is licit since K_0 is supposed to be < 0 at all times. The norm of the Runge-Lenz vector is equal to Le, where e is the eccentricity determined by the relation $G^2 = L^2\,(1-e^2)$. Let $\underset{\sim}{a}$ be a unit vector such that $\underset{\sim}{A} = Le\,\underset{\sim}{a}$. Since $\underset{\sim}{G}.\underset{\sim}{A} = 0$, the direction $\underset{\sim}{a}$ is perpendicular to $\underset{\sim}{n}$, hence it may be decomposed into the sum $\underset{\sim}{a} = \underset{\sim}{\ell} \cos g + \underset{\sim}{n} \times \underset{\sim}{\ell} \sin g$. Simple projections yield the components of the pericenter direction:

$$A_1 = \underset{\sim}{A}.\underset{\sim}{f}_1 = Le\,a_1,$$

$$A_2 = \underset{\sim}{A}.\underset{\sim}{f}_2 = Le\,a_2,$$

$$A_3 = \underset{\sim}{A}.\underset{\sim}{f}_3 = Le\,a_3,$$

$$a_1 = \underset{\sim}{a}\cdot\underset{\sim}{f}_1 = \cos g \cos \nu - \sin g \cos I \sin \nu,$$
$$a_2 = \underset{\sim}{a}\cdot\underset{\sim}{f}_2 = \cos g \sin \nu + \sin g \cos I \cos \nu,$$
$$a_3 = \underset{\sim}{a}\cdot\underset{\sim}{f}_3 = \sin g \sin I.$$

It will be convenient to consider also the direction $\underset{\sim}{b} = \underset{\sim}{n} \times \underset{\sim}{a}$, and its components in the synodic frame

$$b_1 = \underset{\sim}{b}\cdot\underset{\sim}{f}_1 = - \sin g \cos \nu - \cos g \cos I \sin \nu,$$
$$b_2 = \underset{\sim}{b}\cdot\underset{\sim}{f}_2 = - \sin g \sin \nu + \cos g \cos I \cos \nu,$$
$$b_3 = \underset{\sim}{b}\cdot\underset{\sim}{f}_3 = \cos g \sin I.$$

Finally the particle's semi-major axis a and its mean motion n are defined by the relations

$$\mu = n^2 a^3, \qquad K_0 = - \frac{1}{2} n^2 a^2$$

from which it is readily inferred that

$$L = n\, a^2, \quad G = L\, a\, \eta \quad \text{with} \quad \eta = (1 - e^2)^{1/2}.$$

3. The Gyration Frequency

Since the principal part of (4) is an ordinary Keplerian system, a Delaunay normalization is in order [Deprit, 8]. This is a canonical transformation to make the new mean anomaly ℓ'

$$(L, G, N, \ell, g, \nu) \rightarrow (L', G', N', \ell', g', \nu')$$

ignorable in the transformed Hamiltonian, thereby producing the new action L' as an integral -- formally speaking --, and reducing the normalized system to only two degrees of freedom. Since only the first order terms will be retained after averaging, it is adequate to carry out the normalization merely as an infinitesimal contact transformation without deploying the full analytical machinery of a Lie transformation [Deprit, 7]. So the elimination reduces to finding a generating function $W(L', G', N', \ell', g', \nu')$ and a first order Hamitlonian K'_1 to satisfy the conditions

$$(K_0;\, W) + K_1 = K'_1, \qquad (K_0;\, K'_1) = 0.$$

The symbol $(\phi;\, \psi)$ denotes the Poisson bracket of the functions ϕ and ψ, a function that may be built indifferently either in the Cartesian variables as the differential expression

$$(\phi;\ \psi) = \frac{\partial\phi}{\partial u}\frac{\partial\psi}{\partial U} - \frac{\partial\phi}{\partial U}\frac{\partial\psi}{\partial u} + \frac{\partial\phi}{\partial v}\frac{\partial\psi}{\partial V} - \frac{\partial\phi}{\partial V}\frac{\partial\psi}{\partial v} + \frac{\partial\phi}{\partial z}\frac{\partial\psi}{\partial Z} - \frac{\partial\phi}{\partial Z}\frac{\partial\psi}{\partial z}$$

or in the Delaunay variables as the expression

$$(\phi;\ \psi) = \frac{\partial\phi}{\partial \ell}\frac{\partial\psi}{\partial L} - \frac{\partial\phi}{\partial L}\frac{\partial\psi}{\partial \ell} + \frac{\partial\phi}{\partial g}\frac{\partial\psi}{\partial G} - \frac{\partial\phi}{\partial G}\frac{\partial\psi}{\partial g} + \frac{\partial\phi}{\partial \nu}\frac{\partial\psi}{\partial N} - \frac{\partial\phi}{\partial N}\frac{\partial\psi}{\partial \nu}\,.$$

The true anomaly being the angle f such that

$$r \cos f = \underset{\sim}{x}.\underset{\sim}{a} \quad\text{and}\quad r \sin f = \underset{\sim}{x}.\underset{\sim}{b},$$

and the eccentric anomaly, the angle E for which

$$r \cos f = a\,(\cos E - e), \quad r \sin f = a\,\eta \sin E,$$

the perturbation Hamiltonian turns out to be the function

$$K_1 = -\,mG_3 - \varepsilon a_1 ae + \varepsilon a\,(a_1 \cos E + b_1\eta \sin E)$$

in the class of perturbation functions for which the Delaunay normalization is most expeditiously handled by <u>eliminating the eccentric anomaly</u> (Deprit, 1983). The algorithm prescribes that the first order term be written in the form

$$K_1 = -\,mG_3 - \varepsilon a_1 ae + \varepsilon\frac{a^2}{r}\,(-\,\frac{1}{2}\,a_1 e + a_1 \cos E - \frac{1}{2}\,a_1 e \cos 2E$$
$$+\,b_1\eta \sin E - \frac{1}{2}\,b_1 e\eta \sin 2E),$$

and that the generator W and the normalized perturbation be determined to satisfy the conditions

$$(K_0;\ W) = -\,n\,\frac{a}{r}\,\frac{\partial W}{\partial E} = K'_1 - K_1, \qquad \frac{\partial K'_1}{\partial E} = 0.$$

Separating the terms periodic in E and performing elementary quadratures, one satisfies the requirements by setting

$$K'_1 = -\,mG'_3 - \frac{3}{2}\,\varepsilon a' a'_1 e',$$
$$W = \varepsilon\,\frac{a'}{n'}(a'_1 \sin E' - \frac{1}{4}\,a'_1 e' \sin 2E'$$
$$-\,b'_1\eta' \cos E' + \frac{1}{4}\,b'_1 e'\eta' \cos 2E').$$

Because this study is concerned exclusively with the long term behavior of the dust particles, the explicit equations of the infinitesimal transformation

$$\ell = \ell' + (\ell'; W), \qquad L = L' + (L'; W),$$

$$g = g' + (g'; W), \qquad G = G' + (G'; W),$$

$$\nu = \nu' + (\nu'; W), \qquad N = N' + (N'; W)$$

will not be developed. Furthermore, for the sake of simplifying the notations, from here onwards accents will be dropped from all symbols designating averaged state functions.

Insight into the long term behavior of the averaged orbits is gained by introducing the frequency

$$\boxed{\omega_L = \frac{3}{2} \varepsilon \frac{a}{L}.} \tag{5}$$

Although, in the sequel, ω_L will be simply noted ω, it must be emphasized however that each level of averaged L determines its own rate ω, or that $\omega: L \to \omega_L$ specifies a <u>spectrum</u> of frequencies. At a given level of energy, major features of the solution are determined by the relative position of the frequencies m and ω_L in that spectrum. In fact the parameters that will turn up most often will be the <u>gyration frequency</u> $k = (m^2 + \omega^2)^{1/2}$ and the phase λ such that

$$m = k \cos \lambda, \qquad \omega = k \sin \lambda \qquad \text{with} \qquad 0 \leqslant \lambda \leqslant \pi/2.$$

The gyration frequency owes its paramount importance to the fact that, at each level of energy, the averaged perturbation Hamiltonian is the linear combination

$$\boxed{K_1 = - m\, G_3 - \omega\, A_1} \tag{6}$$

of the basic frequencies m and ω. Remarkably enough, the coefficients in (6) have immediate <u>intrinsic</u> meanings; to recall, G_3 is the projection of the particle's angular momentum normal to the planet's orbit, and A_1 is the projection of the particle's Runge-Lenz vector in the direction of the radiation pressure.

So far in the literature [3; 5; 6; 13 and 15], the motion has been analyzed in the equations for its Delaunay coordinates

$$\dot{g} = \frac{\partial K_1}{\partial G} = \omega \left(\frac{\eta}{e} a_1 - \frac{e}{\eta} n_3 \sin \nu \sin g \right), \tag{7_1}$$

$$\dot{\nu} = \frac{\partial K_1}{\partial n} = - m + \omega \frac{e}{\eta} \sin \nu \sin g, \tag{7_2}$$

$$\dot{G} = - \frac{\partial K_1}{\partial g} = \omega \text{ Le } b_1, \tag{7_3}$$

$$\dot{N} = - \frac{\partial K_1}{\partial \nu} = - \omega \text{ Le } a_2 \tag{7_4}$$

and their associate elements

$$\dot{e} = - \omega \eta b_1, \tag{7_5}$$

$$\dot{I} = \omega \frac{e}{\eta} n_1 \cos g. \tag{7_6}$$

One might get a handle on the geometry in the problem by following the evolution of the apsidal frame $(\underset{\sim}{n}, \underset{\sim}{a}, \underset{\sim}{b})$ in the equations

$$\dot{n}_1 = m n_2 - \omega \frac{e}{\eta} n_1 b_1, \qquad \dot{a}_1 = m a_2 + \omega \frac{\eta}{e} a_1 b_1,$$

$$\dot{n}_2 = - m n_1 - \omega \frac{e}{\eta} n_1 b_2, \qquad \dot{a}_2 = - m a_1 + \omega \frac{\eta}{e} a_1 b_2,$$

$$\dot{n}_3 = - \omega \frac{e}{\eta} n_1 b_3, \qquad \dot{a}_3 = \omega \frac{\eta}{e} a_1 b_3,$$

in fact equivalent to the vector differential equations

$$\dot{\underset{\sim}{n}} = \underset{\sim}{\omega} \times \underset{\sim}{n}, \qquad \dot{\underset{\sim}{a}} = \underset{\sim}{\omega} \times \underset{\sim}{a}, \qquad \dot{\underset{\sim}{b}} = \underset{\sim}{\omega} \times \underset{\sim}{b}$$

with the Darboux vector

$$\underset{\sim}{\omega} = - m \underset{\sim}{f}_3 + \omega \left(\frac{e}{\eta} n_1 \underset{\sim}{a} + \frac{\eta}{e} a_1 \underset{\sim}{n}\right).$$

Symmetric in form as they are, these equations are still quadratic in the components of the basis, which creates many analytical complications which obscure the intrinsic simplicity of the problem.

A few classes of particular solutions are derived immediately from equations (7). For example, if there is an instant at which I mod π = 0 and $\dot{I}$ = 0, then, by virtue of the uniqueness theorem applied to (7_6), I mod π and $\dot{I}$ will be equal to 0 at all times; in words, the particle will stay in the plane of the planet. Coplanar orbits, as these solutions are called, have been thoroughly discussed by Mignard, [13]. Furthermore the system admits at least four singular solutions:

I. -- $e = \sin\lambda$, $I = 0$, $\nu + g = 0$;

II. -- $e = \cos\lambda$, $I = \pi/2$, $\nu = g = \pi/2$;

III. -- $e = \cos\lambda$, $I = \pi/2$, $\nu = g = -\pi/2$;

IV. -- $e = \sin\lambda$, $I = \pi$, $\nu - g = 0$.

These are solutions for which the right hand members of equations (7) vanish simultaneously. It will be shown in Section 6 that there are only four critical solutions, and that they are all non-degenerate.

4. Linear and Quadratic Integrals

System (7) has two degrees of freedom, and it admits an integral, which is the Hamiltonian K_1 itself. Naturally one should seek for a second integral. To this end, one will recall the well-known Poisson brackets

$$\begin{aligned}(G_i;\, G_j) &= \sum_{1\leqslant k\leqslant 3} \varepsilon_{i,j,k}\, G_k \qquad (1 \leqslant i,\, j \leqslant 3),\\ (G_i;\, A_j) &= \sum_{1\leqslant k\leqslant 3} \varepsilon_{i,j,k}\, A_k \qquad (1 \leqslant i,\, j \leqslant 3),\\ (A_i;\, A_j) &= \sum_{1\leqslant k\leqslant 3} \varepsilon_{i,j,k}\, G_k \qquad (1 \leqslant i,\, j \leqslant 3);\end{aligned} \tag{8}$$

$\varepsilon_{i,j,k}$ is the Levi-Civita symbol on the permutations of the set (1, 2, 3), it is equal to 1 if the permutation (i, j, k) is even, it is -1 if the transposition is odd, and is zero otherwise. The quantities

$$S_i = \frac{1}{2}(G_i + A_i) \qquad \text{and} \qquad D_i = \frac{1}{2}(G_i - A_i) \qquad (1\leqslant i\leqslant 3)$$

Give rise to an even simpler set of Poisson brackets

$$\begin{aligned}(S_i;\, S_j) &= \sum_{1\leqslant k\leqslant 3} \varepsilon_{i,j,k}\, S_k \qquad (1 \leqslant i,\, j \leqslant 3),\\ (D_i;\, D_j) &= \sum_{1\leqslant k\leqslant 3} \varepsilon_{i,j,k}\, D_k \qquad (1 \leqslant i,\, j \leqslant 3),\\ (S_i;\, D_j) &= 0 \qquad\qquad\qquad\quad (1 \leqslant i,\, j \leqslant 3).\end{aligned} \tag{9}$$

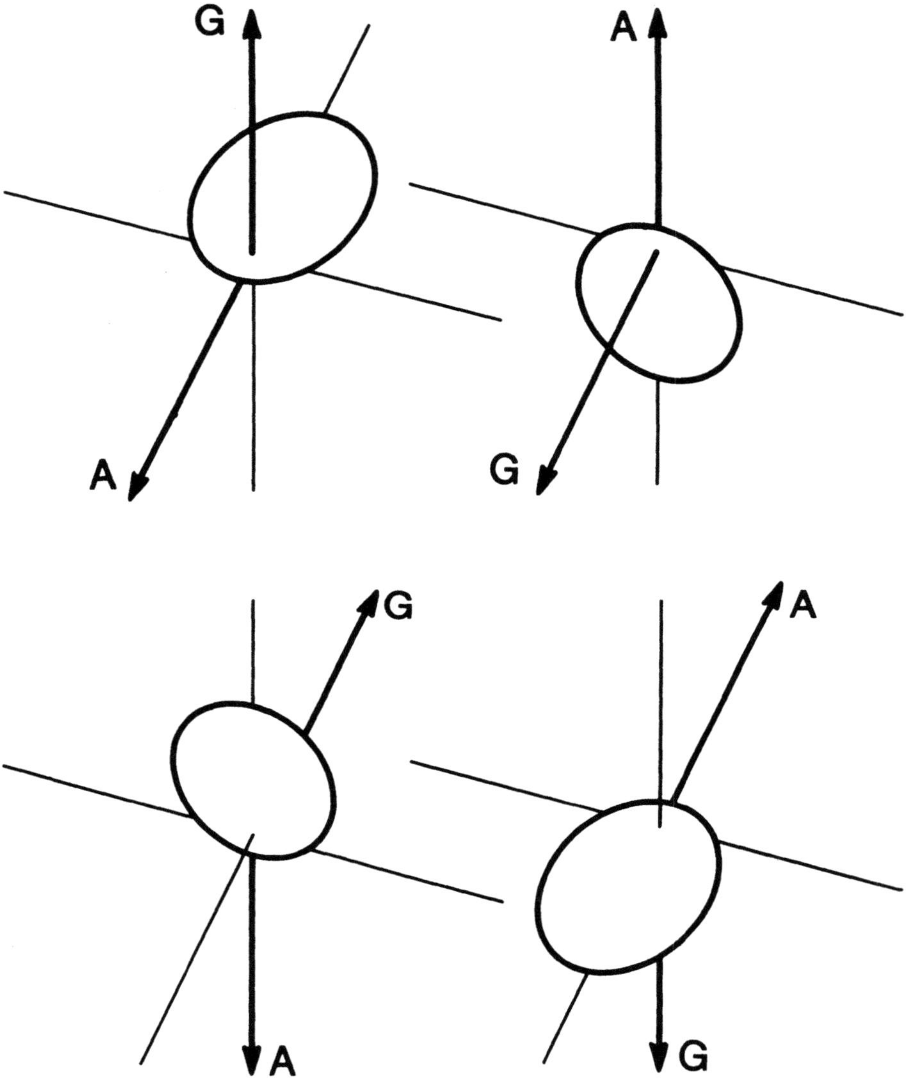

Figure 1. The four singular solutions.

In the phase space $(S_1, S_2, S_3, D_1, D_2, D_3)$,

$$K_1 = -(\omega S_1 + m S_3) + (\omega D_1 - m D_3) \tag{10}$$

In reaching for a second integral, one could begin by enquiring under what conditions the linear combinations with real coefficients

$$J = \sum_{1 \leq i \leq 3} (\alpha_i S_i + \beta_i D_i)$$

satisfy the relation $(J; K_1) = 0$. But a quick calculation yields that

$$(J; K_1) = -m\,\alpha_2 S_1 + (m\alpha_1 - \omega\alpha_3) S_2 + \omega\,\alpha_2 S_3$$
$$- m\,\beta_2 D_1 + (m\beta_1 + \omega\beta_3) D_2 - \omega\,\beta_2 D_3.$$

For J to be an integral, it is therefore necessary to take $\alpha_2 = \beta_2 = 0$, and to choose the remaining four coefficients so as to satisfy the conditions

$$m\,\alpha_1 - \omega\,\alpha_3 = 0 \qquad \text{and} \qquad m\,\beta_1 + \omega\,\beta_3 = 0,$$

or equivalently the conditions

$$\alpha_1 \cos\lambda - \alpha_3 \sin\lambda = 0 \qquad \text{and} \qquad \beta_1 \cos\lambda + \beta_3 \sin\lambda = 0.$$

These relations are satisfied in terms of two arbitrary parameters by taking

$$\alpha_1 = \gamma \sin\lambda, \qquad \beta_1 = -\delta \sin\lambda,$$

$$\alpha_3 = \gamma \cos\lambda, \qquad \beta_3 = \delta \cos\lambda.$$

Thus all integrals of system (7) linear in the actions S_i and D_i reduce to the linear combination $J = \gamma\,\Phi + \delta\,\Psi$ of the two action integrals

$$\boxed{\begin{aligned} \Phi &= S_3 \cos\lambda + S_1 \sin\lambda, \\ \Psi &= D_3 \cos\lambda - D_1 \sin\lambda. \end{aligned}}$$

Needless to say, among the linear integrals, one recovers for $\gamma = \delta = -k$ the perturbation Hamiltonian

$$K_1 = -k\,(\Phi + \Psi). \tag{12}$$

More integrals of the problem are found among the quadratic expressions

$$\Gamma = \sum_{1\leq i,j\leq 3} (\alpha_{i,j}\, S_i\, S_j + \beta_{i,j}\, S_i\, D_j + \gamma_{i,j}\, D_i\, D_j$$

where $\alpha_{i,j} = \alpha_{j,i}$ and $\gamma_{i,j} = \gamma_{j,i}$. Indeed a rather long but straightforward evaluation of Poisson brackets yields that

$$\begin{aligned}
(Q;\, K_1) = &- 2\, \alpha_{1,2} \cos\lambda\, S_1^2\\
&- 2\, [(\alpha_{1,1} - \alpha_{2,2}) \cos\lambda - \alpha_{1,3} \sin\lambda]\, S_1\, S_2\\
&+ 2\, (\alpha_{2,3} \cos\lambda - \alpha_{1,2} \sin\lambda)\, S_1\, S_3\\
&+ 2\, (\alpha_{2,3} \sin\lambda - \alpha_{1,2} \cos\lambda)\, S_2^2\\
&+ 2\, [(\alpha_{3,3} - \alpha_{2,2}) \sin\lambda - \alpha_{1,3} \cos\lambda]\, S_2\, S_3\\
&- 2\, \alpha_{2,3} \sin\lambda\, S_3^2\\
&- (\beta_{1,2} + \beta_{2,1}) \cos\lambda\, S_1\, D_1\\
&+ [(\beta_{1,1} - \beta_{2,2}) \cos\lambda + \beta_{1,3} \sin\lambda]\, S_1\, D_2\\
&- (\beta_{2,3} \cos\lambda + \beta_{1,2} \sin\lambda)\, S_1\, D_3\\
&+ [(\beta_{1,1} - \beta_{2,2}) \cos\lambda - \beta_{3,1} \sin\lambda]\, S_2\, D_1\\
&+ [(\beta_{1,2} + \beta_{2,1}) \cos\lambda + (\beta_{2,3} - \beta_{3,2}) \sin\lambda]\, S_2\, D_2\\
&+ [\beta_{1,3} \cos\lambda - (\beta_{2,2} + \beta_{3,3}) \sin\lambda]\, S_2\, D_3\\
&- (\beta_{3,2} - \beta_{2,1} \sin\lambda)\, S_3\, D_1\\
&+ [\beta_{3,1} \cos\lambda + (\beta_{2,2} + \beta_{3,3}) \sin\lambda]\, S_3\, D_2\\
&- (\beta_{2,3} + \beta_{3,2}) \sin\lambda\, S_3\, D_3\\
&- 2\, \gamma_{1,2} \cos\lambda\, S_1^2\\
&- 2\, [(\gamma_{1,1} - \gamma_{2,2}) \cos\lambda + \gamma_{1,3} \sin\lambda]\, S_1\, S_2\\
&+ 2\, (\gamma_{2,3} \cos\lambda + \gamma_{1,2} \sin\lambda)\, S_1\, S_3\\
&- 2\, (\gamma_{2,3} \sin\lambda + \gamma_{1,2} \cos\lambda)\, S_2^2\\
&- 2\, [(\gamma_{3,3} - \gamma_{2,2}) \sin\lambda + \gamma_{1,3} \cos\lambda]\, S_2\, S_3\\
&+ 2\, \gamma_{2,3} \sin\lambda\, S_3^2
\end{aligned}$$

The conditions for Q to be an integral are thus that

$$\alpha_{1,2} = \alpha_{2,3} = 0,$$

$$(\alpha_{1,1} - \alpha_{2,2}) \cos\lambda = \alpha_{1,3} \sin\lambda,$$

$$(\alpha_{3,3} - \alpha_{2,2}) \sin\lambda = \alpha_{1,3} \cos\lambda,$$

$$\beta_{1,2} = - \beta_{2,1},$$

$$\beta_{2,3} = \beta_{3,2},$$

$$\beta_{1,2} \cos\lambda + \beta_{2,3} \sin\lambda = \beta_{3,2} \sin\lambda - \beta_{2,1} \cos\lambda,$$

$$\beta_{1,2} \sin\lambda + \beta_{2,3} \cos\lambda = 0,$$

$$\beta_{2,1} \sin\lambda - \beta_{3,2} \cos\lambda = 0,$$

$$(\gamma_{1,1} - \gamma_{2,2}) \cos\lambda = - \gamma_{1,3} \sin\lambda,$$

$$\gamma_{1,2} = \gamma_{2,3} = 0,$$

$$(\gamma_{3,3} - \gamma_{2,2}) \sin\lambda = - \gamma_{1,3} \cos\lambda,$$

and they are satisfied by introducing parameters ε_1, ε_2, ε_3, ε_4, ε_5, ε_6, ε_7 so that

$$\alpha_{1,1} = \varepsilon_1 - \varepsilon_2 \sin^2\lambda,$$

$$\alpha_{2,2} = \varepsilon_1,$$

$$\alpha_{3,3} = \varepsilon_1 + \varepsilon_2 \cos^2\lambda,$$

$$\beta_{1,2} = - \beta_{2,1} = \varepsilon_3 \cos\lambda,$$

$$\beta_{2,3} = \beta_{3,2} = - \varepsilon_3 \cos\lambda,$$

$$\beta_{2,2} = \varepsilon_4,$$

$$\beta_{1,1} = \varepsilon_4 \cos^2\lambda - \varepsilon_5 \sin^2\lambda,$$

$$\beta_{1,3} = - \beta_{3,1} = (\varepsilon_4 + \varepsilon_5) \cos\lambda \sin\lambda,$$

$$\beta_{3,3} = \varepsilon_5 \cos^2\lambda - \varepsilon_4 \sin^2\lambda,$$

$$\gamma_{1,1} = \varepsilon_6 - \varepsilon_7 \sin^2\lambda,$$

$$\gamma_{2,2} = \varepsilon_6,$$

$$\gamma_{3,3} = \varepsilon_6 + \varepsilon_7 \cos^2\lambda,$$

Eventually Q appears as the sum

$$Q(\varepsilon_1, \varepsilon_2, \varepsilon_3, \varepsilon_4, \varepsilon_5, \varepsilon_6, \varepsilon_7) \equiv \sum_{1\leqslant j\leqslant 7} \varepsilon_j Q_j$$

of the quadratic integrals $Q_2 = \Phi^2$, $Q_5 = \Phi\,\Psi$, $Q_7 = \Psi^2$, and

$$\boxed{\begin{aligned} Q_1 &= S_1^2 + S_2^2 + S_3^2, \\ Q_3 &= (S_1 \cos\lambda - S_3 \sin\lambda)\, D_2 - (D_1 \cos\lambda + D_3 \sin\lambda)\, S_2, \\ Q_4 &= (S_1 \cos\lambda - S_3 \sin\lambda)\,(D_1 \cos\lambda + D_3 \sin\lambda) + S_2\, D_2, \\ Q_6 &= D_1^2 + D_2^2 + D_3^2. \end{aligned}} \qquad (13)$$

Integrals Q_1 and Q_6 express the well known fact that the orbit space of the averaged problem is the product of two spheres $S^2(R^3)$, each of radius L/2; together with the linear integrals Φ and Ψ, they will lead to an intuitive solution of the entire problem.

5. Angles and Actions

In the KSK construction [19] the functional $J \equiv J(\gamma, \delta)$ acts as a co-moment mapping, from which characterization the Kirillov-Souriau-Kostant theorem deduces that system (6) is integrable. But the KSK construction does not provide a method for defining a symplectic map on the reduced manifold. So it remains to build a canonical transformation

$$(G, N, g, \nu) \rightarrow (\Phi, \Psi, \phi, \psi)$$

that will convert Hamiltonian (6) to its normalized form (12). In the ordinary treatment by classical mechanics, the crucial consideration is not that the system is symmetric for the group of rotations SO(4), but that $(\Phi; \Psi) = 0$, i.e. that the integrals are in involution [12] .

To the integrals Φ and Ψ are adjoined the state functions

$$\Phi' = S_1 \cos\lambda - S_3 \sin\lambda \quad \text{and} \quad \Psi' = D_1 \cos\lambda + D_3 \sin\lambda,$$

and a quick calculation yields the Poisson brackets

$$(S_2;\ \Phi) = \Phi', \qquad (\Phi;\ \Phi') = S_2, \qquad (\Phi';\ S_2) = \Phi,$$

$$(D_2;\ \Psi) = \Psi', \qquad (\Psi;\ \Psi') = D_2, \qquad (\Psi';\ D_2) = \Psi.$$

They confer a geometric meaning to the functions involved, for they prove that both triples (Φ, Φ', S_2) and (Ψ, Ψ', D_2) are bases of Lie algebras (defined by Poisson brackets) isomorphic to so(3), viz the Lie algebra of the group SO(3) of finite rotations in a three - dimensional Euclidean space. In other words, Φ, Φ' and S_2 are generators of infinitesimal rotations of the sphere $Q_1 = L^2/4$. Likewise, Ψ, Ψ' and D_2 are generators of infinitesimal rotations $Q_6 = L^2/4$. But Φ and Ψ are integrals of system (7). Therefore it can already be concluded that the orbits of (7) on the sphere $Q_1 = L^2/4$ consist exclusively of displacements on the small circles at the intersection of the sphere by the planes $S_3 \cos\lambda + S_1 \sin\lambda + \Phi = 0$; likewise, on the sphere $Q_6 = L^2/4$, the orbits of (7) result solely from displacements on the small circles at the intersection of the sphere by the planes $D_3 \cos\lambda - D_1 \sin\lambda - \Psi = 0$. In Figure 2 are depicted the problem's orbits on the S - and D - spheres from the extreme case where $\omega = 0$ for pure Keplerian systems to the other extreme where $m = 0$ for pure Stark effects caused by a non-rotating field of force with constant magnitude.

On the S-sphere, the points

$$(\tfrac{1}{2} L \sin\lambda,\ 0,\ \tfrac{1}{2} L \cos\lambda) \quad \text{and} \quad (-\tfrac{1}{2} L \sin\lambda,\ 0,\ -\tfrac{1}{2} L \cos\lambda)$$

at the extremities of the diameter

$$(S_2 = 0, S_1 \cos\lambda - S_3 \sin\lambda = 0)$$

are fixed; furthermore, the meridian plane containing the vector (S_1, S_2, S_3) rotates around the S_3 - axis when $|\Phi| < (L/2)\cos\lambda$, but librates when $(L/2)\cos\lambda < |\Phi| < L/2$. There is a parallel classification of orbits on the D - sphere. The points

$$(-\tfrac{1}{2} L \sin\lambda,\ 0,\ \tfrac{1}{2} L \cos\lambda) \text{ and } (\tfrac{1}{2} L \sin\lambda,\ 0,\ -\tfrac{1}{2} L \cos\lambda)$$

where the diameter $(D_2 = 0,\ D_1 \cos\lambda + D_3 \sin\lambda = 0)$ intersects around the D_3 - axis when $|\Psi| < (L/2)\cos\lambda$, and librates when $L/2)\cos\lambda < |\Psi| < L/2$.

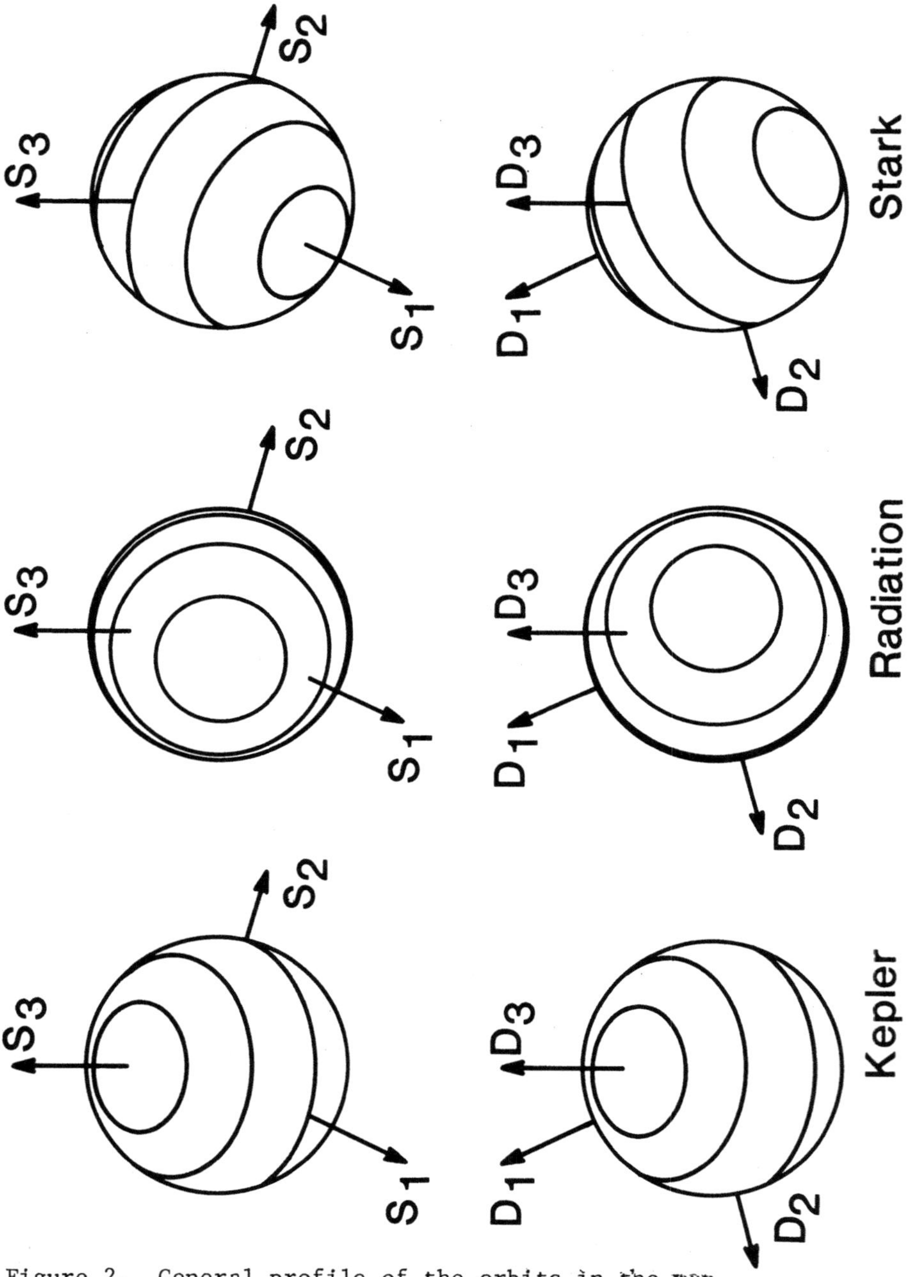

Figure 2. General profile of the orbits in the map (S_1, S_2, S_3, D_1, D_2, D_3).

As λ progresses from 0 to $\pi/2$, the axis of rotation on the S - sphere tilts forward in the meridian plane $S_2 = 0$, starting at the S_3 - axis and ending on the S_1 - axis. The same evolution occurs on the D - sphere except that the tilt is backward from the positive D_3 - axis to the negative D_1 - axis. What is remarkable is that the change is continuous, and free from catastrophic bifurcations.

On account of the relations

$$\Phi^2 + \Phi'^2 + S_2^2 = \Psi^2 + \Psi'^2 + D_2^2 = \frac{1}{4} L^2,$$

it is consistent to parametrize the small circles by angles ϕ and ψ such that

$$\Phi' = (\tfrac{1}{4} L^2 - \Phi^2)^{1/2} \cos\phi, \qquad \Psi' = (\tfrac{1}{4} L^2 - \Psi^2)^{1/2} \cos\psi,$$

$$S_2 = (\tfrac{1}{4} L^2 - \Phi^2)^{1/2} \sin\phi, \qquad D_2 = (\tfrac{1}{4} L^2 - \Psi^2)^{1/2} \sin\psi.$$

The claim is that the transformation

$$(G, N, g, \nu) \rightarrow (\Phi, \Psi, \phi, \psi)$$

defined implicitly by the equations

$$S_3 \cos\lambda + S_1 \sin\lambda = \Phi, \qquad (14_1)$$

$$S_2 = (\tfrac{1}{4} L^2 - \Phi^2)^{1/2} \sin\phi, \qquad (14_2)$$

$$D_3 \cos\lambda - D_1 \sin\lambda = \Psi, \qquad (14_3)$$

$$D_2 = (\tfrac{1}{4} L^2 - \Psi^2)^{1/2} \sin\psi, \qquad (14_4)$$

$$S_1 \cos\lambda - S_3 \sin\lambda = (\tfrac{1}{4} L^2 - \Phi^2)^{1/2} \cos\phi, \qquad (14_5)$$

$$D_1 \cos\lambda + D_3 \sin\lambda = (\tfrac{1}{4} L^2 - \Psi^2)^{1/2} \cos\psi \qquad (14_6)$$

is canonical. In fact it is sufficient to prove that

$$(\phi; \psi) = 0,$$

$$(\phi; \Phi) = 1, \qquad (\psi; \Phi) = 0,$$

$$(\phi; \Psi) = 0, \qquad (\psi; \Psi) = 1, \qquad (\Phi; \Psi) = 0.$$

These identities result from elementary properties of Poisson brackets. First

$$\Phi' = (S_2;\ \Phi) = (\tfrac{1}{4}\ L^2 - \Phi^2)^{1/2}\ \ (\sin\phi;\ \Phi) = (\phi;\ \Phi)\ \Phi',$$

which proves that $(\phi;\ \Phi) = 1$; it is shown in the same manner that $(\psi;\ \Psi) = 1$. Then from the relations

$$0 = (S_2;\ \Psi) = (\tfrac{1}{4}\ L^2 - \Phi^2)^{1/2}\ \ (\sin\phi;\ \Psi) = (\phi;\ \Psi)\ \Phi',$$

$$0 = (\Phi';\ \Psi) = -\ (\phi;\ \Psi)\ S_2,$$

it is deduced straightaway that $(\phi;\ \Psi) = 0$, hence, by analogy, that $(\psi;\ \Phi) = 0$. Next, the relation $(\Phi;\ \Psi) = 0$ follows immediately from (9) and (10). Finally

$$(S_2;\ D_2) = \Phi'\ \Psi'\ (\phi;\ \psi),$$

which shows that $(\phi;\ \psi) = 0$.

In the canonical variables $(\Phi,\ \Psi,\ \phi,\ \psi)$, according to (12), the equations of motion are trivial:

$$\dot{\phi} = \frac{\partial K_1}{\partial \Phi} = -\ k, \quad \dot{\psi} = \frac{\partial K_1}{\partial \Psi} = -\ k, \quad \dot{\Phi} = -\ \frac{\partial K_1}{\partial \phi} = 0, \quad \dot{\Psi} = -\ \frac{\partial K_1}{\partial \psi} = 0.$$

They validate Mignard's contention [13, p. 362] that the system admits two angle variables whose frequencies are equal, the common period being $2\pi/k$.

The meaning of integrals Q_3 and Q_4 may now be elucidated. Since $\dot{\phi} = \dot{\psi} = -\ k$, the phase angle $\phi - \psi$ remains constant along an orbit, and so do the functions

$$(\tfrac{1}{4}\ L^2 - \Phi^2)^{1/2}\ (\tfrac{1}{4}\ L^2 - \Psi^2)^{1/2}\ \cos\ (\phi - \psi) = Q_4.$$

$$(\tfrac{1}{4}\ L^2 - \Phi^2)^{1/2}\ (\tfrac{1}{4}\ L^2 - \Psi^2)^{1/2}\ \sin\ (\phi - \psi) = Q_3.$$

The momenta $(\Phi + \Psi)/2$ and $(\Phi + \Psi)/2$ correspond precisely to the actions J_1 and J_2 which Kramers [11] obtained by contour integrals in his analysis of the Stark effect. The calculation is reproduced in Born [2]. As a reminder of this connection, we propose to name the transformation $(G,\ N,\ g,\ \nu) \to (\Phi,\ \Psi,\ \phi,\ \psi)$ after Kramers. One should note however that, within the context of the Bohr-Sommerfeld quantum mechanics, Kramers was not interested in defining the angles $(\phi,\ \psi)$ associated with the actions $(\Phi,\ \Psi)$, even less in producing formulas of the type (14) to related the angles and actions to the Delaunay orbital elements.

6. The Singular Orbits

The extrema of the function K_1 on the sphere $S^3(R^4) = \{ (S_1, S_2, S_3, D_1, D_2, D_3) \mid S_1^2 + S_2^2 + S_3^2 = L^2/4$ and $D_1^2 + D_2^2 + D_3^2 = L^2/4 \}$ correspond to the singular orbits in the problem. By reason of the constraints

$$S = S_1^2 + S_2^2 + S_3^2 - L^2/4 = 0$$

and

$$D = D_1^2 + D_2^2 + D_3^2 - L^2/4 = 0$$

one introduces the Lagrange multipliers σ and δ, and look for the extrema of the function $F = K_1 + \sigma S + \delta D$ with a view of determining the multipliers so as to satisfy the constraints. Thus, for the equations

$$\frac{\partial F}{\partial S_1} = - \omega + 2 \sigma S_1 = 0, \qquad \frac{\partial F}{\partial D_1} = \omega + 2 \delta D_1 = 0,$$

$$\frac{\partial D}{\partial S_2} = 2 \sigma S_2 = 0, \qquad \frac{\partial F}{\partial D_2} = 2 \delta D_2 = 0,$$

$$\frac{\partial F}{\partial S_3} = - m + 2 \sigma S_3 = 0, \qquad \frac{\partial F}{\partial D_3} = - m + 2 \delta D_3 = 0,$$

the solutions

$$S_1 = \frac{\omega}{2\sigma}, \qquad S_2 = 0, \qquad S_3 = \frac{m}{2\sigma},$$

$$D_1 = -\frac{\omega}{2\delta}, \qquad D_2 = 0, \qquad D_3 = \frac{m}{2\delta},$$

are substituted back into the constraints, and the Lagrange multipliers identifying the extrema of F which happen to be located on the sphere S^4 are such that

$$L^2 \sigma^2 = L^2 \delta^2 = k^2.$$

So there are thus four and only four critical orbits in this problem. At any critical point of F corresponding to the Lagrange multipliers σ and δ, the Hessian of F is

$$\text{Hess } F(S_1, S_2, S_3, D_1, D_2, D_3) = \begin{bmatrix} 2\sigma & 0 & 0 & 0 & 0 & 0 \\ 0 & 2\sigma & 0 & 0 & 0 & 0 \\ 0 & 0 & 2\sigma & 0 & 0 & 0 \\ 0 & 0 & 0 & 2\delta & 0 & 0 \\ 0 & 0 & 0 & 0 & 2\delta & 0 \\ 0 & 0 & 0 & 0 & 0 & 2\delta \end{bmatrix} \{\ker ((S_1, S_2, S_3, D_1, D_2, D_3))\}^{-1}$$

But, at any of the singular points,

$$\ker \{(S_1, S_2, S_3, D_1, D_2, D_3)\}^{-1} = \{ (\cos\lambda, 0, -\sin\lambda), (0, 1, 0)\}$$

where the angle brackets denote the subspace spanned by the enclosed vectors. Thus the Hessian restricted to the plane tangent to S^4 at any critical point is the diagonal matrix

$$\text{Hess } F = \begin{vmatrix} 2\sigma & 0 & 0 & 0 \\ 0 & 2\sigma & 0 & 0 \\ 0 & 0 & 2\delta & 0 \\ 0 & 0 & 0 & 2\delta \end{vmatrix}$$

which of course is non degenerate. It remains to show that the critical points correspond to the four singular orbits mentioned at the end of Section 3. As an illustration, the case when $\Phi = \Psi = L/2$ will be analyzed in detail. The relations

$$S_1 \sin\lambda + S_3 \cos\lambda = \frac{1}{2} L, \qquad S_1 \cos\lambda - S_3 \sin\lambda = 0,$$

$$D_3 \cos\lambda - D_1 \sin\lambda = \frac{1}{2} L, \qquad D_3 \sin\lambda + D_1 \cos\lambda = 0$$

imply that

$$S_3 = D_3 = \frac{1}{2} \cos\lambda, \qquad S_1 = D_1 = -\frac{1}{2} \sin\lambda,$$

from which there results that

$$G_1 = G_2 = 0, \qquad G_3 = L \cos\lambda,$$

$$A_1 = L \sin\lambda, \qquad A_2 = A_3 = 0.$$

Consequently $L^2 \eta^2 \sin^2 I = 0$, hence $G^2 = L^2 \eta^2 = L^2 \cos^2 \lambda$, which implies that $\eta = \cos\lambda$ and $e = \sin\lambda$.But then $\cos I = 1$ and $\sin I = 0$, which makes $I = 0$. Finally $A_1 = L \sin\lambda$ and $A_2 = 0$ mean that $a_1 = \cos(\nu + g) = 1$ and $a_2 = \sin(\nu + g) = 0$, hence that $\nu + g = 0$.

Table I presents the characteristics of the singular solutions. In the Keplerian case ($\omega = 0$), orbits I and IV are circular while II and III are linear or collision orbits. This is exactly the opposite of what happens in the Stark effect ($m = 0$): I and IV in the plane of the planet's orbit are linear whereas II and III in the plane normal to the radiation pressure are circular.

Table 1. The Four Critical Solutions at a Given Energy

	I	II	III	IV
S_1	$\frac{1}{2}$ L sin λ	$\frac{1}{2}$ L sin λ	$-\frac{1}{2}$ L sin λ	$-\frac{1}{2}$ L sin λ
S_2	0	0	0	0
S_3	$\frac{1}{2}$ L cos λ	$\frac{1}{2}$ L cos λ	$-\frac{1}{2}$ L cos λ	$-\frac{1}{2}$ L cos λ
D_1	$-\frac{1}{2}$ L sin λ	$\frac{1}{2}$ L sin λ	$-\frac{1}{2}$ L sin λ	$\frac{1}{2}$ L sin λ
D_2	0	0	0	0
D_3	$\frac{1}{2}$ L cos λ	$-\frac{1}{2}$ L cos λ	$\frac{1}{2}$ L cos λ	$-\frac{1}{2}$ L cos λ
e	sin λ	cos λ	cos λ	sin λ
ν	(?)	$\frac{\pi}{2}$	$-\frac{\pi}{2}$	(?)
g	$\nu + g = 0$	$\frac{\pi}{2}$	$-\frac{\pi}{2}$	$\nu - g = \pi$
I	0	$\frac{\pi}{2}$	$\frac{\pi}{2}$	π
	direct	direct	retrograde	retrograde
K'_1	$-$ k L	0	0	k L
Φ	$\frac{1}{2}$ L	$\frac{1}{2}$ L	$-\frac{1}{2}$ L	$-\frac{1}{2}$ L
Ψ	$\frac{1}{2}$ L	$-\frac{1}{2}$ L	$\frac{1}{2}$ L	$-\frac{1}{2}$ L

7. Initial Value Problem

Most astronomers would rather discuss the particle's motion in the Keplerian elements. As a matter of fact, transposition from the geometric variables (S_1, S_2, S_3, D_1, D_2, D_3) to the traditional coordinates presents no major problem. The components of the angular momentum and of the Runge-Lenz vector are solutions of two separate systems, each made of three linear homogeneous equations with constant coefficients. The first group,

$$\dot{G}_1 = (G_1; K_1) = m\, G_2, \tag{15_1}$$

$$\dot{G}_2 = (G_2; K_1) = -\, m\, G_1 + \omega\, A_3, \tag{15_2}$$

$$\dot{A}_3 = (A_3; K_1) = -\, \omega\, G_2, \tag{15_3}$$

concerns essentially the motion of the node of the particle's orbital plane in the plane of the planet. Its solutions are curves drawn on the sphere

$$G_1^2 + G_2^2 + A_3^2 = 2\,(\tfrac{1}{4}\,L^2 + Q_4 - \Phi\,\Psi)$$

$$= L^2 \sin^2 I\,(1 - e^2 \cos^2 g)$$

whose radius is evidently an integral. Notice that, for coplanar orbits ($I \bmod \pi = 0$), the sphere collapses onto its center. The integral

$$A_3 \cos\lambda + G_1 \sin\lambda = \Phi - \Psi$$

$$= L \sin I\,(e \sin g \cos\lambda + \eta \sin\nu \sin\lambda)$$

restricts the motions on the sphere to being rotations around the fixed diameter of equations $A_3 \sin\lambda = G_1 \cos\lambda$ in the the meridian plane $G_2 = 0$. Indeed, introducing the radius

$$\Gamma = L \sin I\,(1 - e^2 \cos^2 g)^{1/2}$$

and the spherical coordinates (α, δ) such that

$$A_3 \cos\lambda + G_1 \sin\lambda = \Gamma \sin\gamma,$$

$$G_1 \cos\lambda - A_3 \sin\lambda = \Gamma \cos\gamma \cos\alpha,$$

$$G_2 = \Gamma \cos\gamma \sin\alpha,$$

one finds by substitution in the node equations (15) that the elevation γ is constant, and that the azimuth α precesses at the constant rate $\alpha = -k$.

Likewise, for the second system,

$$\dot{A}_1 = (A_1;\ K_1) = m\ A_2, \qquad (16_1)$$
$$\dot{A}_2 = (A_2;\ K_1) = -\ m\ A_1 + \omega\ G_3, \qquad (16_2)$$
$$\dot{G}_3 = (G_3;\ K_1) = -\ \omega\ A_2, \qquad (16_3)$$

which deals basically with the motion of the pericenter, the solutions are curves on the sphere

$$A_1^2 + A_2^2 + G_3^2 = 2\ (\tfrac{1}{4}\ L^2 - Q_4 + \Phi\ \Psi).$$

Note that the sphere collapses onto its center when the focal axis of the particle's averaged ellipse is normal to the orbital plane of the planet. Exactly as was done on the node-sphere, one recognizes that

$$G_3 \cos\lambda + A_1 \sin\lambda = \Phi + \Psi = -\ K_1/k$$

is an integral, hence that the orbits on the pericenter sphere consist of small circles in the planes defined by the integral. In the spherical coordinates such that

$$A_3 \cos\lambda + G_1 \sin\lambda = \Delta \sin\delta,$$
$$G_1 \cos\lambda - A_3 \sin\lambda = \Delta \cos\delta \cos\beta,$$
$$G_2 = \Delta \cos\delta \sin\beta,$$
$$\Delta = [\ 2\ (\tfrac{1}{4}\ L^2 - Q_4 + \Phi\ \Psi)\ ]^{1/2},$$

the elevation δ is constant, and the azimuth β precesses at the constant angular velocity $\dot{\beta} = -k$.

From the global behavior of the orbits as depicted in Figure 3, one gathers the general evolution from $\omega = 0$ to $m = 0$ along a circle of fixed gyration frequency k. In the purely Keplerian case, the ascending node and the projection of the pericenter in the plane of the planet undergo circulations exclusively. As soon as λ depart from 0, librations start appearing. Eventually, in the Stark case, all circulations have disappeared; node and pericenter's projection undergo exclusively librations about the direction of the homogeneous field.

8. Conclusions

In a Whittaker map $(R, \Theta, N, r, \theta, \nu)$ or in a Delaunay map (L, G, N, ℓ, g, ν), Keplerian systems in the presence of a homogeneous field of force are most awkward to handle. The

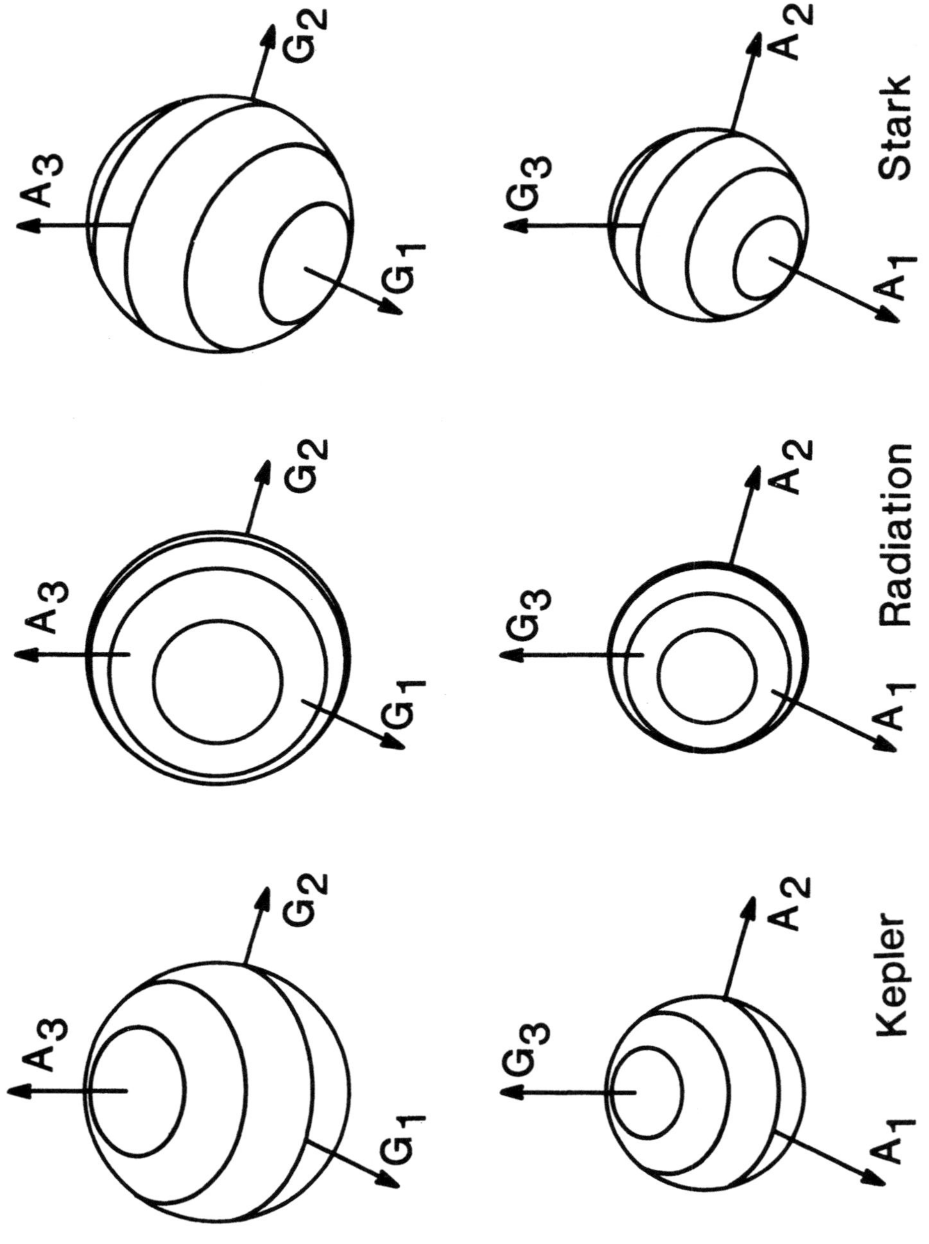

Figure 3. General Profile of the Orbits in the Map (G_1, G_2, G_3, A_1, A_2, A_3).

geometric simplicity of the model comes out most naturally in the map determined by the Cartesian components of the angular momentum and of the Runge-Lenz vector.

Beyond the particular problem put by Bertaux and Blamont which turns out to be an extension of the Stark problem, this Note reaches an element of generality. For it introduces a one-parameter family of canonical transformations from the Delaunay elements (G, N, g, ν) to a new set of phase variables (Φ, Ψ, ϕ, ψ) applicable in principle to any perturbed Keplerian system in two or three dimensions.

For the main problem in the theory of artificial satellites, Kramers' elements provide naturally a symplectic map on the two-dimensional sphere that is the orbit space obtained by eliminating the ascending node from the system after a Delaunay normalization. In particular, they explain most clearly the nature of the so-called critical inclinations. There may be other perturbed Keplerian systems to which the Kramers transformation could be applied, including the theory of minor planets modelled after the three-dimensional restricted problem of three bodies. Especially in resonance situations, Kramers' actions and angles present a definite advantage: they offer the possibility of normalizing the system about any critical frequency by routine techniques beyond the first order.

ACKNOWLEDGEMENTS

The Author thanks Mrs. Elaine Rhodes for the skill with which she typed the manuscript to suit the publisher's requirements.

References

[1] Bertaux, J. L., and Blamont, J. E.: 1973, J. Geophys. Res. **78**, 80 - 91.

[2] Born, M.: 1925, Vorlesungen ueber Atommechanik, Springer Verlag Berlin.

[3] Burns, J. A., Lamy, P. L., and Soter, S.: 1979, Icarus **40**, 1 - 48.

[4] Cartan, E. : 1922, Lecons sur les invariants intégraux, Hermann, Paris.

[5] Chamberlain, J. W. : 1979, Icarus **39**, 286 - 294.

[6] Chamberlain, J. W. : 1980, Icarus **44**, 651 - 656.

[7] Deprit, A. : 1969, Celes. Mech. **1**, 12 - 30.

[8] Deprit, A. : 1982, Celes. Mech. **26**, 9 - 21.

[9] Deprit, A. : 1983, Celes. Mech. **29**, 229 - 247.

[10] Epstein, P. S. : 1916, Ann. Physik. Series 4, **50**, 489 - 520.

[11] Kramers, H. A. : 1920, Z. Physik **3**, 199 - 223.

[12] Meyer, K. R. : 1972, in Dynamical Systems (ed. M. Peixoto), Academic Press, New York, 259 - 272.

[13] Mignard, F. : 1982, Icarus **49**, 347 - 366.

[14] Moser, J. : 1970, Comm. Pure Appl. Math. **23**, 609 - 636.

[15] Peale, S. J. : 1966, J. Geophys. Res. **71**, 911 - 933.

[16] Pauwels, T. : 1983, Celes. Mech. **30**, 229 - 247.

[17] Schwartzschild, K. : 1916, Sitzber. Konigl. Preuss. Akad. Wiss. Berlin (no vol. nr) 548 - 568.

[18] Souriau, J. M. : 1974, Symp. Math. **14**, 343 - 360.

[19] Woodhouse, N. : 1980, Geometric Quantization, Clarendon Press Oxford.

NON GRAVITATIONAL FORCES IN THE EVOLUTION OF THE SOLAR SYSTEM

J. KOVALEVSKY

CERGA, Grasse, France

ABSTRACT

Among the numerous cases of resonnances encountered in the solar system, many of the satellite to satellite and spin-orbit couplings can be explained as the result of an evolution driven by tidal forces. Models of such evolution are described and it is shown that they all lead to a similar type of reduced equations.

It is proposed to solve these equations in form of perturbation of a simpler "restricted tidal problem". Such an approach gives a simple picture of the capture as well as conditions for a permanent capture into a resonant situation, as it is the case of the major satellite resonances and the Mercury rotation. In some other cases, involving higher order resonances, it is possible to have only temporary captures. This has probably happened many times to the Moon, inducing a lengthening of the evolutionary history of the lunar orbit as given by the classical secular deceleration theory.

I. INTRODUCTION

The contributions of Georges Lemaître to Celestial Mechanics concerned essentially the domain of the classical three body problem. This problem - which is far from being fully investigated - is the key problem in the Solar System and more generally in the Universe when only a few bodies are interacting. If, in addition, the masses are all small but one and if the distances between them are large, then the three body problem is used as the basis of the solution for relativistic Celestial Mechanics. Indeed, the effects

A. Berger (ed.), The Big Bang and Georges Lemaitre, 181–194.

of General Relativity can be modelled in the frame of the perturbation theory and they do not disrupt the overall dynamical picture of the system. So, one can say that the basic assomptions of the Celestial Mechanics to which Georges Lemaître has so brilliantly contributed is the ones that allows the very refined description of the present motions of the bodies of the Solar System.

However, this statement is true only if one is allowed to neglect all other forces that inevitably act on them. This is the case to very high accuracy - or, we should better say, to the present accuracy of the observations of their positions - for planets and satellites. In that case, Newtonian classical Mechanics, based on the theoretical studies of the three body problem and slightly modified by the introduction of relativistic corrections, are effectively used for ephemerides and all astronomical and astronautical applications.

There are exceptions. The major exception is the Moon. For more than a century it is known that it is not possible to account exactly for the observations through a purely gravitational theory. An empirical secular term is added to take care of the tidal friction that produces a secular deceleration of its motion in longitude of about $25''/\text{century}^2$. Actually all the bodies are undergoing other forces than direct gravitation: radiation pressure, tidal friction (although it has a gravitational origin) aerodynamical drag, encounters with solid masses, magnetic torques, etc.. The importance of these phenomena in the physics of the Solar System is significant only if the effects produced are directly observable or if one can somehow recognize their consequences. These situations can be summarized as follows :

1. The accelerations produced by these forces are not negligible in comparison of the gravitational forces. The most striking example is the Pointing-Robertson effect on very small particles. The tidal friction on the lunar motion and the solar radiation pressure and mass loss effects on the comets are two other well known examples.

2. The effect of the acceleration is negligible during the time span corresponding to observations but the infinitesimal effects build up and produce a secular evolution that transform, in a long term, the structure of the sub-system. This is the case we shall now concentrate upon.

II. GENERAL DESCRIPTION

The large number of resonances found in the Solar System has always been recognized: the coupling by pairs of Saturn satellites, the three first galilean satellites or the existence of gaps and families in the asteroidal belt have been studied for many years and the resonance theory in Celestial Mechanics has gained considerably from the research driven by these real cases. However, it is Roy and Ovenden (1954) who have first demonstrated that the mean motion commensurabilities in the Solar System are much more frequent than they would be if the periods were randomly distributed. This non-random distribution could be due to some specific initial conditions in the early history of the Solar System and the physical interactions that would have then taken place. It can also be interpreted as the result of a later evolution. In both cases, one should also explain the stability of such situations.

If, in the case of minor planets, many authors have shown that purely gravitational effects are responsible for the gaps and for the Hirayama families (see, for instance, Henrard and Lemaître, 1984), this approach does not hold for satellites. This was first studied in detail by Goldreich (1965) and then developed by him and many other authors (see Peale, 1976, for bibliography). Let us assume, for the sake of a qualitative description of the phenomenon, that originally two satellites of a planet revolve with the mean motions n_0 and n'_0. The well known tidal friction due to the planet increases continously their semi major axes and decreases the mean motions. The evolutions of n and n' are linearly independent, since they are proportional to the 8th power of the semi major axes. So, the ratio n'/n varies with time and, at some stages of the evolution they become commensurable, so that one has

$$in + i'n' = 0 \tag{1}$$

where i and i' are small integers.

It is well known that some of such resonant situations are stable and other are not. Among the stable conditions, the more important are those for which i and i' are small. This is the case of the i=1, i'=2 relationship for the couples Mimas-Tethys and Enceladus-Dione in the Saturn system or Io-Europa and Europa-Ganymede in Jupiter. It is also the case of i=3, i'=4 for Titan-Hyperion. If the resonant situation is such that the interaction between satellites is sufficiently strong to impose a sharing of the angular momentum transferred by the tides, then the ratio n'/n will remain constant and the semi major axes will increase in a fashion governed by the constancy of this ratio. This was called by Goldreich "tidal stability". Similar descriptions can be made also of spin-orbit couplings and, more generally, whenever a resonant situation appears in the course of the evolution of the main periods describing the motions. It is necessary, however, to understand why some situations indeed turn out to lock a sub-system in a stable resonant configuration, while other resonances encountered during the independent evolutionary stage did not lock the system and were bypassed. Only partial answers to this question exist while practically nothing has been said on the evolutionary perturbations that may be induced by such weak resonances.

III. EQUATIONS OF THE PROBLEM

Let us first show how the various cases lead to similar types of equations.

1. Orbit/Orbit coupling

Let us consider first the gravitational equations of motion of a system composed of two satellites S and S' revolving around a planet P. It is well known that the equations of motion can be written in the form (see e.g. Kovalevsky, 1967)

$$\left.\begin{aligned} m_j &= \frac{d^2x_j}{dt^2} = \frac{\partial U}{\partial x_j} \\ m_j &= \frac{d^2y_j}{dt^2} = \frac{\partial U}{\partial y_j} \\ m_j &= \frac{d^2z_j}{dt^2} = \frac{\partial U}{\partial z_j} \end{aligned}\right\} \tag{2}$$

Here, j=1 and 2, and U is the force function. These equations can also be written in a Hamiltonian formulation

$$\frac{dp_i}{dt} = \frac{\partial H}{\partial q_i} \; ; \; \frac{dq_i}{dt} = - \frac{\partial H}{\partial p_i} \quad i = 1,2\ldots6 \tag{3}$$

where, in the Hamiltonian H=T-V one introduces the kinetic energy T.

In solving these equations of motion H is expressed in terms of the metric variables p_i and the angular variables q_i in the form of a multiperiodic trigonometric expansion

$$H = A(p_i) + \sum_i \varepsilon B_i(p_i) \cos(i_1q_1+i_2q_2+\ldots i_6q_6) \tag{4}$$

where $i=(i_1,i_2\ldots i_6)$ and $\varepsilon B_i(p_i)$ is at least of the order 1 of a small parameter ε. Furthermore, the leading part of A is T and is consequently a quadratic function of the p_i. The methods most commonly used to solve equations (3) use infinitesimal contact transformations where the new variables are some kinds of averages of the old ones, so that periodic terms are successively eliminated. In the general case, finally, only terms independant of the q_i remain. If we denote with primes the final variables, the equations take the form :

$$\frac{dp'_j}{dt} = 0 \quad ; \quad \frac{dq'_j}{dt} = A'_j(p'_{(i)}) = n_j(p'_{(i)}) \tag{5}$$

the solution of which is

$$p'_j = P^\circ_j \quad \text{and} \quad q'_j = n_j(P^\circ_{(i)})t+Q^\circ_j$$
$$= n_j t+Q^\circ_j$$

The quantities n_j are the angular velocities corresponding to the proper frequencies of the problem. Among them, we find n and n', the mean motions of the satellites.

This procedure does not apply if there exists some term involving an argument $\phi=\sum_{j=1}^{6} i_j q_j$ that has an almost zero mean motion $n_\phi=i_1n_1+i_2n_2+..i_6n_6$. We are in resonant conditions that occur when n is of the order of the square root of the small parameter ε of the Hamiltonian ($n_\phi \approx k\sqrt{\varepsilon}$). This term in θ cannot be eliminated as previously. By a simple linear change of variables, one can replace one of the q'_i (say q') by ϕ. Let us call X the corresponding conjugate variable. The reduced Hamiltonian is :

$$H' = a(X,p'_i)+ \Sigma b_j(X,p'_i)\cos j\phi \tag{6}$$

so that the equations take the form:

$$\left.\begin{aligned} \frac{dX}{dt} &= - \sum_j jb_j \sin j\phi \\ \frac{d\phi}{dt} &= - \frac{\partial a}{\partial X} - \sum_j \frac{\partial b_j}{\partial X}\cos j\phi \end{aligned}\right\} \tag{7}$$

and $$\frac{dp'_i}{dt} = 0 \quad ; \quad \frac{dq'_i}{dt} = - \frac{\partial a}{\partial q'_i} - \sum_j j\frac{\partial b}{\partial q'_i}\cos j\phi$$

The p'_i being constants, the variables are totally separated and the solution reduces to the resolution of equations (7), the typical form of the resonance problem. The orders of magnitude with respect to the small quantity $\mu=\sqrt{\varepsilon}$ being those indicated above, and replacing the p'_i by constants, one finally reduces the Hamiltonian to :

$$H = A + \mu^2 \sum_j B_j \cos j\phi \tag{8}$$

A and B being functions of X. This form is generalized by Garfinkel (1976) what he calls the "Ideal Resonance Problem" in which

$$H = B(y) + 2\mu^2 A(y) f(x)$$

Let us now assume that an additional non-conservative force of tidal origin exists. Its essential effect is an acceleration of the mean longitude. The dynamical situation may be represented by replacing the semi major axes a and a' and the mean motions n and n' by quantities slowly varying with time $a=a_0+a_1t+..;...$ The equations still hold but in order to proceed with the elimination of variables one has to add a pair of conjugate variables K, and k, where k=t. Formally, the Hamiltonian that one obtains after the elimination of all the resonant terms is :

$$H=A_o+A_1k+A_2k^2+...+\mu^2\Sigma(B_j^{\,0}+B_j^{\,1}k+..)\sin j\phi \tag{9}$$

and since all but one argument entering into ϕ are now linear functions of time and consequently can be expressed in function of ϕ, we have another equivalent formulation of H as

$$H'=A'_o+A'_1\phi+A'_2\phi^2+\mu^2\Sigma(B_j^{\,10}+B_j^{\,11}+..)\sin j\phi$$

The resulting equations are at the first order in t :

$$\left.\begin{aligned}\frac{dX}{dt} &= A'_1+A'_2\phi-\mu^2\Sigma(B'^{\,0}_j+B'^{\,1}_j t)\sin j\phi \\ \frac{d\phi}{dt} &= -\frac{\partial A'_0}{\partial X}-\frac{\partial A'_1}{\partial X}\phi-\mu^2\Sigma\left(\frac{\partial B'^{\,0}_j}{\partial X}+\frac{\partial B'^{\,1}_j}{\partial X}t\right)\cos j\phi\end{aligned}\right\} \tag{10}$$

2. Spin/orbit coupling

When there is a possible interaction between the rotation and the orbital motion of a body (Moon, planets), the basic equation is that of the translational-rotational motion as described by Duboshin (1963). It depends upon a force function and consequently a Hamiltonian formulation exists. The variables are the usual conjugate variables for the orbital motion and Andoyer angles θ^*, ϕ^* and ψ^* with their conjugate variables: L, $L\cos\phi^*$ and $L\cos\theta$ where $\vec{L}$ is the angular momentum of the rotation and θ is one of the Eulerian angles.

The solution of the equations can proceed as in the preceding case. The six independent angular variables are transformed into linear functions of time among which the transformed variable derived from θ^* represents the rotation of the body, the direction of its axis of rotation being given by ϕ^* and ψ^*.

In the case of resonance, there will exist a commensurability between the time coefficient of θ^* and the mean motion n of the body $in_\theta - i'n = o$. So if we set $\phi = i\theta^* - i'l$, the reduced Hamiltonian will take the form (6).

The tidal friction will show in a slow secular decrease of L, so again, as in the preceding case, the coefficients of the Hamiltonian will slowly vary with time as in (9), and we shall end up with equations of form (10).

3. Higher order resonances

In the two examples just given, the resonant argument ϕ was a linear integer combination of mean orbital or rotational motions corresponding to well identified physical situations. However, this is not required by the mathematical formulation we gave. The argument ϕ may a priori be any of the arguments created in the development of the Hamiltonian or in further transformations performed during the solution. Let us give an example in the lunar theory for which, presently, there is no such critical argument either in the main problem or in the planetary perturbations. But when the lunar orbit evolves with time, the periods change. Although it is impossible that any combination of lunar proper periods is zero for the normally used arguments in the lunar theory, this is not true when planetary perturbations are included since the planetary mean motions do not change significantly.

The main problem of lunar theory is a Hamiltonian problem with three degrees of liberty with a time depending external argument, the mean anomaly l' of the Earth's orbit. The general form of the arguments in the Hamiltonian is :

$$\phi = i_1 l + i_2 g + i_3 h + i_4 l$$

where l, g and h are the mean anomaly, the argument of perigee and the longitude of the nodes. If the planetary terms are also searched for, the disturbing function contains, in addition the mean longitudes l_j of the planets, so that one has :

$$\phi = i_1 l + i_2 g + i_3 h + i_4 l' + \sum_{j=1}^{n} i_j l_j$$

The treatment of such a Hamiltonian system is analogous to the normal case, with the use of the already defined additional variables k and K. Consequently, it is possible to follow again exactly the procedure described in the first case and reduce the resonance problem to equations (7) and, when tidal terms are includes, to (9) and (10), since n and a are then slowly varying functions of time.

So, for all the cases, one many reduce the problem to the same type of equations. For the sake of a further simplification, we shall follow all the authors that have studied the subject and restrict the trigonometric series in ϕ to their first term. Then, we remark that the transformations do not modify the 0-th order part of the Hamiltonian, so that A_o is essentially T expressed in the new variables, and is consequently a quadratic form in X. Finally, we shall change X in $X=X_o+x$, where x is the variation of X around the equilibrium value X_o. Changing notations and taking into account these assumption, one gets the following set of equations :

$$\left.\begin{aligned} \frac{dx}{dt} &= A-Bx+(Q+Q't)\sin\theta \\ \frac{d\theta}{dt} &= 2Gx \end{aligned}\right\} \qquad (11)$$

Assuming Q'=o, this formulation corresponds to the equations used by Burns for the Mercury rotation problem (Burns, 1979) or for the study of the transient resonance in lunar theory (Kovalevsky, 1983). On the contrary, if one neglects B, one obtains :

$$\frac{d^2\theta}{dt^2} = 2GA + (Q+Q't)\sin\theta$$

This is the equations used by Sinclair (1972) for his study of Saturn satellites.

So the formulation (11) is common to all the actual tidal resonance problems and can be considered as typical.

IV - REPRESENTATION OF THE SOLUTION

Several methods have been proposed to study the evolution described by equations (11). In particular, Henrard (1982) has applied adiabatic invariants to a more general Hamiltonian and applied it to actual astronomical cases. We have proposed (Kovalevsky, 1983), a more analytical approach.

Let us consider first the case B=Q'=0 as a basic "restricted case". The equations are :

$$\frac{dx}{dt} = A + Q\sin\theta \quad ; \quad \frac{d\theta}{dt} = 2Gx \qquad (12)$$

There exists a Hamiltonian integral

$$A\theta - Q\cos\theta - Gx^2 = C \qquad (13)$$

The trajectories can be studied in the θ-x^2 or θ-x planes. Two cases occur :

- If $|A|<|Q|$, one may have libration orbits, symmetrical circulation orbits or a limiting asymptotic case (fig. 1).

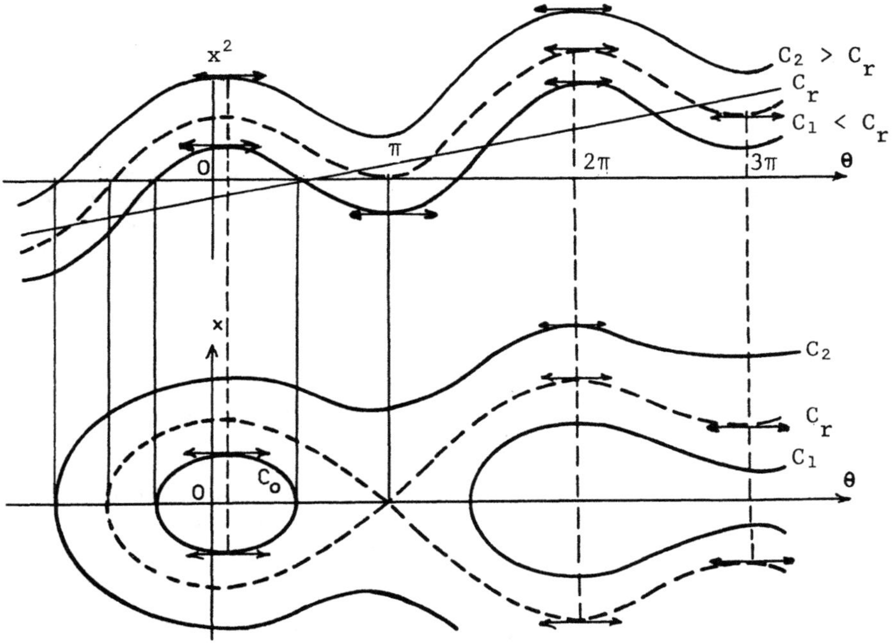

FIGURE 1

- If $|A|>|Q|$, the $x^2=f(\theta)$ curve has no horizontal tangent and only circulation orbits exist (fig. 2).

Let us now consider the case when $B\neq 0$. The equations are :

$$\frac{dx}{dt} = A + Bx + Q\sin\theta \quad ; \quad \frac{d\theta}{dt} = 2Gx \qquad (14)$$

In this case, there is no more Hamiltonian integral. But one can still construct an expression similar to (13), namely :

$$C(t) = (A-Bx)\theta - Q\cos\theta - Gx^2 \qquad (15)$$

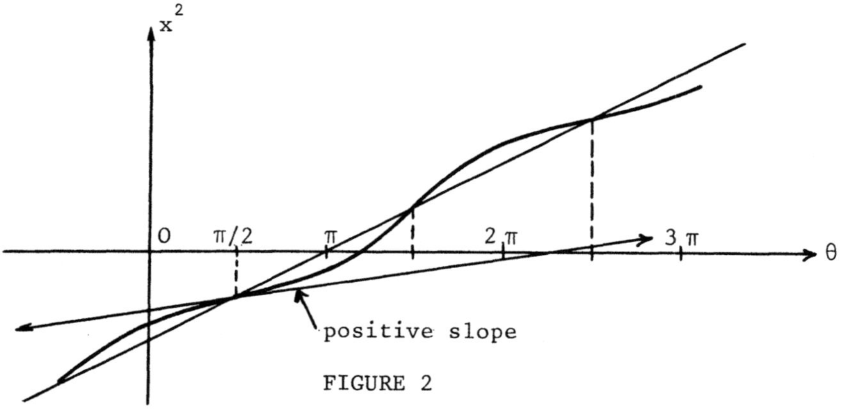

FIGURE 2

This quantity is a function of t. However, Bx is a very small, very slowly varying quantity, and it is legimimate to consider that, for some finite interval of time Δt, x is a constant equal to x_o. Then, during this particular interval of time, the solution behaves like a solution of (12) where one replaces A by $A-Bx_o$ and during that time C(t) can be considered as a constant.

The "restricted case" plays the role of an osculating orbit for the general solution. Consequently it can be described as a continuously varying orbit in a family of orbits represented in figure 1.

Let us assume, for instance, that $|A-Bx|<Q$ and that when A-Bx varies with time, $dc/dt<0$. The evolution of the solutions can then be described in the following manner.

1. Before entering the resonance region, while $|\frac{d\theta}{dt}|=|2Gx|>k\mu$, one can eliminate θ from the equations and the evolution is a non resonant one, with continuous variations of a and n.

2. In the vicinity of the resonance region, one may use equation (15) and the path follows an ascending branch of an open curve of figure 1 shifting slowly from one to another, while C decreases.

3. When x tends to zero, the limit of C(t) is Co. If Co>Cr, then the motion continues close to an ascending branch of an open curve of figure 1. If Co>Cr, then there is a trapping into resonance.

In this situation, the capture probability (or better, following Kyner, a capture measure) as introduced by Goldreich and Peale (1966) is directly applicable. We shall not repeat the definition, but only illustrate in figure 3 the capture and no capture cases. The curve Σ represents in the θ-x^2 plane a common os-

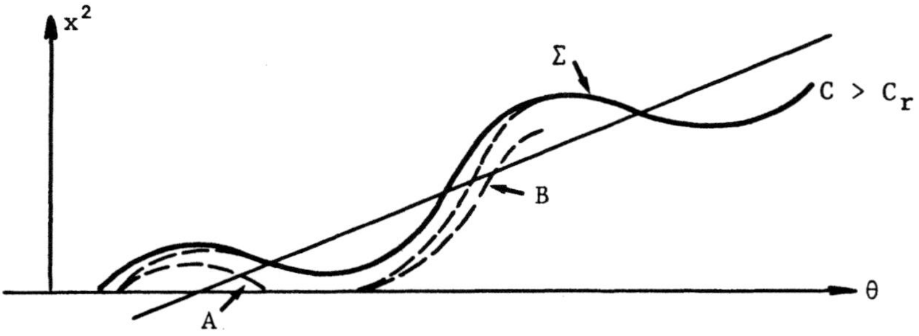

FIGURE 3

culating path. Curve A is a perturbed path that leads to capture. Curve B is another perturbed path that does not lead to capture.

4. If a capture has occured and if the sign of $\frac{dC}{dt}$ does not change, the motion is proceeding inside the resonant region inside the curve C=Cr in figure 1 and the path whirls into it as shown by Murdock (1978) so that the resonant situation is definitively established: this is the tidal stability. If, on the contrary, dC/dt changes its sign, the path within the resonant region expands, and when C=Cr, it excapes, and follows again an open path in the $x^2=\theta$ plane.

5. If no trapping has occured, the path along open curves of figure 1 is followed until $|2Gk|>k\mu$, when the normal evolutionary scheme takes place again. We have however shown by numerical simulations that the time evolution has increased in comparison with what it would have been if there was no resonant term in the equations, especially if $|Q|$ is close to its critical value $|A|$.

V - APPLICATIONS

The cases already studied all enter in the scheme described above.

1. Orbit/Orbit coupling

For planets, the tidal forces are too small to have a sizeable effect. There should be no sign of non-gravitational evolution in the planetary system. The structure of the asteroidal belt is consequently governed by gravitational properties of the solar system.

For the couples of satellite systems locked into resonance, the tidal origin has been assessed for Mimas-Thetys and Enceladus-Dione (Sinclair, 1972; Yoder, 1979). For Titan-Hyperion, this is not sure because a large value of the tidal coefficient would be necessary. For the Galilean satellites, it may be necessary to introduce tidal dissipation in the satellites themselves (Sinclair, 1975).

2. Spin/Orbit coupling

The tidal origin has been assessed by Goldreich and Peale(1966) for Mercury. For Venus, an interesting hypothesis was given by Lago and Cazenave (1979) who suggested that while the solid tidal torque despins the planet, a thermally driven atmospheric tidal torque could have changed the direction of the pole by 180° and then stibilize it at the present position.

3. Earth-Moon system

The co-rotation of the Moon has been proven to be of tidal origin (as well as for practically all satellites). However, there remains the gap between the computed and the actual value of the time necessary to drive the Moon at its present positions. The difference corresponds to a factor larger than 2. Littleton (1980) has suggested to explain this by a large secular variation of the Earth's moments of inertia. However, it is necessary to investigate other purely dynamical causes. One of them is the occurence of transient high order resonances involving temporary trapping in a resonant region. Just for the sake of giving an example, let us consider the lunar perturbation term due to Venus whose argument is

$$\theta = 2V - l' + 2D - 2F = n_\theta (t-to)$$

The present value of n_θ is 0.00772. But when the ratio $m=n/n'$ varies, $2n_D-2n_F$ varies. One has, with the current notations of lunar theory :

$$2(n_D - n_F) = -2m - 1.5m^2 - 0.5625m^3 + 3m'\gamma^2 + 2.25m^2 e'^2 + 1.5m^2 e^2 - 1.3828m^4$$

When m had 0.9 times its present value, n_θ was of the order of -0.018 so that θ crossed the resonant value $n_\theta=0$ in between. Furthermore, the planetary perturbations of the Earth's orbit induce large long periodic variations of the eccentricity e'. This suggests that during the crossing of the resonant region, dn_θ/dt has varied greatly and changed sign several times. It means that, when the system was in the case 4 described in section IV, dC/dt may have changed sign several times and temporary trapping in resonance may have occurred one or several times. During the trapping periods, the semi-major axis of the Moon does not change and the angular momentum lost by the Earth is gained by the planetary orbits involved . I am presently studying this process which might explain at least partly the discrepancy just quoted.

V - CONCLUSION

In the past 20 years, the investigations on the evolution of the dynamical behaviour of bodies in the Solar System were very efficient in explaining resonances and most of the rotation properties. Much is still left to be done, particularly on the evolution outside the resonant region and the time scales. A more rigourous approach, following the papers by Kyner (1970), Murdock (1978) and partly of Yoder (1979), should be pursued since the theory is far from having the completeness and the unity of the non dissipative Celestial Mechanics. Such efforts are very important since, while the classical gravitational approach is quite adequate for the present situation of the Solar System, the non-gravitational effects are fundamental in studying its dynamical evolution, a domain that could also be called "Paleo-Celestial Mechanics".

REFERENCES

Burns, T.J., 1979, Celestial Mechanics, vol 19, p 297.

Duboshin, G.N., 1963, "Celestial Mechanics: basic problems and methods", Fiz.Mat.ed., Moscow, in russian.

Garfinkel, B., 1976, Celestial Mechanics, vol 13, p 229.

Goldreich, P., 1965, Monthly Not.R.A.S., vol 130, p 159.

Goldreich, P. and Peale,S.J., 1966, Astron.J., vol 71, p 425.

Henrard, J., 1982, Celestial Mechanics, vol 27, p 3.

Henrard, J. and Lemaître, A., 1984, presented at the International Symposium Georges Lemaître, Louvain-la-Neuve, october 1983, this volume.

Kovalevsky, J., 1967, "Introduction to Celestial Mechanics", D. Reidel.Publ.Co., Dordrecht, p 7.

Kovalevsky, J., 1983, in "Dynamical trapping and Evolution in the Solar System", IAU Coll.N°74, V.V. Markellos and Y. Kozaï ed., p 3.

Kyner, W.T., 1970, in "Periodic orbits, Stability and Resonances", G.E.O. Giacaglia ed., Reidel Publ. Co., Dordrecht, p 501.

Lago, B. and Cazenave, A., 1979, The Moon and the Planets, vol 21, p 127.

Littleton, R.A., 1980, The Moon and the Planets, vol 22, p 13.

Murdock, J.A., 1978, Celestial Mechanics, vol 18, p 237.

Peale, S.J., 1976, Annual Rev. of Astron. and Astroph., vol 14, p 215.

Roy, A.E. and Ovenden, M.W., 1954, Monthly Not. R.A.S., vol 114, p 232.

Sinclair, A.T., 1972, Monthly Not. R.A.S., vol 160, p 169.

Sinclair, A.T., 1975, Monthly Not. R.A.S., vol 171, p 59.

Yoder, C.F., 1979, Celestial Mechanics, vol 29, p 3.

GENERALIZATIONS OF THE RESTRICTED PROBLEM OF THREE BODIES

Victor Szebehely, Arthur L. Whipple

The University of Texas at Austin

ABSTRACT. This paper offers several generalizations of the restricted problem of three bodies from an analytical and dynamical point of view. First, a short review of the classical restricted problem is offered which is followed by the most general reformulation of the problem. In this most general formulation we consider a dynamical system consisting of several large bodies and of several smaller masses. The influence of the large masses on the small ones can be arbitrary but in most practical cases we consider gravitational forces only. We also allow forces acting between the small bodies influencing their motion. The restriction comes in that we follow the basic idea of the restricted problem of three bodies and do not allow any influence of the small bodies on the motion of the large ones. This complete generalization is then followed with some special situations such as having two primary masses and two smaller masses. In this case we also establish a new Jacobian Integral which might be considered the generalization of the classical well known Jacobian Integral.

INTRODUCTION

It is a great honor for us to partake in Professor George Lemaitre's International Conference which is dedicated to his numerous discoveries in the field of Celestial Mechanics and of Cosmology.

Professor Lemaitre's transformation applied to the classical restricted problem produces a global regularization of the problem and indeed this is a unique result since his global

A. Berger (ed.), The Big Bang and Georges Lemaitre, 195–205.

regularization also is a rationalization of the equations of motion. By this we mean that other global regularizations of the classical restricted problem utilized transformations which when applied result in differential equations containing square roots or non-integer powers of certain expressions while Professor Lemaitre's transformation yields all rational functions without square roots.

It is our pleasure and honour to present the following generalization of the restricted problem of three bodies in these proceedings since the restricted problem of three bodies was one of Professor Lemaitre's great interests. We are offering a physical generalization as opposed to a simple mathematical exercise and we believe that this is in what Professor Lemaitre would be most interested in. We are presenting a new integral of this new problem which is analogous and similar to the classical Jacobian integral.

1. SHORT REVIEW OF THE CLASSICAL RESTRICTED PROBLEM OF THREE BODIES

We are considering two primaries of masses M_1 and M_2 and one small third mass m_1. The relation between these masses is given by the inequality $m_1 << M_2 \leq M_1$. In the restricted problem we also consider only gravitational forces as well as we assume circular orbits for the primaries. As mentioned in the abstract and in the well known literature there is an effect of M_1 and M_2 on m_1 but m_1 is not influencing the motion of the primaries. Since the motion of the primaries is given the problem is to determine the motion of the small body m_1. This classical formulation of the restricted problem of three bodies is one of the famous non integrable dynamical problems. Figure 1 represents the system in an x, y coordinate system which is rotating around the center of mass of the primaries so that M_1 and M_2 are fixed on the x axis. The distances between the primaries and the small mass are r_1 and r_2. The equations of motion are given by

$$\begin{aligned} \ddot{x} - 2\dot{y} &= \Omega_x \\ \ddot{y} + 2\dot{x} &= \Omega_y \end{aligned} \tag{1}$$

where

$$\Omega = \frac{1}{2}[(1-\mu)r_1^2 + \mu r_2^2] + \frac{1-\mu}{r_1} + \frac{\mu}{r_2} \tag{2}$$

and

$$\mu = \frac{M_2}{M_1 + M_2}$$

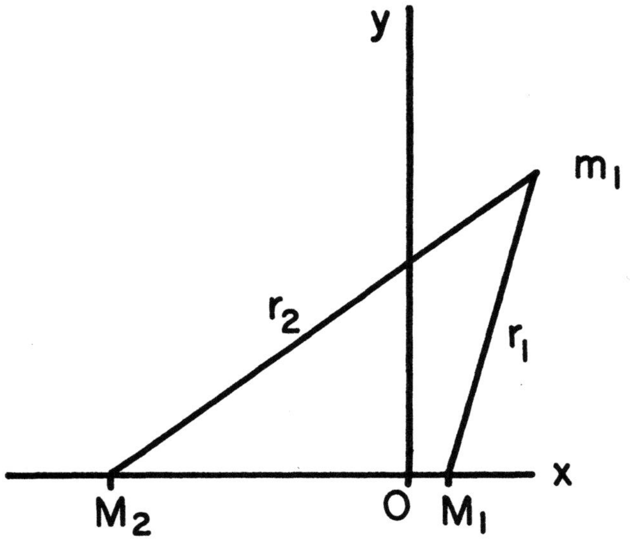

Fig. 1. The classical circular restricted problem of three bodies.

This system of equations is written in the two dimensional form in order to simplify the results but the three dimensional approach is equally possible. The famous Jacobian Integral [1] is given as

$$v^2 = 2\Omega(x,y) - C \tag{3}$$

where C is the Jacobian constant and v is the velocity.

It is mentioned that this dynamical system has five equilibrium points in the rotating (synodic) coordinate system [2].

2. GENERALIZATION

We now consider n primaries M_1, M_2, ... M_i, ... M_n. These primaries are under the influence of each other and they are influencing the motion of the small masses: m_1, m_2, ... m_α, ... m_ν. We note that any of the small masses are much smaller than the primaries so in general we have the inequality $m_\alpha \ll M_i$. This inequality is true for any values of α and i, which satisfy the inequalities $1 \leq \alpha \leq \nu$

and $1 \leq i \leq n$. In the most general formulation the forces acting between the primaries are arbitrary. Also the forces acting between the small bodies are arbitrary but all forces are given. First we determine the orbits of the primaries since they are not influenced by the small orbits. The orbits of the primaries might be determined using the given forces which depend on the masses, on the position, on the velocities, and on the time in general. Consequently we have

$$\ddot{\bar{r}}_i = \bar{F}_i \, (M_i \, \ldots \, M_n; \, \bar{r}_i \, \ldots \, \bar{r}_n; \, \dot{\bar{r}}_i \, \ldots \, \dot{\bar{r}}_n; \, t) \tag{4}$$

Let's assume that the solution of Equations (4) are

$$\bar{r}_i = \bar{r}_i \, (t) \tag{5}$$

Since the primaries are not influenced by the small bodies part of the solution is represented by Equation (5).

The orbits of the small bodies are determined by their mutual interactions and by the effects of the primaries. This might be written as

$$\ddot{\bar{\rho}}_\alpha = \bar{G}_\alpha \, [M_1 \, \ldots \, M_n \, ; \, m_1 \, \ldots \, m_\nu \, ; \, \bar{r}_1(t) \, \ldots \, \bar{r}_n(t) ; \, \bar{\rho}_1 \, \ldots \, \bar{\rho}_\nu \, ; \, \dot{\bar{\rho}}_1 \, \ldots \, \dot{\bar{\rho}}_\nu \, ; \, t] \tag{6}$$

As we can see, the motion of the small bodies is determined by the masses of the primaries, the masses of the small bodies, the motion of the primaries, the location of the small bodies, the velocities of the small bodies, and of the time. In this Equation (6) we are supposed to know the function $\bar{G}_\alpha$, we are supposed to know the location of the primaries, we are supposed to know all the masses, and we have to solve all these equations for the vectors $\bar{\rho}_\alpha$. The solution of Equation (6) will be

$$\bar{\rho}_\alpha = \bar{\rho}_\alpha \, (t) \tag{7}$$

and this Equation (7) is the solution of the generalized restricted problem of $n+\nu$ bodies.

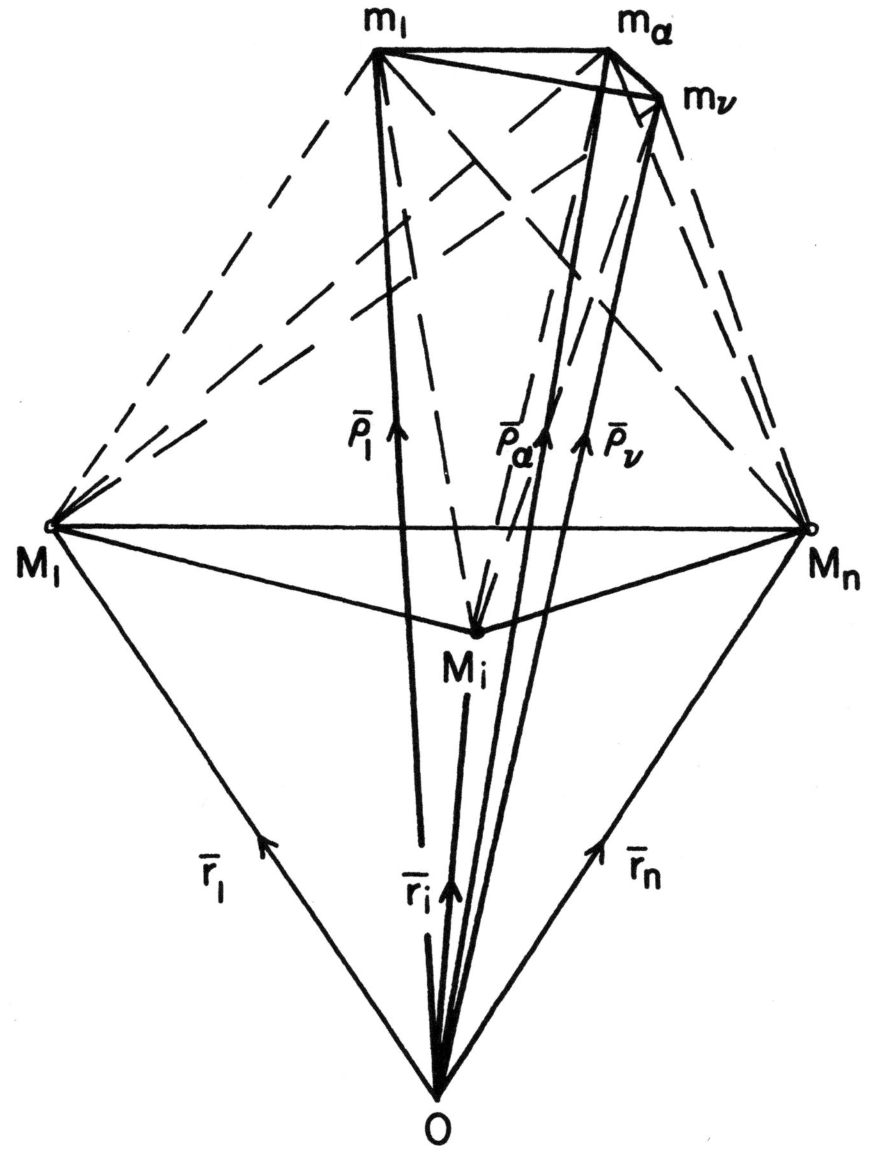

Figure 2 shows the general arrangement. The letters M_i represent the primaries and the figure shows only the first, the ith, and the nth primaries. The position vectors are $\bar{r}_1$, $\bar{r}_i$, and $\bar{r}_n$. The figure also shows the small bodies m_1, m_α, and m_ν. Their locations are given by $\bar{\rho}_1$, $\bar{\rho}_\alpha$ and

$\bar{\rho}_\nu$. The dashed lines in the figure show the forces interacting between the primaries and the small bodies. The solid lines connecting the primaries on one hand and the small bodies on the other side, represent the forces acting between the primaries and the forces acting between the small bodies. We remind the reader that this figure represents the generalized restricted problem of $n+\nu$ bodies, where n represents the number of the primaries and ν represents the number of the small bodies.

3. THE GENERALIZED GRAVITATIONAL RESTRICTED PROBLEM OF n+ν BODIES

The previously discussed completely general situation will now be restricted to gravitational forces only. The equations of motion of the primaries may be written as

$$\ddot{\bar{r}}_i = - G \sum_{\substack{j=1 \\ i \neq j}}^{n} \frac{M_j}{|\bar{r}_{ij}|^3} \bar{r}_{ij} \tag{8}$$

where $\bar{r}_{ij} = \bar{r}_i - \bar{r}_j$ and $1 \leq i, j \leq n,\ i \neq j$. This is of course simply the representation of the many body gravitational problem. The equations of motion of the small bodies are not as simple as the same for the primaries because they affect each other as well as their motion is affected by the primaries. Consequently the equations of motion of the small bodies might be written as

$$\ddot{\bar{\rho}}_\alpha = - G \sum_{j=1}^{n} \frac{M_j}{|\bar{\Delta}_{\alpha j}|^3} \bar{\Delta}_{\alpha j} - G \sum_{\substack{\beta=1 \\ \beta \neq \alpha}}^{\nu} \frac{m_\beta}{|\bar{\rho}_{\alpha\beta}|^3} \bar{\rho}_{\alpha\beta} \tag{9}$$

where

$$\bar{\Delta}_{\alpha j} = \bar{\rho}_\alpha - \bar{r}_j \qquad \bar{\rho}_{\alpha\beta} = \bar{\rho}_{\alpha\beta} - \bar{\rho}_\beta$$

$$1 \leq \alpha \leq \nu \qquad 1 \leq \alpha,\ \beta \leq \nu$$

$$1 \leq j \leq n \qquad \alpha \neq \beta$$

In equation 9 we notice that the first term on the right hand side represents the effect of the primaries on the small bodies and the second term represents the forces acting between the small bodies. As a simple result we might mention that the formulation of the generalized gravitational restricted problem of n+1 bodies is an immediate consequence of Equation (9). If we have only one small body (consequently having $\alpha = 1$) we have the equation of motion of this small body:

$$\ddot{\bar{\rho}} \equiv \ddot{\bar{\rho}}_1 = - G \sum_{j=1}^{n} \frac{M_j}{|\bar{\Delta}_{ij}|^3} \bar{\Delta}_{ij} \tag{10}$$

where

$$\bar{\Delta}_{ij} = \bar{\rho} - \bar{r}_j \quad ; \quad 1 \leq j \leq n$$

If in addition we would assume that we have $n = 2$ and these two primaries would move on circles Equation (10) would reduce to the equation of the classical restricted problem of three bodies.

4. GENERALIZED GRAVITATIONAL RESTRICTED CIRCULAR PROBLEM OF 2+2 BODIES

The geometry of this problem is shown in Figure 3. The primaries M_1 and M_2 are located again on the x axis which is rotating so that M_1 and M_2 describe circular orbits around the origin of the coordinate system. The small bodies are denoted by m_3 and m_4 with their respective coordinates. The distances between the primaries and the small bodies are denoted by $\tilde{r}_{ij}$ and the distance between the small bodies is $\tilde{\Delta}$. All these quantities shown on Figure 3 have a wavy bar representing the fact that these are dimensional coordinates. We note that the previously given Equations (1), (2), and (3) all contain dimensionless coordinates. In order to change our system once again to dimensionless coordinates we introduce x_3, y_3, x_4, y_4, r_{13}, r_{14}, r_{23}, r_{24} and Δ, as dimensionless lengths which are simply obtained by division of the corresponding dimensional quantities by the quantity $(a + b)$. The dimensionless time $t = t^*n$, where t^* is the dimensional time and n is the mean motion of the system. We also introduce dimensionless mass parameters which are

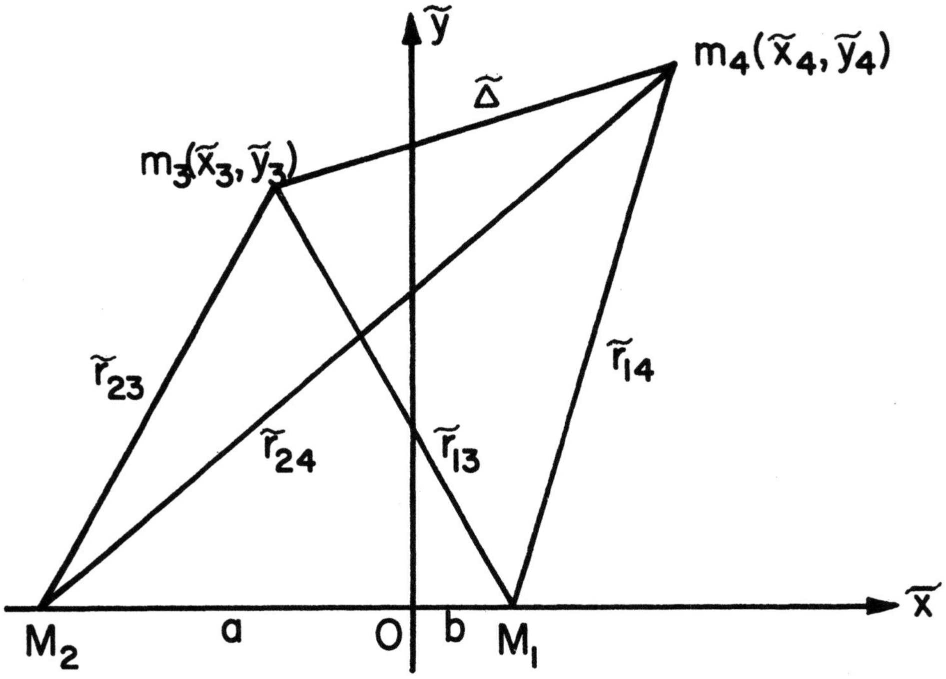

Fig. 3. Synodic dimensional coordinates for 2 primaries and 2 small bodies.

given by Equations (11):

$$\mu = M_2(M_1 + M_2)^{-1}$$
$$\mu_3 = m_3(M_1 + M_2)^{-1}$$
$$\mu_4 = m_4(M_1 + M_2)^{-1} \quad (11)$$

In this system M_1 and M_2 are the primaries and their motions are circular orbits around the origin of the coordinate system. We now write the equations of motion of the small masses as influenced by coriolis forces, centrifugal forces, by the attraction exerted by the primaries, as well as by their mutual interaction. We again restrict ourselves to two dimensions but the 3-dimensional generalization is simple and trivial.

The equations of motion become

$$\frac{d^2x_3}{dt^2} - 2\frac{dy_3}{dt} - x_3 =$$

$$= - \left[\frac{(1-\mu)(x_3-\mu)}{r_{13}^3} + \frac{\mu(x_3+1-\mu)}{r_{23}^3} + \frac{\mu_4(x_3-x_4)}{\Delta^3}\right] \tag{12}$$

$$\frac{d^2y_3}{dt^2} + 2\frac{dx_3}{dt} - y_3 = - y_3\left[\frac{1-\mu}{r_{13}^3} + \frac{\mu}{r_{23}^3}\right] - \mu_4\frac{y_3-y_4}{\Delta^3} \tag{13}$$

$$\frac{d^2x_4}{dt^2} - 2\frac{dy_4}{dt} - x_4 =$$

$$= - \left[\frac{(1-\mu)(x_4-\mu)}{r_{14}^3} + \frac{\mu(x_4+1-\mu)}{r_{24}^3} + \frac{\mu_3(x_4-x_3)}{\Delta^3}\right] \tag{14}$$

$$\frac{d^2y_4}{dt^2} + 2\frac{dx_4}{dt} - y_4 = -y_4\left[\frac{1-\mu}{r_{14}^3} + \frac{\mu}{r_{24}^3}\right] - \mu_3\frac{y_4-y_3}{\Delta^3} \tag{15}$$

In these equations the distances are defined by

$$\Delta = r_{34} = \left[(x_4-x_3)^2 + (y_4-y_3)^2\right]^{1/2} \tag{16}$$

$$\left.\begin{aligned} r_{1i} &= \left[(x_i-\mu)^2 + y_i^2\right]^{1/2} \\ r_{2i} &= \left[(x_i-\mu+1)^2 + y_i^2\right]^{1/2} \end{aligned}\right\} \quad i = 3,4 \tag{17}$$

We may eliminate the interacting terms between the small bodies by writing $\mu_4 = 0$ or $\mu_3 = 0$. In this way the classical restricted problem of three bodies is obtained.

The generalized Jacobian Integral may be obtained by proper multiplication of these equations with certain quantities and by adding the results. This is quite similar with the derivation of Equation (3) as it is obtained from Equation (1) except in the present general case Equations (12) and (13) will have to be multiplied by $\mu_3 dx_3/dt$ and $\mu_3 dy_3/dt$. Equations (14) and (15) are multiplied by $\mu_4 dx_4/dt$ and $\mu_4 dy_4/dt$.

Once these equations are multiplied by the proper quantities and the results are added we obtain the generalized Jacobian Integral in the form

$$\mu_3 v_3^2 + \mu_4 v_4^2 = 2\Omega - C \tag{18}$$

where C is the generalized Jabobian constant, v_3 and v_4 are the velocities, and Ω will be a generalization of the function given by Equation (2). In our case Ω becomes

$$\begin{aligned}\Omega = \frac{1}{2}\mu_3 & \left[(1-\mu)r_{13}^2 + \mu r_{23}^2\right] + \\ & \frac{1}{2}\mu_4\left[(1-\mu)\, r_{14}^2 + \mu\, r_{24}^2\right] + \\ & \mu_3\left[\frac{1-\mu}{r_{13}} + \frac{\mu}{r_{23}}\right] + \\ & \mu_4\left[\frac{1-\mu}{r_{14}} + \frac{\mu}{r_{24}}\right] + \frac{\mu_3\,\mu_4}{r_{34}}\end{aligned} \tag{19}$$

As we can see, the terms corresponding to the effect of the primaries on the small bodies as well as the interaction between the small bodies are all represented. The interaction between the small bodies is represented by the last term of Equation (19). As an interesting exercise we might consider the situation when the mass of one of the small bodies becomes 0. If $\mu_4 = 0$ the Jocobian Integral becomes

$$v_3^2 = 2\Omega - C \tag{20}$$

where now Ω is identical with the function given by Equation (2) except for a slightly different notation:

$$\Omega = \frac{1}{2} (1-\mu) r_{13}^2 + \mu r_{23}^2 + \frac{1-\mu}{r_{13}} + \frac{\mu}{r_{23}} \tag{21}$$

Please note that the generalization of the Jacobian Integral is available for any number of small bodies which are interacting with each other provided that the primaries are still moving on circular orbits. Equation (18) is the final result of this paper and its applications are left for some future publication. Nevertheless, we might mention that if two asteroids are closely interacting with each other and the primaries are the Sun and Jupiter we might be able to treat with Equation (18) the so-called binary asteroid problem mentioned in recent literature. Other applications can be easily thought of, especially interesting are those using the so-called zero-velocity surfaces or Hill curves in the classical restricted problem. As the number of variables with the number of the small bodies increases we have high dimensional surfaces representing the so-called zero velocity regions and the geometry and the topology of the surfaces can become increasingly complicated.

REFERENCES

(1) C.G.J. Jacobi, Compt. Rend. Vol. 3, p. 59, (1836).

(2) V. Szebehely, "Theory of Orbits" Academic Press, p. 133, (1967).

SECULAR PERTURBATIONS OF ASTEROIDS WITH COMMENSURABLE MEAN MOTIONS

Yoshihide KOZAI

Tokyo Astronomical Observatory, Mitaka, Tokyo 181, Japan

ABSTRACT

Secular perturbations of asteroids are derived for mean motion resonance cases under the assumptions that the disturbing planets are moving along circular orbits on the same plane and that critical arguments are fixed at stable equilibrium points. Under these assumptions the equations of motion are reduced to those of one degree of freedom with the energy integral. Then equi-energy-curves in the phase plane are derived with given values of the two parameters, the semi-major axis and the z-component of the angular momentum, and the variations of the eccentricity and the inclination as functions of the argument of perihelion are estimated.

The same method is also applied to Pluto-Neptune system and the results are found to agree with those by a method of numerical integrations and show that the argument of perihelion of Pluto is librating around 90°.

1. INTRODUCTION

Secular perturbations of asteroids are treated in several text books of celestial mechanics by assuming that the eccentricities and the inclinations of both the disturbing planets and the asteroids are very small, namely, by neglecting fourth power terms with respect to the small quantities and squares of masses of the planet in the disturbing function. Then the equations are reduced to two independent sets of linear differential equations, one depending on the eccentricity and the longitude of the perihelion and the other on the inclination and the longitude of the node, each of which is

A. Berger (ed.), The Big Bang and Georges Lemaitre, 207–215.

identical to the equation for pendulums.

However, when the mean motion of the asteroid is commensurable with that of Jupiter and the critical argument is librating, the disturbing function for the secular perturbations cannot be derived by a conventional way of the power series expansion as secular terms are also produced from the critical term which librates.

When the eccentricity and the inclination of the asteroid are not so small the differential equations are not linear any more and the equations of two degrees of freedom should be solved simultaneously even for non-resonant cases. However, if it is assumed that the disturbing planets are moving along circular orbits on the same plane the equations are reduced to those of one degree of freedom with the energy integral after averaging the disturbing function with respect to the mean longitudes of the planets and the asteroid as the z-component of the angular momentum is conserved. Therefore, they can be solved by quadratures(Kozai, 1962).

General properties of the solutions thus derived for high eccentricity and inclination cases are different from those by the linear theory. Namely, the eccentricity and the inclination vary very widely as functions of the argument of perihelion because of $e^2\sin^2 i \cos 2g$ term(g being the argument of perihelion) in the disturbing function which is neglected in the linear theory. If the value of $(1-e^2)^{1/2}\cos i$ which is constant takes a small value below 0.8 the argument of perihelion can librate around 90° or 270° according to the non-linear theory. In fact there are a few asteroids, for which the arguments of perihelion are librating(Kozai, 1979 and 1980).

In this paper it is intended to extend the non-linear theory to resonance cases. However, as the equations cannot be reduced to those of one degree of freedom generally for resonant cases even after averaging the disturbing function with respect to the mean longitudes, it is assumed that one of the critical arguments is fixed at a stable equilibrium point to reduce the degrees of freedom by one. Then the same method as the non-resonant cases can be applied to derive the secular perturbations for resonant cases and the ranges of the variations of the eccentricities and the inclinations of existing asteroids can be estimated. This method is also applied to Pluto-Neptune system under the assumption that the critical argument is fixed at 180° and the results are found to agree with those by a numerical integration method with librating critical argument(Kinoshita and Nakai, 1983).

2. EQUATIONS OF MOTION

The equations of motion for an asteroid is formulated by using

Delaunay canonical variables as follows;

$$dL/dt = \partial F/\partial l, \quad dG/dt = \partial F/\partial g, \quad dH/dt = \partial F/\partial h,$$
$$dl/dt = -\partial F/\partial L, \quad dg/dt = -\partial F/\partial G, \quad dh/dt = -\partial F/\partial H, \qquad (1)$$

where F is the Hamiltonian depending on Delaunay variables of the asteroid, the semi-major axes which are constant and the mean longitudes of the planets which are known linear functions of time.

As it is assumed that the disturbing planets are moving along circular orbits on the same plane, the longitude of the ascending node, h, does not appear in the disturbing function. Therefore, the following integral exists;

$$H = \{a\ (1 - e^2)\}^{1/2} \cos i = \text{const.} \qquad (2)$$

In order to derive the secular part of the Hamiltonian it is averaged with respect to the mean anomaly of the asteroid and the mean longitudes of the disturbing planets by a numerical method. By this way the method can be applied to any orbit including one, for which the heliocentric distance varies across those of the disturbing planets. For non-resonant cases the averaging can be done by changing the angular variables independently. However, for resonant cases the averaging is made under the condition that the critical arguments are fixed at certain stable equilibrium points.

Among resonant asteroids, there are some, for which the critical arguments are making complete revolutions and a few, for which they are librating. However, there is none, for which the critical argument is fixed. Therefore, the assumption made here is unrealistic. However, since the difference between this case and the librating critical argument case is of the order of square root of the disturbing mass in the disturbing function, it is expected that by making this assumption the results are valid within the accuracy of this order. Also any resonance with planets other than Jupiter is not considered.

After the averaging the equations of motion are reduced to those of one degree of freedom with G and g as the two variables. By the assumption made for the critical argument the semi-major axis, or L , is constant. Therefore, the integral (2) can be written as,

$$H/L = (1 - e^2)^{1/2} \cos i = \Theta = \text{const.} \qquad (3)$$

in this secular perturbation theory. And, since the mean longitudes of the disturbing planets have been eliminated the Hamiltonian is now constant, that is, the energy integral exists.

3. AVERAGED VALUES OF THE DISTURBING FUNCTION

As the main term in the Hamiltonian depends on L only and, therefore, is constant, it is omitted here. The averaging is made for the disturbing function which is the sum of the following terms for all the planets;

$$m' / \Delta \tag{4}$$

where Δ is the distance between the asteroid and the planet with mass m' and is given by

$$\Delta^2 = r^2 + r'^2 - 2\, rr' \cos S,$$

$$\cos S = \cos^2(i/2) \cos(f+g-\lambda') + \sin^2(i/2) \cos(f+g+\lambda'). \tag{5}$$

Here r and r' are, respectively, the heliocentric distances of the asteroid and the planet, i is the inclination of the asteroid to the orbital plane of the disturbing planets, f is the true anomaly and λ' is the mean longitude of the planet measured from the ascending node.

The averaged value of the disturbing function, that is, the sum of terms (4), is denoted by R. It is evident that R is a periodic function of $2g$ as the argument of perihelion appears in (4) through (5) only as $g + \lambda$ and $g - \lambda'$.

In this paper the values of R are computed for various values of the argument of perihelion and X defined by

$$X = G/L = (1 - e^2)^{1/2}, \tag{6}$$

with given values of the semi-major axis and Θ and are plotted on $(2g\text{-}X)$ plane and then equi-R value-curves which are trajectories of the solutions for the secular perturbations for various values of R are derived. The value of X is in the range of 1 and Θ. By the relations (3) and (6) it is clear that as X decreases the eccentricity increases and the inclination decreases.

When both the eccentricity and the inclination are very small R depends on X only but not on g as R is analytically expressed as,

$$R = Ae^2 - A\sin^2 i. \tag{7}$$

For a given value of the semi-major axis which determines A R increases as X decreases since for more eccentric orbits the smallest distance to Jupiter is smaller. For this case every equi-R value-curve is parallel to $2g$- axis as the expression (7) shows. And since

$$dg/dt = -\partial R/\partial G, \tag{8}$$

the argument of perihelion progresses.

As the value of Θ decreases, the term $e^2\sin^2 i\cos 2g$ becomes more important in R. For this case also R increases as X decreases and for the same value of X it decreases as g approaches $2g = 180°$, and *vice versa* for the eccentricity.

However, if Θ takes a value below a certain value around 0.8 depending on the semi-major axis the dependence of R on the argument of pelihelion becomes stronger than that on X if X is nearly 1. And, therefore, R takes a minimum value at a point at $2g = 180°$ and nearly $X = 0.8$, and a libration region appears around this point on $(2g-X)$ plane. Inside the region the argument of perihelion librates around $2g = 180°$ and the eccentricity and the inclination take their maximum and minimum values both at $2g = 180°$ as the equi-R value-curves inside the libration region twice cross the line of $2g = 180°$ (Kozai, 1962).

For resonant cases the situations are different. If the critical argument is at a stable equilibrium point or librating around it, more eccentric orbits can avoid very close approach to Jupiter more easily. Therefore, generally speaking, R decreases as the value of X decreases and the argument of perihelion regresses except for Trojan case, for which more eccentric orbits have more chance to approach Jupiter very closely. Also for some cases the effects of other planets, particularly, that of Saturn which is not commensurable with the asteroid tends to cancel out the effects by Jupiter as the dependence of R on X is in opposite sense for the two cases. However, as the masses of the other planets are much smaller than that of Jupiter, their effects are usually very small.

4. TROJAN CASES

For Trojan case it is assumed that the difference of the mean longitudes of the asteroid and Jupiter is 60°. The value of R increases as X decreases and, therefore, the argument of perihelion progresses generally. However, when Θ is less than 0.87, R takes its maximum value at $2g = 120°$ and $X = 0.87$ which corresponds to $e = 0.50$ as the heliocentric distance of the asteroid at $f = \pm 120°$ where the difference between the true and the mean anomalies is nearly 60° is almost equal to that of Jupiter. Therefore, along any equi-R value-curve except that in a libration region around the maximum value of R X takes its maximum value at $2g = 120°$ and its minimum value at $2g = 280°$. In fact when Θ is less than 0.87 a libration region around the maximum value at $2g = 120°$ appears. When Θ becomes less than 0.65 another libration region which is shallow appears around the minimum point at $2g = 280°$.

Including the effects of the other planets does not change

general features of the diagrams except that the equilibrium points are shifted a little and the difference between the maximum and the minimum values of R are reduced.

5. 4:3, 3:2 AND 2:1 CASES

For the three cases it is assumed that the opposition takes place only when the asteroid is at its perihelion, that is, the critical arguments, $4\lambda' - 3\lambda - g$, $3\lambda' - 2\lambda - g$ and $2\lambda' - \lambda - g$, respectively, are always 0°, where λ and λ' are, respectively, the mean longitudes of the asteroid and the resonant disturbing planet.

Under the assumption the critical terms which have a factor e produce additional secular terms which have more important effects than conventional ones having e^2as a factor, and for more eccentric orbits R takes less values as for more eccentric orbits the smallest distance to Jupiter is larger because of the fixed value of the critical argument. Therefore, the eccentricity takes its maximum and the inclination takes its minimum at $2g = 0°$ and *vice versa* at $2g = 180°$. The argument of perihelion regresses and its secular motion is very rapid near $X = 1$.

R takes its minimum value at $2g = 180°$ and $X = 0.96$ for 4:3 case, 0.91 for 3:2 case and 0.74 for 2:1 case. When Θ takes any value below these ones a shallow libration region appears around the minimum value of R. As the value of Θ decreases the libration region which is bounded by a curve through $X = \Theta$ corresponding to $i = 0°$ expands and when Θ is less than 0.96 for 4:3 case, 0.89 for 3:2 case and 0.59 for 2:1 case the libration region touches off the line $X= \Theta$ and are combined with neighboring regions through singular points on $2g = 0°$.

The situation is very complicated, however, for 4:3 case when Θ is below 0.75. In fact there is a sharp maximum of R at $2g = 72°$ and $X = 0.75$, where the asteroid can approach Jupiter very closely. Then a libration region around this maximum point appears.

For numbered asteroids belonging to these resonant cases the eccentricities take their maximum and the inclination take their minima at $2g = 0°$ and *vice versa* at $2g= 180°$.

6. 3:1 CASE

For 3:1 case it is assumed that the critical argument, $3\lambda' - \lambda - 2g$, takes the value of 180° corresponding to a stable configuration, that is, any opposition takes place only when $2g = 180°$ and R takes smaller values at $2g = 0°$ than at $2g = 180°$, for which case the opposition takes place only near the orbital plane of Jupiter.

Although more eccentric orbits usually have smaller values of R, there is a sharp maximum at $X = 0.998$ and $2g = 180°$, around which there is a libration region bounded by a line through $X = 1$(circular orbits).

When Θ is below 0.79 a shallow libration region appears around the minimum point at $2g = 180°$ and $X = 0.75$, and it is bounded by a line through $X = \Theta$. As Θ reduces further the libration region expands and it touches off the line $X = \Theta$ and is combined with neighboring libration regions through singular points on $2g = 0°$. The effects of the other planets are usually small.

7. 2:3 CASE

In order to apply this method to Pluto-Neptune system 2:3 case is also treated here, by assuming that the opposition takes place only when the asteroid is at aphelion, in other words, that the critical argument, $3\lambda - 2\lambda' - g$, is always 180°. For this case also R decreases as X decreases and, therefore, the argument of perihelion regresses. And usually the value of R at $2g = 180°$ is smaller than that at $2g = 0°$. When Θ is below 0.90 a libration region appears around the minimum point at $2g = 180°$ and $X = 0.90$. As Θ is reduced the libration region bounded by a line through $X = \Theta$ touches off the axis $X = \Theta$ and is combined with neighboring libration regions through singular points on $2g = 0°$.

For Pluto the value of Θ is 0.9315, and, therefore, the argument of perihelion cannot be librating as there is no singular point in the diagram corresponding to the value of $\Theta = 0.9315$ if the effects of Neptune only are included. However, if the effects of the other planets, particularly, those of Uranus which has a mass comparable with that of Neptune, are included a libration region appears even when Θ is as large as 0.965. Of course a libration region appears for $\Theta = 0.93$, and it is concluded that as the value of R for Pluto is inside the libration region the argument of perihelion of Pluto is librating between 67° and 113°. It will cease to increase soon and the eccentricity is decreasing and the inclination is increasing. They will change between 0.223 and 0.273 and between 17.°0 and 14.°4, respectively, the present values being 0.248 and 15.°9 with $g = 112°$, where the inclination and the argument of perihelion are referred to the orbital plane of Neptune. The results agree with those by a numerical integration method (Kinoshita and Nakai, 1983) and those by a more exact analytical method including the effect of the librating critical argument by Nacozy and Diel(1974 and 1977) based on the analytical theory by Hori and Giacaglia(1968). In fact according to Kinoshita and Nakai the eccentricity, the inclination and the argument of perihelion change between 0.218 and 0.266, 16.°6 and 14.°6 and 64° and 116°, respectively.

8. CONCLUSIONS

In this paper the secular perturbations of asteroids with commensurable mean motions with Jupiter are derived under the assumption that all the disturbing planets are moving along circular orbits on the same orbital plane and that the critical arguments are fixed at their stable equilibrium points and by drawing equi-energy-curves, that is, trajectories in the phase space which is two-dimensional for 1:1, 4:3, 3:2, 2:1 and 3:1 cases. This method is also applied to Neptune-Pluto case and it is found that the variations of the eccentricity and the inclination as well as the amplitude of the libration of the argument of perihelion for Pluto derived by this method agree with those by a method of numerical integrations.

However, when the secular perturbations are derived for numbered asteroids with commensurable mean motions the amplitudes of the variations of the eccentricites and the inclinations are not so large for most of them and there are only three asteroids of 3:1 case in libration regions. General properties for them are summarized as follows:

For Trojan asteroids the minima of the eccentricities and the maxima of the inclinations take place at $2g = 120°$ and *vice versa* at $2g = 280°$ and the arguments of perihelion move in direct direction. For 4:3, 3:2 and 2:1 cases the maxima of the eccentricities and the minima of the inclinations take place at $2g = 0°$ and *vice versa* at $2g = 180°$ and the arguments of perihelion move in retrograde direction. For 3:1 case the minima of the eccentricities and the maxima of the inclinations take place at $2g = 0°$ and *vice versa* at $2g = 180°$ and the arguments of perihelion move in retrograde direction for most of them. However, for three of them the arguments of perihelion librate around the maximum points of R and the maxima and the minima of the eccentricities and the inclinations take place both at $2g = 180°$. And for other three asteroids the maxima of the eccentricities and the minima of the inclinations take place at $2g = 0°$ and *vice versa* at $2g = 180°$ and the arguments of perihelion move in direct direction as their value of Θ is small and for two of the three asteroids two libration regions appear. In fact their trajectories are far below the high libration region and above the shallow libration region around the minimum value of R if it exists.

However, it has not yet been checked whether the critical arguments for the asteroids treated here are librating or not. If they are it is expected that the results derived here express solutions of good approximation for the secular perturbations for these asteroids.

The computations in this paper were made by using the FACOM 380R computor of the Computing Center of the Tokyo Astronomical Observatory and a computor program provided by Dr. H. Kinoshita.

REFERENCES

Hori, G. and Giacaglia, G.E.O. : 1968, *Researches in Celestial Mechanics and Differential Equations, University of Sao Paulo*, 67-68-01.
Kinoshita, H. and Nakai, H. : 1983, *private communication.*
Kozai, Y. : 1962, *Astron. J.*, 67, pp. 591-598.
Kozai, Y. : 1979, in *Dynamics of the Solar System*, ed. by R.L. Duncombe, pp. 231-237, D. Reidel Publishing Company.
Kozai, Y. : 1980, *Icarus*, 41, pp. 89-95.
Nacozy, P.E. and Diehl, R.E. : 1978a, *Astron. J.*, 83,pp. 522-530.
Nacozy, P.E. and Diehl, R.E. : 1978b, *Celestial Mechanics*, 17, pp. 405-421.

A MECHANISM OF DEPLETION FOR THE KIRKWOOD'S GAPS

Jacques HENRARD and Anne LEMAITRE*

Department of Mathematics, Facultés Universitaires de Namur, B-5000 Namur, Belgium

* : Aspirante au F.N.R.S.

An analytical model describing the effect of a displacement of the Jovian resonances in the asteroid belt is analysed.

This model is based upon a truncated approximation of the averaged circular and planar restricted problem in the vicinity of a resonance. We investigate in details only the resonance 2/1. The model leads to a one degree of freedom Hamiltonian system with a parameter δ .

When the parameter δ decreases slowly with the time (and thus the resonance moves slowly in the belt) the theory of the adiabatic invariant predicts that a truncated uniform density distribution of asteroïds changes into a distribution showing a gap at the location of the resonance. The observed gap at the 2/1 resonance corresponds to a change of the parameter of a few units.

As a possible physical explanation for the decreases of the parameter δ (and thus the displacement of the resonance) we investigate the effect of the removal of an accretion disk in the early stage of the Solar System.

We first identify the important parameter for such an effect. It is the amount of material contained between the orbit of Jupiter and the orbit of the asteroid and not the total amount of material contained in the disk or in the proto-Sun. Models of proto-Sun and accretion disk can thus vary widely with respect to the two last parameters and still produces the same variation of δ and the same gap.

A. Berger (ed.), The Big Bang and Georges Lemaitre, 217–218.

Removal of a disk containing a few percent of the present solar mass between the orbit of the asteroid and the orbit of Jupiter is sufficient to account for the observed Hecuba gap. Such a value is somewhat larger but not incompatible with Weidenshilling estimates.

Details of this investigation can be found in a paper recently published by the authors in Icarus (55, 482-494, 1984).

ON THE STABILITY OF THE SOLAR SYSTEM AS HIERARCHICAL DYNAMICAL SYSTEM.

Andrea Milani and Anna M. Nobili

Dipartimento di Matematica, Università di Pisa, Italia

ABSTRACT - The Stability of the Solar System as hierarchical dynamical system is investigated with analytical methods i) by using the topological stability criterion of the general 3-body problem for 3-body subsystems ii) by exploiting the hierarchical arrangement of the whole system to develop a new perturbation theory containing as smallness parameters both mass and distance ratios. At the first order in these parameters we get, for the inner Solar System, a minimum lifetime of $\sim 10^8$y. Most of the results presented here are discussed in detail in Milani and Nobili (1982,1983a,1983b) which we shall refer to as Paper I, II and III, respectively.

N-body systems existing in nature, with N of the order of 10, tend to arrange themselves in a hierarchical structure, thus suggesting that this kind of arrangement does minimize the mutual perturbations hence making the system more likely to survive. As an example, in the Solar System the planetary orbits (with the exception of the Pluto-Neptune system, that we shall discuss later) do not cross in a time interval much longer than the longest orbital period and close approaches are avoided (even in the Pluto-Neptune case). Moreover, 70% of all the observed 3 and 4-body stellar systems are a close pair with a distant companion or two close pairs at a large distance. The 6-body stellar system Castor also exibits a hierarchical structure. Figures 1 and 2 show a planetary and double-binary hierarchy respectively with their Jacobian radius vectors $\underline{\rho}_i$.

The stability of hierarchical dynamical systems can be investigated by using Jacobian coordinates and decomposing the system into N-1 2-body subsystems coupled by perturbations. Hierarchical

A. Berger (ed.), The Big Bang and Georges Lemaitre, 219–230.

stability is defined as the property of preserving the hierarchical arrangement of these 2-body subsystems in such a way that orbit crossing is avoided.

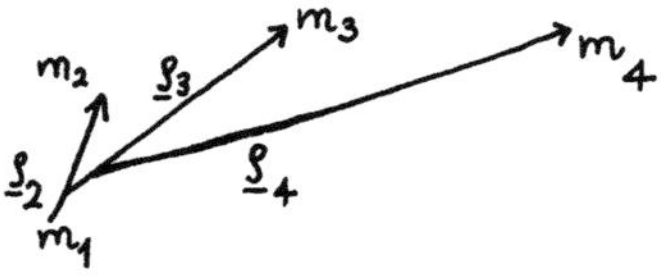

Fig.1. Planetary 4-body hierarchy .Each jacobian radius vector is centred in the center of mass of previous bodies

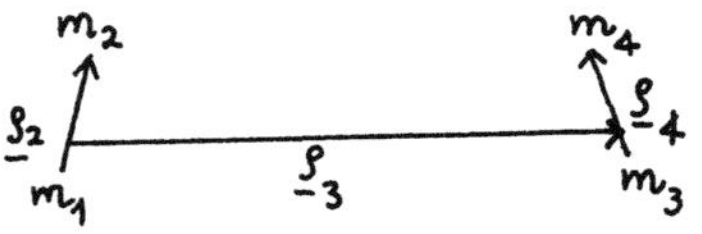

Fig.2. Double binary hierarchy. $\underline{\rho}_3$ goes from the center of mass of m_1, m_2 to the center of mass of m_3, m_4.

Jacobian coordinates have been introduced in the past by the fathers of Celestial Mechanics because of their formal properties. For instance, the total angular momentum and momentum of inertia of the N-body system (with respect to some axis) can be written as sum of the angular momenta and momenta of inertia of the bodies M_i with position vectors $\underline{\rho}_i$, M_i being the reduced masses:

$$M_1 = \sum_{i=1}^{N} m_i$$
$$M_2 = \frac{m_1 m_2}{m_1 + m_2}$$
$$M_3 = \frac{m_3 (m_1 + m_2)}{m_1 + m_2 + m_3} \qquad (1)$$
$$M_4 = \frac{m_4 (m_1 + m_2 + m_3)}{m_1 + m_2 + m_3 + m_4}$$
$$\vdots$$

In recent years Jacobian Coordinates have been re-introduced by Roy (1979) because, besides their properties, they allow to quantify the hierarchical structure of the system, i.e. the fact that the more the system is hierarchical, the smaller are the mutual gravitational perturbations between its 2-body subsystems. The equations of motion are:

$$M_j \ddot{\underline{\rho}}_j = \nabla_{\underline{\rho}_j} U \qquad j = 2, \ldots\ldots, N \qquad (2)$$

where U is the gravitational potential:

$$U = \sum_{i \leq j < k} G \frac{m_i m_k}{r_{ik}} \qquad (3)$$

(G is the universal constant of gravitation). By using Jacobian coordinates U can be expanded as a sum of 2-body terms plus interaction potentials (Paper II):

$$U = \sum_{j=2}^{N} \frac{G M_j N_j}{\rho_j} + \sum_{2 \le i < j < N} R_{ij} \tag{4}$$

N_j is the total mass attached to the jacobian vector $\underline{\rho}_j$ (e.g. for $\underline{\rho}_4$ of Fig.1 $N_4 = m_1 + (m_1 + m_2 + m_3)$; for $\underline{\rho}_4$ of Fig.2 $N_4 = (m_1 + m_2) + (m_3 + m_4)$) and the R_{ij} give the mutual perturbations. They can be written as:

$$R_{ij} \simeq \varepsilon_{ij} \frac{G M_j N_j}{\rho_j} P_2 + \text{higher order terms} \tag{5}$$

where P_2 is the Legendre Polynomial of order 2 and ε_{ij} are the smallness parameters first introduced by Walker et al. (1980) and then generalized for every possible hierarchy in Paper II. It is important to stress that they are a combination of both mass and distance ratios. For instance, in the 3-body planetary case, the smallness of the perturbation by $\underline{\rho}_2$ on $\underline{\rho}_3$ is given by:

$$\varepsilon_{23} = \frac{m_1 m_2}{(m_1 + m_2)^2} \left(\frac{\rho_2}{\rho_3}\right)^2 \tag{6}$$

while the perturbation by $\underline{\rho}_3$ on $\underline{\rho}_2$ is of the order of:

$$\varepsilon_{32} = \frac{m_3}{m_1 + m_2} \left(\frac{\rho_2}{\rho_3}\right)^3 \tag{7}$$

These formulas show that scale ratios as well as mass ratios are relevant to assess the smallness of the perturbations. For instance, in the Earth-Moon-Sun system the perturbation caused by the Sun is of the order of:

$$\varepsilon_{32} \simeq \frac{m_\odot}{m_\oplus + m_\rho} \left(\frac{\rho_{\oplus\rho}}{\rho_{\oplus\odot}}\right)^3 \simeq 3 \times 10^{-3}$$

i.e. it is small, even though the mass ratio is very large ($\sim 3 \times 10^5$), because the system is strongly hierarchical ($\rho_{\oplus\odot} \gg \rho_{\oplus\rho}$). Moreover for fixed values of the mass ratio, the more the system is hierarchical, the smaller are the corresponding ε_{ij}.

During the last 15 years a lot of interesting work has been done on the general 3-body problem so that it is now as deeply understood as the restricted one.

In the restricted circular 3-body problem (2 primaries in circular orbits plus a massless third body; see Szebehely,1967) the level manifolds of the Jacobi integral J in the phase space $(\underline{\rho}_3, \dot{\underline{\rho}}_3)$ are disconnected for:

$$J < J_{CRIT} \tag{8}$$

(J_{CRIT} being the critical value of J corresponding to the equilibrium point L_2) in 3 components. Their projections in the configuration space are also disconnected in a reference frame rotating with the primaries. The so called zero-velocity curves provide different regions of trapped motion and the well known Hill stability criterion (or "J criterion") can be formulated: if $J<J_{CRIT}$ and the third body lies in one of the 3 components (see Fig.3) at t=0, then it will stay there forever.

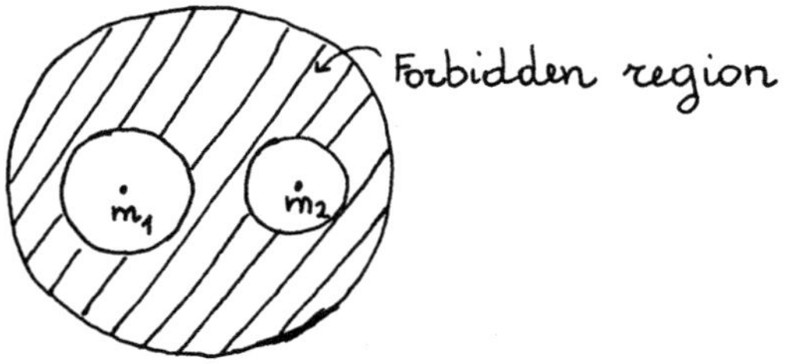

Fig.3. 3 separate regions of motion in the restricted circular 3-body problem ($J<J_{CRIT}$).

On the other hand, if $J>J_{CRIT}$ this does not necessarily mean that the third body will actually leave the region where it was at t=0. It simply means that we can't say anything. In fact numerical experiments do show that there is a region of empirical stability well above the analytical one (Nacozy,1977; Walker and Roy, 1981).

In the general 3-body problem, in which the gravitational action of the third body m_3 is taken into account, there are still 4 integrals (total energy and total angular momentum with respect to the center of mass):

h=const.

$\underline{c}$=const.

It has been recently proved (e.g. Paper I) that the structure of the level manifolds h=const., $\underline{c}$=const. in the phase space ($\underline{\rho}_2, \underline{\rho}_3, \dot{\underline{\rho}}_2, \dot{\underline{\rho}}_3$) depends only on:

$$z = c^2 h \tag{9}$$

The significant role played by this particular combination of integrals of motion can be understood by noting that it is not changed neither by rotations, nor by changes of scale:

$$\underline{\rho} \longmapsto \lambda \underline{\rho} \ , \quad t \longmapsto \tau t \ , \quad \dot{\rho} \longmapsto \frac{\lambda}{\tau} \dot{\rho} \tag{10}$$

that preserve the equations of motion provided that the length factor λ and the time factor τ satisfy a "3rd Kepler law" relationships:

$$\lambda^3 = \tau^2 \tag{11}$$

It is worth stressing that the universal constant of gravitation

G is also invariant under transformation (10) with condition (11). The z integral thus plays in the general 3-body problem the same role as the Jacobi integral in the restricted circular problem. In particular, it has been proved that if

$$z < z_{CRIT} \tag{12}$$

(z_{CRIT} being the critical value of z corresponding to the equilibrium point L_2) the level manifolds in the phase space are disconnected in 3 components. Moreover, their projections in the configuration space are also disconnected in a rotating pulsating reference frame (rotating with the instantaneous angular velocity of the binary and with the unit of length equal to the instantaneous distance between the primaries). In such a reference frame the bounding curves which separate different regions of trapped motion are very much alike the zero-velocity curves of the restricted circular problem and a "z stability criterion" (on the analogy of the J criterion) can be formulated: if $z<z_{CRIT}$ and m_3 lies, at t=0, in one of the 3 disconnected components, then it will stay there forever. However we must point out that the meaning of the connected components is different in the general and in the restricted circular case. In the restricted circular problem a zero velocity curve, enclosing a bounded region of admissible motion,means that the test particle cannot escape. On the contrary, in the general case, the rotating system is pulsating and the (x,y) plot must be multiplied by a variable scale factor, hence escape of one of the bodies is always possible but for $z<z_{CRIT}$ we can ensure that the hierarchy cannot be broken i.e. orbits will not cross. Again, nothing can be said about the stability of the system for $z>z_{CRIT}$.

It is hard to believe that the analogy of the z and J criterion is just a coincidence and in fact they have been thought to be related to one another. The z criterion was first applied to Solar System subsystems like Sun-Jupiter-external (or internal) planet and quite puzzling results were obtained: stability (in the sense of hierarchical stability discussed above) cannot be guaranteed, with this criterion, for Mercury, Mars, Pluto and any of the asteroids. For instance, in the Sun-Mercury-Jupiter system it is not possible to guarantee that Mercury will not cross the orbit of Jupiter. Neither is possible to guarantee that anyone of the asteroids will not cross the orbit of Jupiter. All the other 3-body subsystems can be proved to be hierarchically stable. How is it possible that Sun-Jupiter-Saturn can be proved to be stable while Sun-Mercury-Jupiter can not? After all, Saturn is much more perturbed by Jupiter than Mercury ($\varepsilon_{24}=2.8\times10^{-4}$ $\varepsilon_{2☿}=2.6\times10^{-6}$). Why should the z criterion "fail" just when the third body is a tiny one? When the third body is a tiny one we are probably "nearer" to the restricted case than we are to the full general one, in which the 3 bodies have comparable masses and this suggests that the reply to such paradoxical results is hidden in the

relationship between z and J. But how to compare problems with different dimensionality? The phase space of the general 3-body problem has dimension 12, and we must reduce this number up to dimension 2 of the configuration space in the rotating reference frame of the restricted circular problem where the zero velocity curves are drawn. This is possible in 4 steps (Paper I):
1st: velocities are eliminated by writing Easton inequality

$$IU^2 \geq -2c^2h = -2z \tag{13}$$

with

$$I = \sum_{j=2}^{N} M_j \left| \underline{\rho}_j \times \frac{1}{c} \underline{c} \right|^2$$

and we are reduced to a problem of dimension 6.
2nd: 2 dimensions are eliminated by projection on the invariable plane. Since U increases in doing so, (13) is still satisfied.
3rd: a scaleless configuration is chosen fixing I=1 and the problem is now of dimension 3.
4th: a reference direction is chosen ($\underline{\rho}_2=(1.0)$, elimination of the nodes) and we are eventually in synodic pulsating coordinates (x,y). The two problems are now comparable and the obvious way of making apparent the relationship between the two is to expand the gravitational potential U(x,y) computed in the last step in power series of the mass ratio between the third and the secondary body: $\varepsilon = m_3/m_2$. We get:

$$U(x,y) = G M_2^{3/2} (m_1 + m_2) \left(1 + \frac{1}{1+\mu} \varepsilon \, \Omega(x,y) + O(\varepsilon^2)\right) \tag{14}$$

$(\mu = m_2/(m_1+m_2))$
where $\Omega(x,y)$ is the well known function whose level lines give the zero-velocity curves in the restricted circular problem. Equation (14) says that, at first order in ε, the confinement curves of the general case are essentially given by the same function Ω giving the zero velocity curves in the restricted circular case. The comparison between z and J criterion is now straightforward, and the condition for hierarchical stability is:

$$\Delta z = z - z_{CRIT} = \mu(1-\mu)\left(\frac{e_2^2}{2} + \frac{\varepsilon}{1-\mu}(J - J_{CRIT}) + O(\varepsilon^2) + O(\varepsilon e_2^2)\right) < 0 \tag{15}$$

(e_2 is the osculating eccentricity of the primaries).
By neglecting terms of the order of ε^2 or εe_2^2 an approximate z criterion is obtained:

$$\frac{\delta_z}{\mu(1-\mu)} = \frac{e_2^2}{2} + \frac{\varepsilon}{1-\mu}(J - J_{CRIT}) < 0 \tag{16}$$

It says that if a 3-body system is stable in the restricted circular approximation (i.e. $J-J_{CRIT}\sim -1$), then hierarchical stability can be ensured in the general case (neglecting terms of the order of ε^2 or $e_2^2\varepsilon$) only if the third mass is larger than a minimum value:

$$m_3 > m_{3min} \sim \frac{m_1 m_2}{m_1+m_2}\frac{e_2^2}{2} \tag{17}$$

involving the reduced mass of the binary and its osculating eccentricity squared. In the Sun-Jupiter case (17) gives

$$m_{3min} \simeq 0.4\, m_\oplus$$

which means that the z criterion is unable to ensure stability for masses smaller than $\sim 0.4 m_\oplus$. This is not unexpected, because a tiny third body does contribute very little to the total integrals of the system, hence to z, and therefore a criterion based on z is simply unapplicable in this case. Inequality (17) is important because it quantifies this intuitive statement and defines, for a binary of given reduced mass and eccentricity, the thresold m_{3min} below which the z criterion of the general 3-body problem is unapplicable. No matter how small this thresold is (e.g. no matter how small is e_2), there is a range of values of m_3 between 0 and m_{3min} for which neither the general 3-body problem nor the restricted circular one will be the right framework to investigate the system. We will use then the elliptic restricted model, in which the third body is massless but the eccentricity of the binary is taken into account. It is well known that in this case the Jacobi "integral" is not constant any longer, and therefore confinement curves do no longer exist. In fact, as far as close approaches are avoided the Jacobi function changes slowly and its level lines still provide a boundary. As an example, if an approximation of the Jacobi "integral" is computed for real asteroids, far from Jupiter, the values are found to be below the critical value, with few exceptions that can be accounted for by librational protection mechanisms (Kresák, 1979). This means that regions do exist where the dynamical behaviour is almost the same as in the restricted circular model. On the contrary, when close approaches can happen only numerical integrations in the elliptic restricted model can give an idea of the dynamical behaviour of the system, while the results obtained in the restricted circular model are certainly wrong (Paper III). But numerical experiments, while can show instabilities (such as the ejection of an outer belt asteroid from a chaotic region caused by Jupiter's eccentricity), can never ensure stability forever. They only show stability as long as they are meaningful , i.e. as long as the numerical error is small enough to leave some significant figures, which means in particular that chaotic regions are necessarily excluded.

Let us now consider an N-body system with $N \geqslant 4$. It can be proved (Paper II) that no stability criterion based on the ten classical integrals only and valid for an arbitrary span of time can be given for $N \geqslant 4$. The main confinement condition (Easton inequality (13)) is still valid, but it does no longer provide separate regions of trapped motion. However, the z stability criterion applies to every 3-body subsystem whose z "integral" will of course vary in time because of the other bodies perturbations. In the N=4 case we decompose the system into two 3-body subsystems (Fig.4) whose z functions, z_{23} and z_{34}, at t=0 are smaller than the corresponding critical values z_{23CR}, z_{34CR} so that both the subsystems are initially hierachically stable. Then the hierarchical arrangement of the 4 bodies cannot be broken until either z_{23} or z_{34} is changed by an amount $z_{ijCR} - z_{ij}(0)$; i.e. the whole system is hierarchically stable for a time span not shorter than the minimum between $\Delta t_{23} = (z_{23CR} - z_{23}(0))/\dot{z}_{23}$ and $\Delta t_{34} = (z_{34CR} - z_{34}(0)/\dot{z}_{34}$. We have given up the idea of proving that the Solar System is stable forever; we are trying to prove that it is hierarchically stable for a time span of the order of its present age, i.e. a few times 10^9y.

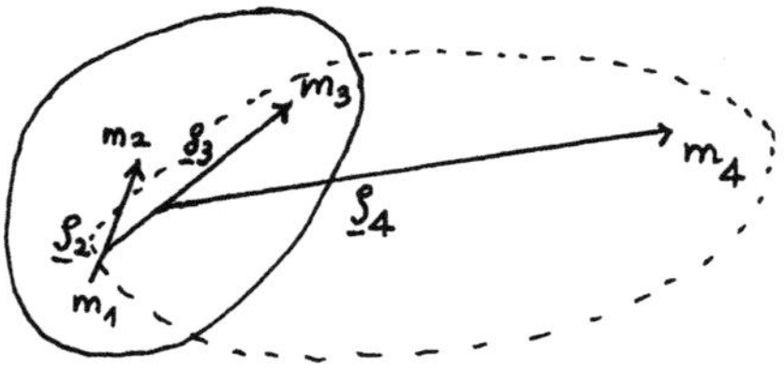

Fig.4. A 4-body system split into two 3-body subsystems. In the $\underline{\rho}_3$, $\underline{\rho}_4$ subsystem the binary m_1, m_2 is concentrated in the center of mass.

A new perturbative theory has been developed (Paper II) to compute the secular time variations of z_{23} and z_{34} in which the smallness parameters are the ε_{ij} instead of the classical mass ratios. Since they contain also the scale ratios the asymptotic expansions are much more rapidly decreasing than the usual expansions in powers of the mass ratios only. Both this new kind of perturbative theory, exploiting the hierarchical arrangement of the system, and the availability of the z stability criterion for 3-body subsystems, make this new approach to the old problem of the stability of the Solar System very promising, despite of the large amount of work that is required. The secular time variation of the z functions is computed by using Poisson bracket formalism. For instance, if a 4-body system with planetary hierarchy is split at t=0 into two stable (according to the z criterion) 3-body subsystems (see Fig.4) we have:

$$z_{23}(0) < z_{23CR} \qquad\qquad z_{34}(0) < z_{34CR}$$

$$z_{23} = c_{23}^2 h_{23} \qquad\qquad z_{34} = c_{34}^2 h_{34}$$

$$\underline{c}_{23} = M_2 \underline{\rho}_2 \times \dot{\underline{\rho}}_2 + M_3 \underline{\rho}_3 \times \dot{\underline{\rho}}_3 = \underline{c}_2 + \underline{c}_3 \qquad \underline{c}_{34} = M_3 \underline{\rho}_3 \times \dot{\underline{\rho}}_3 + M_4 \underline{\rho}_4 \times \dot{\underline{\rho}}_4 = \underline{c}_3 + \underline{c}_4$$

$$h_{23} = \frac{1}{2} M_2 \dot{\varsigma}_2^2 - \frac{G M_2 N_2}{\varsigma_2} + \frac{1}{2} M_3 \dot{\varsigma}_3^2 - \frac{G M_3 N_3}{\varsigma_3} - R_{23} = h_2 + h_3 - R_{23}$$

$$h_{34} = \frac{1}{2} M_3 \dot{\varsigma}_3^2 - \frac{G M_3 N_3}{\varsigma_3} + \frac{1}{2} M_4 \dot{\varsigma}_4^2 - \frac{G M_4 N_4}{\varsigma_4} - R_{34} = h_3 + h_4 - R_{34} \tag{18}$$

(23 and 34 subscripts refer to the two different subsystems). While $\underline{c}_{23}$, $\underline{c}_{34}$, h_{23} and h_{34} change in time, the total energy and angular momentum of the 4-body system are constant:

$$h = h_2 + h_3 + h_4 - R_{23} - R_{34} - R_{24} = \text{const.}$$
$$\underline{c} = \underline{c}_2 + \underline{c}_3 + \underline{c}_4 = \text{const.} \tag{19}$$

We compute the Poisson brackets $\{z_{23}, h\}, \{z_{34}, h\}$ and average over the angular variables. In the 4-body planar case we prove that, at first order in the smallness parameters the z_{ij} do not undergo any secular perturbation because of the interaction with the other subsystem. After the long-period perturbations have been accounted for, and if resonances are avoided, only second order terms have to be considered in the computation of Δt_{23}, Δt_{34}. In the very pessimistic assumption that all the second order terms affect in a secular way the z_{ij} an order-of-magnitude lower bound for the disruption of the hierarchy of the whole system can be obtained. The same techniques can be used in 3 dimensions and the perturbations by other bodies can be included as well, although the computations become more troublesome .

The results obtained in the 4-body planar case can be quite successfully applied to the Sun-Mercury-Venus-Jupiter system. The two 3-body subsystems are Sun-Mercury-Venus and (Sun+Mercury)-Venus-Jupiter, which can both be proved to be hierarchically stable. But how long will Jupiter take to break the hierarchy of the Sun-Mercury-Venus system? (Of course, the perturbations due to the fact that Sun-Mercury is actually a binary and not a point mass does act much more slowly). Our method gives a lower bound of 1.1×10^8 y, which is still 1 order of magnitude shorter than our goal but much longer than previous results. For instance, we can now rule out that Mercury has been a satellite of Venus, at least as far as hundreds of millions of years are concerned. Perturbations by other planets, either internal or external, do not substantially change this result because at first order they simply add up to Jupiter's perturbations and the corresponding ε_{ij} are smaller than $\varepsilon_{4☿}$ that gives the perturbation of Jupiter on Mercury. It is worth stressing that although Mercury is too tiny to apply the z criterion to the Sun-Mercury-Jupiter 3-body system, the very existence of Venus provides a kind of "gravitational screening" effect: when each of the 3-body subsystems is assumed to be isolated, Mercury cannot cross the orbit of Venus, and Venus cannot cross the orbit of Jupiter; this allows to investigate the stability of the whole system with the analytical methods discussed above.

What about Mars, which is too small for the z criterion to be meaningful and is not "screened" by any other big enough planet, as Mercury is by Venus? Since the Sun-Mars-Jupiter 3-body subsystem cannot be proved to be hierarchically stable there is no 4-body system including Mars and Jupiter that can be decomposed into two initially stable 3-body subsystems; therefore our methods are simply unapplicable. The same is true for any of the asteroids, and from this point of view Mars is nothing but a big asteroid. Numerical integrations (either in the elliptic restricted Sun-Jupiter-asteroid model or taking into account the perturbations by the Earth) are the only way of getting, if not a definitive answer at least an indication of the long-term behaviour of these objects. But we must not try to prove stability at any cost. After all, Earth and Mars crossing asteroids do exist; the outer belt has been largerly depleted because of close approaches to Jupiter (Paper III); the 3/1 gap has been found to give rise to strong instabilities (jumps in eccentricity) that seem to explain the lack of asteroids there (Wisdon 1982,1983).On the other hand, dynamical protection mechanisms are known to prevent asteroids from being ejected. This suggests that chaotic regions, as opposed to ordered regions in the phase space must first be spotted (Paper III): orbit crossing is likely to happen in chaotic regions while long-term integrations are worthwile only in ordered regions. Pluto, being a Neptune crosser, is not hierarchically stable. It is known to be protected by a 3/2 libration with Neptune which avoids close approaches in the vicinity of Pluto's perihelion. Moreover, there are reasons to think (Farinella et al., 1980) that Pluto is an escaped satellite of Neptune that ended up in an heliocentric orbit and "adjusted" itself in a protected region: just because it was hierarchically unstable it needed an ad hoc mechanism to survive or, better, it survived only because entered a dynamically protected region . The same probably happened for Apollo-Amor objects, and this suggests that many of them must be dynamically protected.(Janiczek et al., 1972)

Let us now consider the 4 outer planets. Jupiter, Saturn, Uranus and Neptune. In this case all the 3-body subsystems are initially stable according to the z criterion but the gravitational perturbation of Jupiter on Saturn is at least one order of magnitude stronger than any other ε_{ij} ($\varepsilon_{4\hbar} \sim 3\times10^{-4}$) and this means that whenever the two main planets are included the lifetime estimates obtained with our present first order theory are too short to be competitive with numerical integrations already available up to 5×10^6y. Moreover, our estimates being based on a perturbation method, resonances -both in mean motion and secular- must be avoided as long as we want the method to be applicable. But are there exact resonances- with libration of critical arguments - between the main planets of the Solar System? Cohen et al.(1973) published position and velocity of the outer planets every 40000y between -5×10^5y and $+5\times10^5$y. We implemented these data on our program in order to compute the time variation of the z "integrals" of 3-body subsystems. For each subsystem with Jacobian vectors $\underline{\rho}_i$, $\underline{\rho}_j$ the difference $\Delta z_{ij}(t)=z_{ij}(t)-z_{CR}$ was monitored.One expected that Δz_{ij} should remain almost constant over a time span 3 orders of magnitude smaller than the age of the Solar System. On the contrary, in both the Sun-Jupiter-Saturn and Sun-Uranus-Neptune subsystems we have found a clear oscillation with a period of about 10^6y (Fig.5). Moreover, the two oscillations were almost in opposite phase, as if the subsystems were locked to one another exchanging angular momentum (no such long period oscillation appears in semimajor axis, hence in energy). We presently don't know what's the resonance responsible for such "locking", nor what's the critical argument involved.

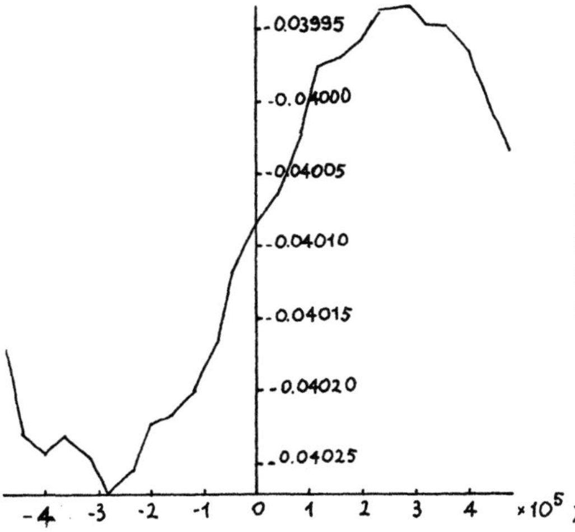

Fig.5. $\Delta z_{♃♄}$ as function of time for Sun-Jupiter-Saturn (normalized as in Paper I). The ephemerides used to compute $\Delta z_{♃♄}$ are from Cohen et al.(1973)

As it might have been expected an attempt of using new powerful techniques to prove the hierarchical stability of the Solar System for a time span comparable to its age has arisen fundamental questions about its structure .

References

Cohen, C.J., Hubbard, E.C. and Oesterwinter, C. (1973), Astr.Pap. Ephem. 22, 1.

Janiczek, P.M., Seidelmann, P.K. and Duncombe, R.L. (1972), A.J. 77, 764.

Kresák, L. (1979), in "Asteroids" (T.Gehrels ed.), Univ. of Arizona, Tucson, p.289.

Farinella, P., Milani, A., Nobili, A.M. and Valsecchi, G.B. (1980), Icarus 44, 810.

Milani, A. and Nobili, A.M. (1983), "On Topological Stability in the General 3-Body Problem", Celestial Mechanics, 31, 213.

Milani, A. and Nobili, A.M. (1983a),"On the Stability of Hierar-

chical 4-Body Systems", Celestial Mechanics, 31, 241.
Milani, A. and Nobili, A.M. (1983b) "The Depletion of the Outer Asteroid Belt", submitted to the Astronomical Journal.
Nacozy, P.E. (1977), Celestial Mechanics, 16, 77.
Roy, A.E. (1979), in "Instabilities in Dynamical Systems" (V.Szebehely ed.), Reidel Pu. Co., Dordrecht, p.177.
Walker, I.W., Gordon Emslie ,A. and Roy, A.E. (1980), Celestial Mechanics 22, 371.
Walker, I.W. and Roy, A.E. (1981), Celestial Mechanics, 24, 195.
Wisdom, J. (1982), A.J. 87, 577.
Wisdom, J. (1983), Icarus 56, 51.
Szebehely, V. (1967), "Theory of Orbits", Academic Press, N.Y. London.

A RELATIVISTIC APPROACH TO THE KEPLER PROBLEM

John G. Bryant

Université Paris VI

ABSTRACT The classical Kepler problem can be modified so that the velocity is always bounded. This is done by using a new time variable. A new Hamiltonian function can also be constructed, and it receives a relativistic interpretation : Newtonian mass and energy are replaced by relativistic (proper) mass and energy. A classical model for the photon can also be given. On the other hand the reformulated Kepler problem can equally well be used in special relativity as a model for a relativistic particle in a gravitational field.

1. THE CLASSICAL KEPLER PROBLEM

1.1 The motion of a particle with mass m in the gravitational field created by a second particle, which we assume fixed at the origin, can be described by the following Hamiltonian function :

$$(1) \qquad H(p_i, q^i) = \sum_{i=1}^{3} \frac{p_i^2}{2m} - \frac{m\alpha}{r}$$

where $r^2 = \sum_{i=1}^{3} (q^i)^2$ and α is a positive constant. The equations of motion are written :

$$(2) \qquad \frac{dq^i}{dt} = \frac{\partial H}{\partial p_i} = \frac{p_i}{m} \quad ; \quad \frac{dp_i}{dt} = -\frac{\partial H}{\partial q^i} = m\alpha \frac{\partial}{\partial q^i}\left(\frac{1}{r}\right)$$

A. Berger (ed.), The Big Bang and Georges Lemaitre, 231–237.

and they lead to Newton's equations when we eliminate the p_i. The Hamiltonian $H(p_i, q^i)$ is a constant of the motion, and the equation H = constant = h yields the following expression for the velocity v of the particle :

$$(3) \qquad v^2 = \sum_{i=1}^{3} \left(\frac{dq^i}{dt}\right)^2 = \sum_{i=1}^{3} \left(\frac{p_i}{m}\right)^2 = \frac{2\alpha}{r} + \frac{2h}{m}$$

This formula shows that in theory v has no upper bound. For a given value of r, v can be made as large as we want by increasing the value of h ; and for a given value of h there are motions for which r can become arbitrarily small (even if we exclude collision orbits).

1.2 This problem can be avoided by adopting the formalism of relativity. However we then lose the simplicity of the classical problem, since ordinary 3-dimensional space is replaced by 4-dimensional space-time and the Kepler orbits are replaced by more complicated ones, as in the case of the Schwarzschild model.

Another way to avoid the problem, without radically changing the formalism, would be to modify the classical equations. The simplest way to do this is to introduce a new time variable. This is the possibility we shall now investigate.

2. THE MODIFICATION OF THE CLASSICAL EQUATIONS

2.1 We introduce a new time variable τ by the following equation :

$$(4) \qquad dt = (1 + \rho(q^i))d\tau$$

where ρ is some function of the q^i only that we wish to determine. The Hamiltonian system (2) now becomes :

$$(5) \qquad \frac{dq^i}{d\tau} = \frac{p_i}{m(1+\rho)} \quad ; \quad \frac{dp_i}{d\tau} = \frac{m\alpha}{1+\rho} \frac{\partial}{\partial q^i}\left(\frac{1}{r}\right)$$

We see that :

a) the trajectories are unchanged
b) if we assume $\rho \ll 1$, then motion in time is practically unchanged
c) $H(p_i, q^i)$ is no longer the Hamiltonian. However, H is still a constant of the motion.

2.2 We would like to choose $\rho(q^i)$ in such a way that

we can define a new Hamiltonian function $E(p_i, q^i)$ for the problem. We even assume the following very special form for E :

$$(6) \qquad E(p_i, q^i) = \psi(p_i)\,\varphi(q^i)$$

i.e. E is the product (and not the sum) of a function of the p_i and a function of the q^i. We must then have :

$$(7) \qquad \frac{dq^i}{d\tau} = \frac{\partial E}{\partial p_i} = \varphi \frac{\partial \psi}{\partial p_i} \quad ; \quad \frac{dp_i}{d\tau} = -\frac{\partial E}{\partial q^i} = -\psi \frac{\partial \varphi}{\partial q^i}$$

Comparison with (5) yields the following equations :

$$(8) \qquad \frac{p_i}{m(1+\rho)} = \varphi \frac{\partial \psi}{\partial p_i} \quad ; \quad \frac{m\alpha}{1+\rho} \frac{\partial}{\partial q^i}\left(\frac{1}{r}\right) = -\psi \frac{\partial \varphi}{\partial q^i}$$

For the second group of equations to have meaning, it becomes necessary to restrict (7) to a given surface E = constant = k (this will in fact be equivalent to H = h as we shall see). We can then replace ψ by k/φ and the second group of equations becomes :

$$(9) \qquad \frac{m\alpha}{1+\rho} \frac{\partial}{\partial q^i}\left(\frac{1}{r}\right) = -\frac{k}{\varphi} \frac{\partial \varphi}{\partial q^i}$$

It is clear from these equations that $\rho(q^i)$ must be some function of $1/r$, and we are led to the following choice :

$$(10) \qquad \rho(q^i) = \frac{2\alpha}{c^2 r}$$

where c is the velocity of light. $\rho(q^i)$ is thus a dimensionless quantity, and we have $\rho \ll 1$ except when r is very small. (9) now becomes :

$$(11) \qquad \frac{mc^2}{1+\rho} \frac{\partial \rho}{\partial q^i} = -\frac{2k}{\varphi} \frac{\partial \varphi}{\partial q^i}$$

and we make the following choice for $\varphi(q^i)$:

$$(12) \qquad \varphi(q^i) = (1+\rho)^{-\frac{1}{2}} = \left(1 + \frac{2\alpha}{c^2 r}\right)^{-\frac{1}{2}}$$

(this will be justified later on). We then have :

$$(13) \qquad k = mc^2$$

In other words, the Hamiltonian system defined by $E(p_i, q^i)$ must be restricted to a <u>single</u> surface $E = mc^2$. In fact it becomes <u>too</u> restricted. We must therefore modify our approach by assuming that E is of the form :

$$(14) \qquad E = \psi(p_i, a)\,\varphi(q^i)$$

where a is some constant. We thus seek to obtain a 1-parameter <u>family</u> of Hamiltonian functions (the parameter being a), each one being restricted to the surface $E = mc^2$.

To determine $\psi(p_i, a)$, we must solve the first group of equations (8) which, when we take account of (12) and (13), has the form :

$$(15) \qquad c^2 p_i = \psi \frac{\partial\psi}{\partial p_i}$$

This leads to the following solution :

$$(16) \qquad \psi(p_i, a) = c\sqrt{a + \sum_{i=1}^{3} p_i^2} \quad ; \quad a = \text{constant}$$

E is therefore written :

$$(17) \qquad E(p_i, q^i) = c\sqrt{\frac{a + \sum_{i=1}^{3} p_i^2}{1 + \frac{2\alpha}{c^2 r}}}$$

and has the desired form.

2.3 The constant a is linked to h and m since we have :

$$a = m^2c^2\left(1 + \frac{2\alpha}{c^2 r}\right) - \sum_{i=1}^{3} p_i^2$$

$$a = m^2c^2 + \frac{2m^2\alpha}{r} - \left(\frac{2m^2\alpha}{r} + 2hm\right)$$

(taking account of H = h) and thus :

$$(18) \qquad a = m^2c^2\left(1 - \frac{2h}{mc^2}\right)$$

$E = mc^2$ is therefore equivalent to $H = h$.

The expression for the new velocity v' of the particle is given by :

$$v'^2 = \sum_{i=1}^{3}\left(\frac{dq^i}{d\tau}\right)^2 = \frac{\sum_{i=1}^{3} p_i^2}{m^2(1+\rho)^2} = \left(\frac{2\alpha}{r} + c^2 - \frac{a}{m^2}\right)\left(1 + \frac{2\alpha}{c^2 r}\right)^{-2}$$

$$(19) \qquad \frac{v'^2}{c^2} = \left(1 + \frac{2\alpha}{c^2 r}\right)^{-1}\left[1 - \frac{a}{m^2c^2}\left(1 + \frac{2\alpha}{c^2 r}\right)^{-1}\right]$$

Thus, for v' to always have an upper bound, we have the following conditions on a and h :

$$(20) \qquad a \geqslant 0 \iff h \leqslant \frac{mc^2}{2}$$

which lead to

$$(21) \qquad v' \leqslant c$$

Also, when $r \simeq 0$, then $v' \simeq 0$.

3. THE RELATIVISTIC INTERPRETATION OF THE MODIFIED EQUATIONS

3.1 We have achieved our goal of obtaining a modified Kepler problem, where the orbits are preserved, but the velocity is always bounded. The equations of motion keep their Hamiltonian form, but there is no globally defined Hamiltonian, only a 1-parameter family of Hamiltonians (note that this situation is similar to the one we obtain when applying the Principle of Maupertuis).

At this stage a very useful comparison can be made with relativity. In the case of a <u>free</u> particle ($\alpha = 0$), the expression of E becomes :

$$(22) \qquad E = c\sqrt{a + \sum_{i=1}^{3} p_i^2} = mc^2$$

If we write

$$a = m_o^2 c^2 \quad ; \quad m_o \geqslant 0$$

then E is <u>identical</u> to the Hamiltonian of a free relativistic particle with proper mass m_o and energy mc^2. Only the <u>interpretation</u> of the parameters is different. In relativity, m_o is a characteristic parameter of the particle and m (= E/c^2) is a constant of the motion, whereas, in the preceding interpretation, m_o (= $\sqrt{a}/c$) is a constant of the motion (linked to h) and m is a characteristic parameter of the particle.

To give global significance to $E(p_i, q^i)$ in this particular case, as well as in the general case ($\alpha \neq 0$), we simply have to adopt the relativistic interpretation of the parameters (this is in fact possible owing to our choice of $\varphi(q^i)$ given by (12)). In practice (i.e. planetary motion) this is not a problem since we generally have $v' \ll c$; $2\alpha/c^2 r \ll 1$, so that, according to (19) :

$$(23) \qquad m \simeq m_o$$

The Hamiltonian function $E(p_i, q^i)$, which we now write :

$$(24) \qquad E(p_i, q^i) = c\sqrt{\frac{m_o^2 c^2 + \sum_{i=1}^{3} p_i^2}{1 + \frac{2\alpha}{c^2 r}}}$$

is no longer restricted to a particular value. We have thus obtained a formulation of the <u>classical</u> Kepler problem in <u>relativistic</u> terms, where :

a) the <u>Kepler orbits</u> are preserved, with, according to (18) :

$h < 0 \iff m < m_o$: Elliptic orbits
$h = 0 \iff m = m_o$: Parabolic orbits
$h > 0 \iff m > m_o$: Hyperbolic orbits

b) motion in time is only slightly modified (at least when r is not too small). In the elliptic case, Kepler's third law becomes :

$$\frac{T^2}{\lambda^3} = \frac{4\pi^2}{\alpha}\left(1 + \frac{2\alpha}{c^2\lambda}\right) \tag{25}$$

with : T : period λ : semi-major axis.

3.2 As an added bonus to this formulation of the Kepler problem, we can now define a <u>classical model for a photon</u> in a gravitational field. This corresponds simply <u>to the</u> case $m_o = 0$. $E(p_i, q^i)$ becomes :

$$E_{photon} = c\sqrt{\frac{\sum_{i=1}^{3} p_i^2}{1 + \frac{2\alpha}{c^2 r}}} \tag{26}$$

We then have $h = mc^2/2 > 0$, so the trajectories are hyperbolas. The "gravitational bending of light rays" is thus a classical phenomenon.

3.3 Finally, if we place ourselves in the context of <u>special</u> relativity, our reformulated Kepler problem can <u>equally</u> well be used as a model for a relativistic particle in a gravitational field. We have in fact a <u>new</u> derivation for the concept of the relativistic particle, which encompasses the <u>free</u> particle.

The model is no <u>longer</u> valid however in <u>general</u> relativity. This is easily seen in the elliptic case where there is <u>no</u> advance of the perihelion since the orbits are <u>closed</u> ; and one can also show that the deflection of photons is too small by half. Nevertheless, it is possible to account for these effects by adopting a "perturbed" form of the Hamiltonian $E(p_i, q^i)$ given by (24), whose orbits are in complete agreement with those of the Schwarzschild model. (See : Bryant J.G. : "Variétés de contact et variétés canoniques en mécanique", Thèse de Doctorat d'Etat, Université Paris VI, 1983)

THE THREE-BODY PROBLEM IN STELLAR DYNAMICS

Piet Hut

Institute for Advanced Study
Princeton, New Jersey 08540
USA

In studies of the three-body problem, bound orbits and especially periodic orbits have traditionally received much attention. Complementary to these, however, are the scattering orbits, where at least one body is unbound initially and (in the generic case) also at least one body escapes after a finite period of close three-body interaction. These scattering orbits play an important role in stellar dynamics. Three-body interactions exchange energy and angular momentum between internal degrees of freedom of binary stars, and external degrees of freedom of both the single stars and the center of mass motion of binaries. In this way three-body scattering can significantly influence the evolution of a star cluster.

The scattering of single stars off tight binary stars on average increases the binding energy of these binaries. The energy released heats up the star cluster, just as nuclear reactions generate energy within a star. To describe these exothermic processes in detail, accurate binary - single star scattering cross sections are needed. Several million numerical orbit integrations have been performed in order to determine differential cross sections and their dependence on the many parameters which characterize individual three-body scattering configurations.

Several general conclusions follow from these detailed investigations. As an important example, quantitative results are presented to illustrate the role of the three-body scattering problem as a heat source in stellar dynamics.

A. Berger (ed.), The Big Bang and Georges Lemaitre, 239–255.

1. INTRODUCTION

Although the three-body problem is natural and simple to formulate, the richness and variety of its solutions is truly overwhelming. Despite its long history, with contributions made by many outstanding mathematicians over the last three centuries, many new developments have occurred during the last three decades. Far from being exhaustive, let me just mention a few areas before restricting myself to astrophysical applications.

Qualitative methods such as developed by Kolmogorov, Arnol'd and Moser have resulted in a better understanding of the stability of the three-body problem (cf. Arnol'd, 1978). Specific examples of an unexpectedly rich spectrum of solutions to a very simple sub-class of initial conditions were found by Sitnikov and Alekseev, who studied the behavior of a single star oscillating along the symmetry axis perpendicular to the orbit of a double star (cf. Alekseev, 1981). Another recent development is the study of orbits near triple collisions. McGehee (1974) showed how a triple collision manifold could be defined and pasted onto the boundary of the regular energy manifold, and that the equations of motion induced a gradient-like flow on this boundary, significantly simpler than the regular flow (cf. Simó, 1981 for more recent extensions).

The use of electronic computers has also stimulated the renewed study of classic problems, by providing powerful ways to search for periodic solutions, for example. A completely different class of solutions is formed by the scattering orbits, where a binary star and/or initially unbound single star(s) meet each other in a close encounter. In the generic case the period of genuine three-body interaction is relatively short, and the system disperses in three unbound stars, or a binary and an unbound single star. There is a rich variety of qualitatively different types of scattering orbits, including exchange, temporary and permanent capture, total dispersion and oscillatory orbits (Alekseev, 1981).

The topology of the set of all possible scattering orbits shows a bewildering variety, including an infinite nesting of orbits of qualitatively different outcome within a finite region of the space of parameters describing the initial conditions. I have reported an initial exploration of some of this variety, together with some more mathematical aspects of three-body scattering, elsewhere (Hut, 1983c). In the present paper I will concentrate on more astrophysical applications.

In Sect. 2.1 I will review our current understanding of the simplest type of the general N-body problem: the long-term evolution of large spherically symmetric systems of many point-particles. This seemingly simple mathematical physics problem is, of course, far removed from real star clusters, where effects of stellar evolution (*e.g.* mass loss), external fields (from the galaxy) and non-gravitational effects (*e.g.* tidal energy dissipation and physical collisions) can be important. However, many fundamental questions have not yet been answered even in this simple case, and it seems wise to try and understand this isolated problem first.

In Sect. 2.2, I will emphasize the important role in stellar dynamics of binaries, which can be described as particles with internal degrees of freedom. In more advanced stages of star cluster evolution, the central star density quickly grows very large, leading to the formation of new binaries. Three-body scattering of single stars off these binaries provides an energy source for subsequent star cluster evolution, in a way analogous to nuclear energy generation which powers stellar evolution. This energy production might prevent further contraction of the central regions, balancing the continuing energy loss in the form of heat conduction into the outer regions which is caused by two-body interactions.

In Sect. 3.1, I will give an outline of an extensive project of numerical orbit integrations to determine cross sections of scattering processes between single stars and binaries (for more details, see Hut and Bahcall, 1983; Hut, 1983a,b). The fundamental law of binary dynamics, that hard binaries tend to become harder while soft binaries tend to become softer, is illustrated in Sect. 3.2 with new quantitative results. A discussion of applications to the evolution of star clusters is presented in Sect. 3.3.

2. EVOLUTION OF STAR CLUSTERS

Take a large number of point masses, sprinkle them at random inside a limited volume of space, and let them interact via Newton's law of gravity. Here we have a recipe with simple ingredients and a simple prescription. Still, the evolution of such a simple system was very poorly understood before the advent of fast computers. And even today, the qualitative behavior of the later stages of evolution of such a system is the subject of considerable debate.

From an analytic point of view, a major stumbling block is the long-range attractive character of gravitational forces, which precludes all standard statistical physics treatment. For example, each star undergoes encounters with many other stars simultaneously, with no intermediate nearly free motion. Furthermore, there is an infinite amount of phase space available since stars can escape from the bound system. And worse, in very close encounters velocities can grow without bound. These last two aspects of self-gravitating systems lead to a thermal distribution of binaries which formally diverges both for very wide and for very tight binaries! Yet another aspect of these problems shows up when we study deviations from equilibrium, which cannot be made arbitrarily small with an independent tuning parameter, because a self-gravitating system has no real equilibrium state. Instead, small deviations from pseudo-equilibrium are governed by the same coupling constant (Newtons gravitational constant) which determines the overall characteristics of the pseudo-equilibrium state in the first place. The situation is therefore inherently more complicated than that in the laboratory for a gas with short range forces where, *e.g.*, an arbitrarily small temperature difference can be imposed in order to study heat conduction.

More fundamentally, gravity cannot be treated as a thermodynamic system in the usual sense, because no extensive quantities can be defined; adding extra particles changes the overall pseudo-equilibrium in a strongly non-linear way. Still, when we look up at the night sky, equipped with a modest-sized telescope, we can find many globular clusters, aggregates of some 10^5 - 10^6 stars, in which the stars seem to be distributed in a nicely smooth spherical way. Moreover, these systems have an age of at least 10^{10} years! If statistical mechanics tells us that these systems cannot be in true equilibrium, they certainly present to us the challenge of trying to describe their pseudo-equilibrium, which seems so well-behaved, all theoretical objections notwithstanding.

2.1 Evolution Towards Core Collapse

To come back to our original recipe, let us sprinkle N point-like stars at random within a limited volume of space, for simplicity giving them all equal masses. What will happen? A number of partial answers have emerged from many different approaches, analytical as well as numerical, in the last quarter century. It is impossible to give a complete description here, and I will merely point out some of the general results, together with a few references to reviews on particular aspects of the general problem.

For small systems, $N < 100$, evolution proceeds on a dynamical time scale. In a few crossing times the density increases significantly in the center, while stars escape steadily from the outskirts of the system. Soon the central density becomes so high that at least one tight binary is formed in the center, from a simultaneous encounter of three single stars followed by the escape of one star which carries away the necessary amount of kinetic energy to leave the other two in a tightly bound orbit. Shortly thereafter the tightest binary grows even tighter by subsequent encounters with single stars, and quickly acquires more than half the binding energy of the whole system. The energy released leads to a slow expansion of the system, and the evolution slows down considerably. But single stars keep escaping through two- and three-body encounters, until the whole system is dispersed in single independent units, mostly single stars and binaries with occasional (meta-)stable triples or even higher hierarchical objects.

For somewhat larger systems, $100 \leq N \leq 1000$, computer experiments in which all $\frac{1}{2} N(N-1)$ forces are computed directly become very costly, and only few of them have been reported in the literature. The overall behavior seems to be similar to that of smaller systems, albeit on a somewhat longer time scale. Again one central hard binary is formed which plays a major role in heating up its surroundings. Here and in the following a hard (soft) binary is defined as having a binding energy much larger (smaller) than the typical kinetic energy of the field stars. More details and references can be found in the review by Aarseth and Lecar (1975), while a comparison with the observations of open clusters is reviewed by King (1980).

Of great interest, from a theoretical point of view, are much larger systems for which $N \sim 10^5 - 10^6$. Not only do these correspond to observed systems such as globular clusters, they also allow a clear distinction of time-scales. The two-body relaxation time is defined as the time scale on which the motion of a typical star is changed significantly through repeated (generally weak) interactions with individual other stars. Two-body relaxation takes place on a time scale much larger than a crossing time, and the system stays relatively close to an equilibrium state. In the simplest case of a spherically symmetric star cluster the equilibrium distribution function of stars in phase space is a function only of energy and angular momentum. Two-body interactions prevent such an equilibrium to be reached exactly, and these small fluctuating forces cause some stars to become more energetic and therefore to move away from the center. The remaining stars move slower which causes the core to contract. The resulting higher density causes a decrease in the two-body relaxation time-scale, and the whole process accelerates more and more.

These large - N systems have been studied only indirectly, in a variety of elegant approximations. The methods used generally involve a scheme to solve numerically an appropriate Fokker-Planck equation for the distribution function of the stars under certain simplifying assumptions, using either Monte Carlo methods (*e.g.* Hénon, 1972; Spitzer, 1975) or direct integration (Cohn, 1979, 1980). The general picture emerging from these works is that of an accelerated increase in central density, driven by two-body relaxation processes, resulting in a nearly homologous core collapse, seemingly to infinite density in a finite time. A simple and surprisingly accurate description of this development was given Lynden-Bell and Eggleton (1980, following earlier work by Hachisu *et al.*, 1978), who used a gaseous model similar to that used to model the evolution of the fluid elements in a single star. A review of earlier work with applications to globular clusters is given by Lightman and Shapiro (1978).

2.2 Evolution Beyond Core Collapse

The end state resulting from core collapse is not really singular, or course. Although the density of the core keeps increasing, the number of particles in the core decreases. When this number falls below a hundred or so, statistical fluctuations become dominant and a description in terms of average quantities is not reliable anymore. A detailed numerical modeling of this situation is very difficult, because of the large dynamical range in densities and time scales.

The pioneer in this field is Hénon (1961, 1965, 1975), who was the first to construct simple analytical and numerical models to describe the evolution of a large N-body system after core collapse. He conjectured that some unspecified energy source, presumably in the form of binaries, would appear to replenish the energy lost continuously through conduction to the outer layers, and eventually lost by escaping stars.

This picture was substantiated in more detailed investigations by Stodółkiewicz (1982), who found the reaction of the N-body system to be rather insensitive to the precise nature of the central energy source. Very recently, a number of investigations, using different approximations have greatly expanded our understanding of post-core collapse evolution (Hénon, 1961, 1975; Stodółkiewicz, 1982; Inagaki and Lynden-Bell, 1983; Heggie, 1983; Goodman, 1983, Sugimoto and Bettwieser, 1983; Bettwieser and Sugimoto, 1983), but also raised several controversies, and many questions still remain to be answered.

For systems with 10^5 - 10^6 stars, it is unlikely that one tight binary will play as dominant a role as for smaller systems, even when the finite size of the stars is neglected. It seems more plausible that several binaries form in the core, each giving off heat to the system until becoming so hard that a single encounter with a field star can give rise to a recoil velocity large enough to remove the binary into the halo, or even out of the cluster. To give a detailed statistical description of these effects, and to improve on the preliminary investigations mentioned above, accurate binary-single star cross sections are needed which describe, *e.g.*, rates of excahnge of energy, momentun and angular momentum. This is the subject of the next section. An alternative approach to study the behavior immediately around the time of core collapse has been followed by McMillan (1983) who developed a hybrid N-body / Fokker Planck computer code to handle extremely large density contrasts.

3. BINARIES AS A CENTRAL ENERGY SOURCE IN STAR CLUSTERS

There is an analogy between stars, powered by nuclear reactions, and star clusters, powered by gravitational reactions, in the form of binary-single star scattering. This analogy is helpful in understanding the overall evolution of star clusters, although there are important qualitative differences. For example, nuclear reaction rates increase at higher density, and the same is true for gravitational three-body reactions. However, nuclear reactions rates increase steeply with increasing temperature, whereas binary reactions show the opposite behavior, releasing less energy with higher dispersion velocities in the star system.

Even without considering binaries, there are important differences between stellar dynamical systems and the gaseous interior of a single star. For example, increasing the density at constant temperature will make a star more opaque, thus lowering the conductivity. Increasing the density in a stellar system, while holding the velocity dispersion fixed, will increase the rate of two-body relaxation effects, and increase the effective conductivity of energy through the system.

With these cautions in mind, I will describe below the approach I have taken to tackle these problems at a fundamental level. This approach consists of three steps, starting from a microscopic description of three-body scattering, I have derived local statistical quantities to describe the effects of binaries on their immediate surroundings, from which I hope to arrive at a global statistical description of the evolution of the whole system. Results from the first two steps are described in the next two subsections, while the last subsection summarizes work in progress on the last step.

3.1 Three-body Reactions In The Laboratory

An extensive project of numerical orbit integrations has been initiated by Hut and Bahcall (1983; paper I) to determine cross sections of scattering processes between single stars and binaries. All scattering experiments are carried out over a full range of initial conditions, in which all parameters are allowed to change independently of each other. This exploration of the full dimensionality of parameter space can be carried out in practice only by a Monte Carlo sampling of many different initial conditions, followed by a deterministic orbit calculation of the scattering process in each case. In this way several million numerical scattering experiments were carried out, exceeding by more than an order of magnitude the total number of experiments reported in the literature (the following papers describe more than a thousand experiments each: Saslaw *et al.*, 1974; Hills, 1975; Valtonen (1975); Valtonen and Aarseth, 1977; Valtonen and Heggie, 1979; Hills and Fullerton, 1980; Fullerton and Hills, 1982).

Although the results of previous numerical investigations are interesting, especially in transition regions between different domains of validity of analytic approximations, their accuracy is difficult to estimate. All authors mentioned above adopted some form of constraint on the initial conditions in their scattering experiments, such as zero impact parameter and zero eccentricity, or incoming velocity at infinity either zero or rather large. The significance of the large number of experiments conducted in the present project lies not only in the smaller statistical spread in the resulting cross sections, and the resulting possibility to exhibit the detailed dependence of (differential) cross sections on individual parameters, but more importantly in the *guarantee* that the error bars are only determined by statistical noise and not by systematic trends caused by non-uniform sampling.

It is interesting that here classical mechanics causes computational difficulties in an area where quantum mechanics makes life easy. In classical mechanics there do not exist simple highly symmetric low-lying energy levels (let alone a spherically symmetric ground state). For any choice of star cluster parameters, all binary orbital parameters have to be taken into account; the only consistent simplifying choice allowed is that of equal masses for all stars. The effect is an all-or-nothing situation. If one wants to obtain reliable and accurate quantitative results in answer to any *specific* question, it is necessary to first build up a large data base containing the outcome of many individual experiments. This is the only way to acquire a statistical resolution large enough to simultaneously discriminate between the dependence on the different parameters of interest. But from that point on *many other* questions, specific as well as general, can be answered using the same data base.

Detailed results for the equal mass case are presented in paper I, together with an outline and background of the project. These results, in the form of a variety of total and differential cross sections, tend to asymptotic limits which are in excellent agreement with analytic estimates by Heggie (1975; for a review: 1980), extended where necessary by Hut (1983a; paper II). The only limitation left in paper I concerns the case of resonance scattering in which the incoming star forms a bound system with the two binary members. A resonant bound state is unstable, but only weakly so: it often will take thousands of original binary periods before one of the stars escapes. Since such complex orbits require orders of magnitude more computer time, the first few million orbit calculations were halted whenever the occurrence of a resonance was established. This still made possible the determination of total cross sections for resonance scattering in paper I, but no information was obtained about the amount of energy exchange since the outcome was left undetermined. Full resonance scattering calculations have been carried out subsequently, and will be reported in paper III (Hut, 1984), which will provide the first unrestricted and detailed description of equal mass binary--single star scattering. The case of unequal masses poses no extra computational difficulty, and in subsequent papers several other mass ratios will be explored.

All experiments mentioned above involved a Monte Carlo sampling of initial conditions, in order to obtain a physicist's description of gravitational three-body scattering in terms of cross sections. From the point of view of a mathematician interested in the three-body system as a dynamical system, additional insight into the extremely rich microscopic structure of the space of orbits has been obtained from a series of experiments for a grid of initial conditions, determined by stepwide varying several parameters independently while keeping the other parameters fixed (Hut, 1983c). Specific astrophysical applications, such as binary--single star exchange scattering as a formation mechanism for cataclysmic variables in globular clusters, are discussed by Hut and Verbunt (1983a,b).

3.2 Three-body Reactions In the Field

3.2.1 The Fundamental Law of Binary Dynamics

Energy exchange between external and internal degrees of freedom is the most important feature of binary--single star scattering. Hard binaries, with an orbital velocity much larger than typical field star velocities, behave differently from soft binaries, for which the orbital velocity is much lower than that of the field stars. A fundamental law of three-body stellar dynamics is: hard binaries tend to become harder while soft binaries tend to become softer. This can be described heuristically by the following equipartition argument.

A fast star moving past a slowly revolving binary will on average lose some energy to the binary. However, trying to speed up the binary members will put them in wider orbits with an actually *lower* velocity (loosely speaking a Kepler orbit seems to have a negative 'specific heat', a general phenomenon for gravitational interactions, cf. Lynden-Bell, 1973). Hard binaries, on the other hand, can capture a slowly incoming field star under formation of a bound triple system. After some time, generally orders of magnitude longer than the initial binary period, one of the stars is ejected more or less stochastically. The velocity with which it reaches infinity is typically of order of the internal binary velocities, and therefore much larger than the initial field star velocity. The binary has to increase its binding energy in order to give off this energy, thereby shrinking and increasing its orbital velocity.

Of course, not all hard binaries harden during each encounter with a field star, nor do soft binaries loosen up monotonically; both processes take place for both types of binaries, but the net energy balance has a different sign. On the whole, soft binaries do not play an important role for the energy budget of a star cluster, since the kinetic energy of their center of mass motion already exceeds their binding energy. Even the complete dissolution of many soft binaries will hardly affect the temperature (i.e. the velocity dispersion) of the cluster. Hard binaries cause much more dramatic effects, as is discussed in Sect. 3.3.

3.2.2 Cross Sections and Reaction Rates: Definitions

The differential cross section $\frac{d\sigma}{d\Delta}(\Delta;\, v,\, e,\, m_1,\, m_2,\, m_3)$ gives the probability distribution for different amounts of energy exchange in a binary--single star scattering. Here Δ is the relative change in binary binding energy during one scattering event:

$$\Delta = \frac{E_{bin}(t\to+\infty) - E_{bin}(t\to-\infty)}{E_{bin}(t\to-\infty)}\,, \tag{1}$$

where *bin* stands for *bin*ary *bin*ding energy, and e is the initial binary orbital eccentricity. As in Paper I v denotes the velocity of the incoming field star with respect to the center of mass of the binary in units of the critical velocity v_c, given by

$$v_c^{\,2} = G\,\frac{m_1 m_2 (m_1 + m_2 + m_3)}{m_3 (m_1 + m_2)}\,\frac{1}{a}\,, \tag{2}$$

for which the total energy of the three-body system vanishes. G is the gravitational constant, a is the initial semimajor axis of the binary, m_3 is the mass of the incoming field star, and m_1, m_2 are the masses of the binary components.

In three-body scattering, three types of reactions can occur: 1) a single star incident on a binary breaks up the binary, leaving all three stars unbound (ionization); 2) a binary emerges from a binary--single star scattering event with a different energy and possibly different stars (fly-by and exchange); 3) three unbound stars can interact to form a new binary (creation). In the following we will discuss only the second process of energy *exchange; destruction* (ionization) rates are given in paper I, which can be used to calculate *creation* rates via detailed balance relations. The definition of the average energy change is formally

$$\langle\Delta\rangle = \lim_{\varepsilon\to 0} \frac{\int_{-1}^{-\varepsilon} \frac{d\sigma}{d\Delta}\,\Delta\, d\Delta + \int_{\varepsilon}^{\infty} \frac{d\sigma}{d\Delta}\,\Delta\, d\Delta}{\int_{-1}^{-\varepsilon} \frac{d\sigma}{d\Delta}\, d\Delta + \int_{\varepsilon}^{\infty} \frac{d\sigma}{d\Delta}\, d\Delta}\,, \tag{3}$$

where the lower limit of integration $\Delta = -1$ is chosen to exclude ionization events.

Unfortunately, both terms in the denominator in Eq. (3) diverge for $\varepsilon \rightarrow 0$: there is an infinitely little energy exchange to take place. This type of infrared singularity *always* appears in classical mechanics, no matter how fast the interaction strengths drops as a function of impact parameter (even for an exponential drop off), as long as the interaction is not identically zero. The familiar fact that short range forces lead to finite total cross sections is a consequence of quantum mechanics. This follows directly from a partial wave analysis, and can be understood physically from the uncertainty principle (for a lucid description, see Landau and Lifshitz, 1965).

The two terms in the numerator in Eq. (3), however, do approach a finite limit as ε vanishes. The reason is that a field star passing the binary at large distance interacts with the internal binary degrees of freedom only via a tidal force, which nearly cancels out when averaged over a complete binary orbital period. It can be shown that the binding energy of the binary is an adiabatic invariant in this limit of very wide encounters (cf. Heggie, 1975).

A useful expression with a finite limit for $\varepsilon \rightarrow 0$ is the product of the cross section weighted by the amount of energy exchange:

$$< \sigma\Delta > = \lim_{\varepsilon \rightarrow 0} \left\{ \int_{-1}^{-\varepsilon} \frac{d\sigma}{d\Delta} \Delta \, d\Delta + \int_{+\varepsilon}^{\infty} \frac{d\sigma}{d\Delta} \Delta \, d\Delta \right\} . \quad (4)$$

For simplicity we will still refer to $< \sigma\Delta >$ as the average energy exchange, although it really is a quantity which enters in the average energy exchange *rate* $n < \sigma\Delta > v$, where n is the density of field stars.

3.2.3 Some Illustrative Results

The average energy exchange in scattering processes between binaries and single stars of equal masses is plotted in Figs. 1 and 2 as a function of binary hardness/softness, measured by the velocity of the incoming field star. Figs. 1*a* and 2*a* show the results for initially circular binary orbits, while Figs. 1*b* and 2*b* are for the opposite extreme case, where the original binary orbit has eccentricity $e = 0.99$. The results for both cases are remarkably similar, and show that the dependency on eccentricity is relatively small, a situation which greatly simplifies application to realistic star clusters (cf. paper I). In practice, it often suffices to simply take the root-mean-square (r.m.s) value of a thermal eccentricity distribution, $e = 0.7$ (Hut, 1983b). A large range of binary energies is explored in Figs. 1 and 2: the ratio of initial binary binding energy versus incoming kinetic

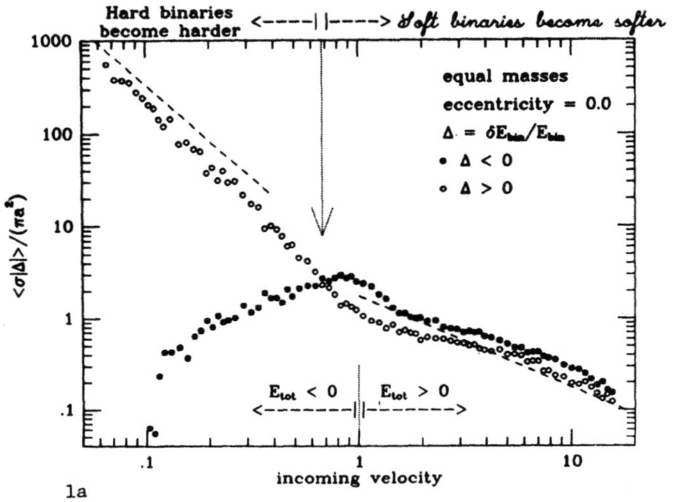

1a

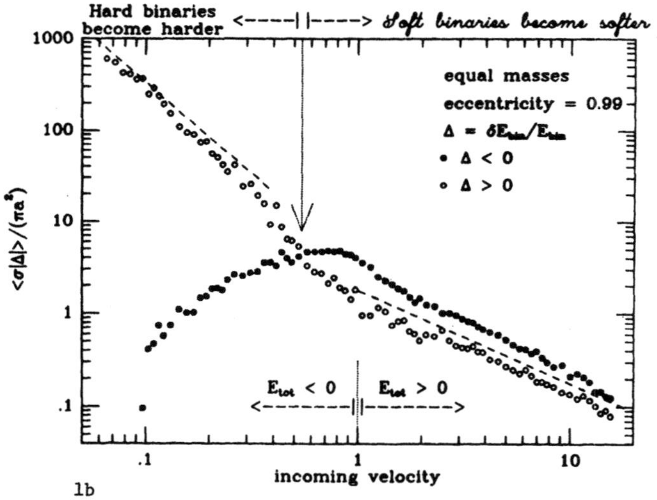

1b

Fig. 1 The average change δE_{bin} in binary binding energy is plotted separately for the scattering events with a binding energy increase ($\Delta>0$) and for those with a binding energy decrease ($\Delta<0$), for the two extreme eccentricity values e=0 and e=0.99. The watershed velocity, where the average energy increase and decrease are equal, is indicated by the dotted arrow. The incoming velocity v is given in units of the critical velocity (Eq.2), for which the total energy vanishes; in these units the r.m.s. binary orbital velocity is $v_{orb} = \frac{1}{2}\sqrt{3} \simeq 0.87$.

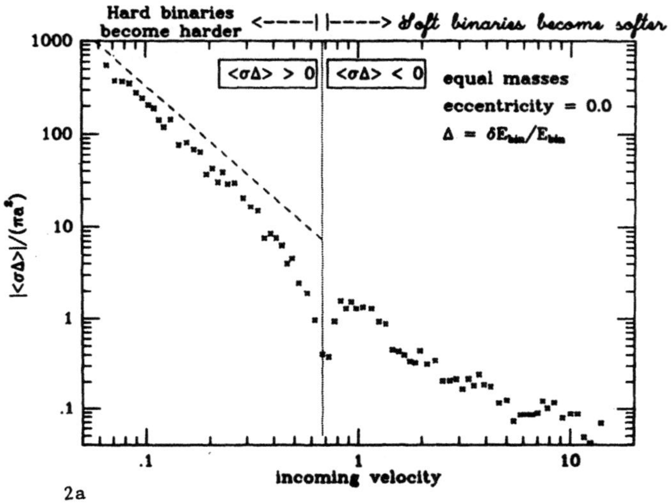

2a

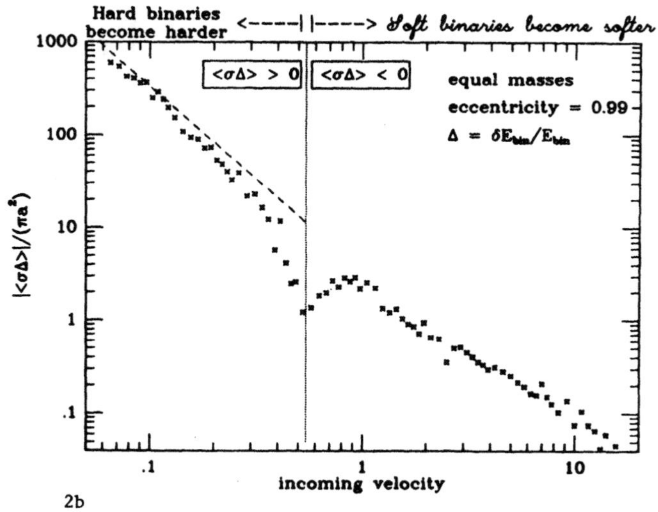

2b

Fig. 2 The net average change in binary binding energy. Hard binaries, at the left side of the watershed, will on average gain binding energy, thus moving to the left and becoming harder. Soft binaries, at the right, on average loose binding energy, move to the right and become softer.

energy, equal to unity at the center, drops to 1/256 at the right side and climbs to 256 at the left side, spanning sixteen octaves or nearly five orders of magnitude.

A comparison with analytic approximations by Heggie, given by the dashed lines in Figs. 1 and 2, shows remarkable agreement. For soft binaries this has been noticed already in paper I (cf. Figs. 9-12 therein). For hard binaries Heggie's analytic approximations for close encounters only are plotted, which give the dominant contribution to the average energy exchange. When the analytic contribution from wide encounters is taken into account, it seems that the total analytic result is somewhat too large, but by less than a factor two. A comparison of the differential cross section curves $\frac{d\sigma}{d\Delta}(\Delta)$ for hard binaries (not shown here) helps to account for this. The numerical cross sections agree well in shape with Heggie's prediction for large Δ, but are as a whole displaced to somewhat lower values. Details of this comparison and a similar discussion for small Δ are presented in paper III. The main conclusion is that Heggie's assumption of an effective stochastic mixing (loss of memory of the initial state) for resonance scattering is a good approximation.

The numerical results presented here are most useful in providing an accurate determination of the shape of the energy exchange curves near the watershed between hard and soft binaries. By definition the expectation values for energy gain and loss are equal at this point, indicated by the arrows in Fig. 1a,b. Experiments at different eccentricity values give slightly different watershed velocities, and a good fit is $v = 0.67-0.13e \pm 0.02$, with a 1σ estimated error. Fig. 2 indicates that the transition between the two asymptotic regimes of hard and soft binaries is very rapid: the total width of the transition zone spans only half an order of magnitude in incoming velocity for a given binary (equivalently: an order of magnitude in binary binding energy for given incoming velocity). This transition zone determines the formation rate of hard binaries through hardening of a fraction of the soft binaries, via repeated encounters with single stars.

Note that the present definition of watershed is a *local* one: it is concerned with the average energy change in the next scattering event only. An alternative *global* definition might be more appropriate, which evaluates the combined result of all subsequent scattering events, leading ultimately to either a complete breakup (ionization) of the binary or a continuing hardening. However, such a global definition requires the specification of the (in general time-dependent) distribution of field stars. In paper III values for a global watershed are presented for the simple case of a time independent Maxwellian background field.

3.3 Three-body Reactions and the Evolution of a Large Star Cluster

The cross sections discussed in Sect. 3.1 can be used as input data in a statistical treatment of large N-body systems, where direct integration of the equations of motion becomes impractical (*i.e.* N > 1000). The most widely used approach is the Fokker-Planck treatment in which the effects of strong two-body scattering is neglected, since these become relatively less important with respect to those of weak encounters as N increases.

A pioneering study in which three-body effects were added in a Monte Carlo Fokker-Planck treatment was carried out by Spitzer and Mathieu (1980). They used as input data analytical approximations by Heggie (1975) and numerical cross sections by Hills (1975). Starting with an initial fraction of the stars in the form of moderately hard binaries, they followed the initial stages of cluster evolution leading to the onset of core collapse. They found that the presence of binaries did slow down the rate of central contraction, but could not prevent core collapse.

The logical next step is to try to follow the evolution through and beyond core collapse, starting for simplicity with a system of single stars only. Such a system will evolve until the core has contracted so much that it only contains few (30 - 100, say) stars, at which time the density will have become high enough to cause an appreciable rate of binary formation, directly from three-body reactions between unbound single stars.

The simplest way to mimic the formation and subsequent energy generation of central binaries is to include an appropriate source term in a gaseous model for a star cluster (Inagaki and Lynden-Bell, 1983; Heggie, 1983; Goodman, 1983, Sugimoto and Bettwieser, 1983; Bettwieser and Sugimoto, 1983). Including energy generating terms in Monte Carlo Fokker Planck codes (Hénon, 1975; Stodółkiewicz, 1982) is more realistic, but the dynamic range in star density is not as large in a Monte Carlo approach with a finite number of shells.

A much better treatment could be given in a direct integration of the Fokker-Planck equation, as done by Cohn (1979, 1980) for the pre-core collapse phase for single stars, using a two-fluid approach for binaries and single stars. The interaction between the center of mass motion can be modelled with the standard Fokker-Planck terms describing the cumulative effect of many distant two-body encounters. The interaction between external degrees of freedom and internal degrees of freedom of the binaries can be described statistically using the data mentioned in Sect. 3.1 (Ultimately one would need binary--binary cross sections as well as binary--single star cross sections, but the gaseous models hint that few binaries will coexist in the core, at least in the first phase after core collapse; a first rough numerical determination of some binary - binary cross sections is given by Mikkola, 1983).

A simpler approach which still allows a very large dynamic range is the use of a direct Fokker-Planck code with only one fluid of single stars, modelling the effects of binary generation and hardening with the average reaction rates discussed in the previous subsections. Work along these lines is in progress, and the results will combine the large dynamical range (as in the gaseous models) with a more realistic stellar dynamical treatment (as in the Monte Carlo Fokker-Planck models).

Acknowledgements

I thank Jeremy Goodman and Robert Lupton for helpful comments on the manuscript. This work was supported in part by the National Science Foundation through grant PHY-8217352.

REFERENCES

Aarseth, S. J. 1972, in *Gravitational N-Body Problem,* IAU colloq. 10, ed. M. Lecar (Dordrecht: Reidel), p. 88.
Aarseth, S. J. and Lecar, M. 1975 in *Ann. Rev. Astron. Astrophys.* 13, p. 1.
Alekseev, V. M. 1981, *Amer. Math. Soc. Transl. (2)* 116, p. 97. [Translation of *Ninth Math. Summer School* (Kaciveli, 1971, Izdanie Inst. Mat. Akad. Nauk Ukrain. SSR, Kiev, 1972, pp. 212-341]
Arnol'd, V. I. 1978, *Mathematical Methods of Classical Mechanics* (Springer).
Bettwieser, E. and Sugimoto, D. 1983, preprint, to be published in *Mon. Not. R. astr. Soc.*
Cohn, H. 1979, *Astrophys. J.* 234, p. 1036.
Cohn, H. 1980, *Astrophys. J.* 242, p. 765.
Fullerton, L. W., and Hills, J. G. 1982, *Astron. J.* 87, p. 175.
Goodman, J. 1983, preprint, to be published in *Astrophys. J.*
Heggie, D. C. 1975, *Mon. Not. R. astr. Soc.* 173, p. 729.
Hachisu, I., Nakada, Y., Nomoto, K. and Sugimoto, D. 1978, *Prog. Theor. Phys.* 60, p. 393.
Heggie, D. C. 1980, in *Globular Clusters,* ed. D. Hanes and B. Madore, p. 281 (Cambridge, U.K.: University Press).
Heggie, D. C. 1983, preprint, to be published in *Mon. Not. R. astr. Soc.*
Hénon, M. 1961, *Ann. Astrophys.* 24, p. 369.
Hénon, M. 1965, *Ann. Astrophys.* 28, p. 62.
Hénon, M. 1972, in *Gravitational N-Body Problem,* IAU colloq. 10, ed. M. Lecar (Dordrecht: Reidel), p. 406.
Henon, M. 1975, in *Dynamics of Stellar System,* IAU Symp. 69, ed. A. Hayli (Dordrecht: Reidel), p. 133.
Hills, J. G. 1975, *Astron. J.* 80, p. 809.
Hills, J. G., and Fullerton, L. W. 1980, *Astron. J.* 85, p. 1281.
Hut, P. and Bahcall, J. N. 1983, *Astrophys. J.* 268, p. 319 (Paper I).
Hut, P. 1983a, *Astrophys. J.* 268, p. 342 (Paper II).
Hut, P. 1983b, *Astrophys. J. Lett.* 272, L29.
Hut, P. 1983c, *Astron. J.* 88, p. 1549.
Hut, P. 1984, in preparation (Paper III).
Hut, P. and Verbunt, F. 1983a, *Nature* 301, 587.
Hut, P. and Verbunt, F. 1983b, in *Cataclysmic Variables and Low-Mass X-Ray Binaries,* Seventh North American Workshop,

ed. D. Q. Lamb and J. Patterson (Dordrecht: Reidel)
Inagaki, S. and Lynden-Bell, D. 1983, preprint, to be published in *Mon. Not. R. astr. Soc.*
King, I. R. 1980, in *Star Clusters*, IAU Symp. 85, J. E. Hesser (Dordrecht: Reidel), p. 139.
Landau, L. D. and Lifshitz, E. M. 1965, *Course of Theoretical Physics*, Vol. 3, *Quantum Mechanics* (Pergamon), second edition, p. 473 (third edition, 1977, p. 507).
Lightman, A. P. and Shapiro, S. L. 1978, *Rev. Mod. Phys.* 50, p. 437.
Lynden-Bell, D. 1973, in *Dynamical Structure and Evolution of Stellar Systems*, ed. L. Martinet and M. Mayor, Third Advanced Course of the Swiss Society of Astronomy and Astrophysics at Saas Fee (Geneva Observatory, Sauverny, Switzerland).
Lynden-Bell, D. and Eggleton, P. P. 1980, *Mon. Not. R. astr. Soc.* 191, 483.
McMillan, S. L. W. 1983, PhD thesis, Harvard.
McGehee, R. 1974, *Invent. Math.* 27, 191.
Mikkola, S. 1983, *Mon. Not. R. astr. Soc.* 203, p. 1107.
Monaghan, J. J. 1976a, *Mon. Not. R. astr. Soc.* 176, p. 63.
Monaghan, J. J. 1976b, *Mon. Not. R. astr. Soc.* 177, p. 583.
Monaghan, J. J. 1977, *Mon. Not. R. astr. Soc.* 179, p. 31.
Nash, P. E. and Monaghan, J. J. 1978, *Mon. Not. R. astr. Soc.* 184, p. 119.
Saslaw, W. C., Valtonen, M. J., and Aarseth, S. J. 1974, *Astrophys. J.* 190, p. 253.
Simó, C. 1981, in *Classical Mechanics and Dynamical Systems*, ed. R. L. Devaney and Z. H. Nitecki (Dekker, New York).
Spitzer, L. 1975, in *Dynamics of Stellar Systems*, IAU Symp. 69, ed. A. Hayli (Dordrecht: Reidel), p. 3.
Spitzer, L. and Mathieu, R. D. 1980, *Astrophys. J.* 241, p. 618.
Stodółkiewicz, J. S. 1982, *Acta Astron.* 32, p. 63.
Sugimoto, D. and Bettwieser, E. 1983, *Mon. Not. R. astr. Soc.* 204, 19p.
Valtonen, M. J. 1975, *Mem. R. astr. Soc.* 80, 61; p. 77.
Valtonen, M. J., and Aarseth, S. J. 1977, *Rev. Mex. Astron. Astrophys.* 3, p. 163.
Valtonen, M. J., and Heggie, D. C. 1979, *Celest. Mech.* 19, p. 53.

THE CRITICAL PERIODIC ORBITS IN THE STÖRMER PROBLEM

R.A. Broucke

University of Texas, Austin, TX, 78712, USA

Abstract

We study some of the periodic orbits, among the families f_0, f_1, f_2, f_3, f_4 and f_5 that were previously found by Goudas and Markellos. We concentrate on the periodic orbits which have a special value (= +2, -2, -1 or 0) of the stability index, because, for these orbits, the period of the periodic solutions of the variational equations is an integer multiple (1, 2, 3 or 4) of the original period. We classify several such solutions according to the Hénon type or the Contopoulos Resonant type. We also computed several bifurcations or trifurcations of new families out of these critical orbits, in order to illustrate the extreme complexity of the phase space, even in the quasi-periodic regions.

1. The Störmer Problem

The problem of the motion of a charged particle in the magnetic field of a dipole has been studied by C. Störmer (1907) since the turn of the century, especially because of its interest in relation to the Northern Light. From the 1930's until well in the fifties, these studies have been continued by G. Lemaitre and his group at Louvain, because of the connec-

A. Berger (ed.), The Big Bang and Georges Lemaitre, 257–267.

tions of the problem with cosmic rays, the eventual remnants of the big Bang, (Lemaitre, 1934, de Vogelaere, 1958, Bouckaert, 1934 and Godart, 1938).

The interest in the problem has very much remained alive in the last two decades especially in the Greek schools with C. Goudas and V. Markellos (1976). The aspect that has been studied in most detail is the classification of the periodic orbits in the rotating meridian plane. Some detailed mathematical studies of the problem have also been published in the last few years (Braun, 1970, 1970, 1979, 1981) (Dragt, 1935). The problem can be considered a typical non-integrable dynamical system with two degrees of freedom. We use it here as a model to illustrate the complex structure even in the regions of the stable motions. We show that many complex bifurcation phenomena actually occur.

2. The Stability of the Periodic Orbits

In the present work, we concentrate on the stability properties of the periodic orbits of the problem. We especially study the critical periodic orbits which are at the transition between stability and instability and which are at the origin of the bifurcations between families of periodic orbits as well as new families of periodic orbits.

In order to determine the stability we used two difference methods: the standard variational equations in rectangular coordinates, leading to a 4 by 4-monodromy matrix R, and the Hill method with two normal variations, leading to a 2 by 2-matrix H, called Hill or Hénon matrix, (Hénon 1965). The trace of Hill matrix is what we call the stability index k, the sum of its two eigenvalues, (Deprit and Henrard, 1967):

$$k = \lambda + \lambda^{-1}$$

If the monodromy matrix R is used, there are two additional unit eigenvalues, so that the stability index is obtained by, (Deprit and Price, 1965):

$$k = \lambda + \lambda^{-1} = \text{Trace}(R)-2.$$

The interval $-2<k<+2$ corresponds to stability. This is the case with two eigenvalues on the unit-circle $\{\lambda,\lambda^{-1}\} = \cos\beta \pm i\sin\beta$. The exceptional cases occur when $k = +2$ or -2 or, more generally when $\beta = 2\pi/n$, for n integer. The integers n = 1, 2, 3, 4, which are the only ones that we studied, corresponding respectively to k = +2, -2, -1 and 0. They correspond to eventual bifurcations to new periodic orbits with the same, double, triple or quadruple period:

Several classifications of the exceptional periodic orbits have been presented in the litterature, especially by Hénon and Contopoulos. Hénon has four types of periodic orbits with stability index k = 2 (Hénon 1965) =

Type One: Extremum of Energy; no bifurcation.

Type Two: Bifurcation between two symmetric families of periodic orbits, but no extremum of the energy.

Type Three: Bifurcation between two symmetric families; extremum of the energy on one family.

Type Four: Bifurcation between a symmetric and a non-symmetric family of orbits.

Contopoulos has classified some additional types: his five resonant types, (Contopoulos 1970). These classifications are essentially based on the eigenvalues, the rank and the Jordan form of the matrices H and R or R-I. More precisely, the following statements may be made in relation to these classifications:

- The rank of (R-I) is usually only 3, due to the fact that R has a (double) eigenvalue.

- The Rank of (R-I) is 2 on the critical periodic orbits with k = +2, where bifurcations between families may occur, (these are the 4 Hénon types).

- The Rank of (R-I) may be 1 on some orbits that we call super-critical. These are the five Contopoulos resonant types. Bifurcations and trifurcations between 2 or 3 families of periodic orbits are possible. In the case of trifurcations, one of the families always consists of non-symmetric periodic orbits. We will now describe different examples of these special cases of exceptional periodic orbits.

3. Stability of the Six Basic Families

The periodic orbits of the six basic families $f_0, f_1, f_2, f_3, f_4, f_5$ have many properties in common. First of all, the families f_o, f_2, and f_4 contain all open orbits: they have a zero velocity point at each end and they cross the equator at a right angle. The other 3 families f_1, f_3, f_5 are branches out of the three previous families, but they contain only closed orbits, without any zero-velocity points. We computed many members of each of the six families, in order to establish an accurate stability diagram (x_o, k) for each one of them. It turns out that the stability behavior for all the open orbits f_o, f_2, and f_4 is fairly similar. Also, the stability behavior of the closed orbits f_1, f_3, f_5 is very similar for each of the three families. We reproduce below the sketch of the stability diagram for the families f_o and f_1 only (the four others being similar).

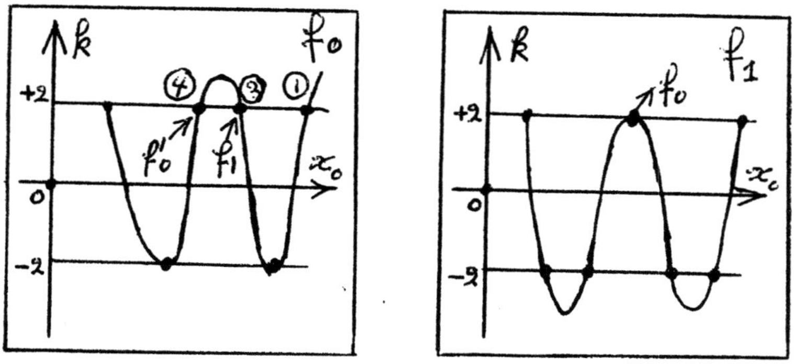

Some of the important features visible on these two stability diagrams are as follows. As for f_o there are two tangent points to the stability limit k = -2. There are also four important intersection points with the other stability limit +2. These 4 points k=+2 have the following meaning, (from left to right on the x_o - axis):

- The first point is the beginning of the family.

- The next point has the type 4, in the Hénon classification and is at the origin of a new family of non-symmetric periodic orbits. We call this family f'_0 (or f'_2, f'_4) and we give a short description below.

- The next point is of the Hénon type 3 and is at the origin of the family f_1, of closed symmetric periodic orbits.

- The next point k = +2, is of the Hénon type 1. It corresponds to the maximum Energy point of the family and there is no bifurcation with another new family. Again, everything that was just said for the families f_0, f'_0, f_1 also holds for f_2, f'_2, f_3 and f_4, f'_4, f_5 . We will now describe these features in more detail.

4. The Three Maximum Energy Periodic Orbits

We mention here the first of the important common characteristics of the three families f_o, f_2 and f_4 : it is that each one of these families has a Maximum Energy Orbit which is a critical periodic orbit with k=+2 and Rank (R-I)=2. In the Hénon classification this is the type one.

We now give the summary of the most important numerical data of these three periodic orbits.

	f_o	f_2	f_4
x_o =	1.11261769	1.18703859	1.19723845
$\dot{y}_o$ =	0.39162092	0.24839729	0.20534800
E =	0.0808216	0.0396606	0.0305513
T/2 =	4.738	8.703	10.451
H =	$\begin{bmatrix} 1 & 12.18 \\ 0 & 1 \end{bmatrix}$	$\begin{bmatrix} 1 & 46.00 \\ 0 & 1 \end{bmatrix}$	$\begin{bmatrix} 1 & 75.87 \\ 0 & 1 \end{bmatrix}$

5. The Type-3 Orbit of f_0 and the Bifurcation to f_1

As we have said before, each of the three families f_o, f_2 and f_4 of periodic orbits has a special member where the stability index k is +2, the Hénon type is 3, the Hénon matrix H has a zero element (H_{21}=c=0) and two units on the diagonal. Also for these orbits, the rank of R-I (where R is the monodromy matrix) is +2. In fact R has a quadruple unit eigenvalue. These points are at the origin of the new symmetric families of closed periodic orbits, f_1, f_3, f_5. We give below the numerical data for all three critical periodic orbits.

	f_o-f_1	f_2-f_3	f_4-f_5
x_o =	0.993861565	1.13598137	1.1697032
$\dot{y}_o$ =	0.377117485	0.259065122	0.2130875
E =	0.07112811	0.0391093	0.0303953
T/2 =	4.33380	8.28603	10.2052
H =	$\begin{bmatrix} 1 & -6.66 \\ 0 & 1 \end{bmatrix}$	$\begin{bmatrix} 1 & -36.17 \\ 0 & 1 \end{bmatrix}$	$\begin{bmatrix} 1 & -68.42 \\ 0 & 1 \end{bmatrix}$

We finally mention another property which is common to all three bifurcation points: At this point, one of the families (f_1, f_3 or f_5) has an extremum of the Energy Constant. In fact this extremum is a maximum in all three cases. On the other three families, the energy is not a extremum at this point.

6. Bifurcations to Non-Symmetric Periodic Orbits

We will now describe a few examples of bifurcation points corresponding to branches with new families of non-symmetric periodic orbits. These non-symmetric families can and do arise in two different situations.

- At the stability points k = +2 with the Hénon type = +4, there is normally a branch of non-symmetric periodic orbits. At these points, the Rank of the matrix (R-I) is +2. We describe three examples below.

- At the stability points with k = -2, -1, 0 (=2 cos 2 π /n, where n is integer), where the matrix R^n-I is of rank +1, we can have a trifurcation rather than a bifurcation: two new failies of periodic orbits are born: one of them is symmetric and the other is non-symmetric. We will describe two examples later. We call these bifurcation or trifurcation points with a rank = 1, supercritical orbits.

a. The Three Non-Symmetric Families f'_0, f'_2, f'_4

The three basic families f_o, f_2, f_4 each have a member with stability index k = +2 and a Hénon type of 4. These are thus each at the origin of a family of a non-symmetric periodic solutions with the same period, at the bifurcation point.

We begin by giving the initial conditions of the three. symmetric periodic orbits of Type 4, (with $y_o = \dot{x}_o = 0$)

	f_o	f_2	f_4
x_o =	0.890576	0.9301048	0.94808426
$\dot{y}_o$ =	0.132572	0.1243051	0.107710767
E =	0.01830494	0.01098977	0.074687445
T/2	6.283445	10.723763	14.1962995
H =	$\begin{bmatrix} 1 & 0 \\ -11 & 1 \end{bmatrix}$	$\begin{bmatrix} 1 & 0 \\ -28 & 1 \end{bmatrix}$	$\begin{bmatrix} 1 & 0 \\ -40 & 1 \end{bmatrix}$

On each one of the three non-symmetric families, about 20 periodic orbits have been computed. The table below summarizes the initial conditions of one member of each of the families, with $y_o = 0$:

	f'_0	f'_2	f'_4
x_o =	0.97769140	0.97615105	0.979279
$\dot{x}_o$ =	0.17425781	0.105692	0.073453
$\dot{y}_o$ =	0.11057837	0.11479595	0.102234
E =	0.02156902	0.01248059	0.008157
T =	13.233376	21.459386	28.015419
k =	-5.8391	-3.7433	-2.5549

The three non-symmetric families f'_o, f'_2, f'_4 have completely similar stability characteristics: all three families are stable, at last they start of with stable orbits. The stability index k begins at 2 and decreases. It eventually crosses the value -2 and the orbits become then odd-unstable.

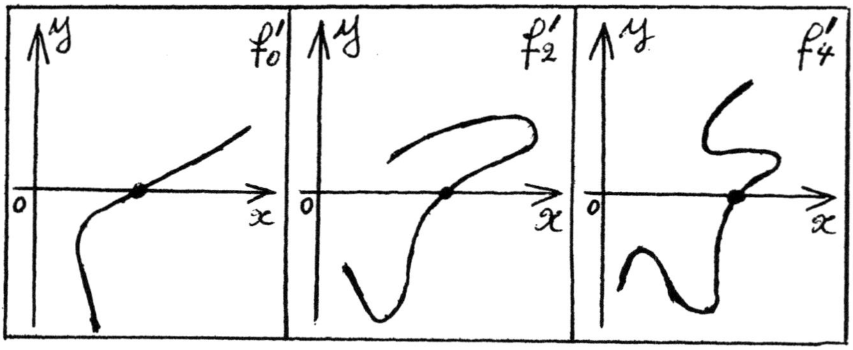

b. A Triple Period Non-Symmetric Periodic Orbit

This is a family of Non-Symmetric Periodic Orbits that begins at a supercritical symmetric periodic orbit of the basic family f_o, at the value -1 of the stability index k and the argument $\beta = 120^o$. The initial conditions of this symmetric orbit are

$$(1.0203243, 0, 0, 0.38825176)$$
$$E = 0.07556028 \; ; \; T/2 = 4.370994.$$

The Hill or Hénon matrix of this periodic orbit is

$$H = \begin{bmatrix} -0.5 & -3.9711 \\ 0.18886 & -0.5 \end{bmatrix}, \quad H^3 = \begin{bmatrix} 1 & 0 \\ 0 & 1 \end{bmatrix}.$$

The new non-symmetric family will begin at this bifurcation point with three times the above period. One of the members has the initial conditions:

$$(1.02328 \, , 0, .0008, 0.388369)$$
$$E = 0.075663 \, , \; T = 26.238.$$

The Family is unstable

c. A Quadruple Period Non-Symmetric Periodic Orbits

This family of non-symmetric periodic orbits originates at a supercritical periodic orbit of faily f_o with stability index k = 0 and argument $\beta = 90^o$ with initial conditions:

$$(1.04853119, 0, 0, 0.3938845),$$
$$E = 0.078546796; \; T/2 = 4.4522.$$

The Hill or Hénon matrix of this periodic orbit is of the form

$$H = \begin{bmatrix} 0 & -1/\alpha \\ \alpha & 0 \end{bmatrix} ; \; H^4 = \begin{bmatrix} 1 & 0 \\ 0 & 1 \end{bmatrix}.$$

The new non-symmetric family begins thus at this point with a quadruple period. One of the members of this non-symmetric family has the initial conditions

(1.0486579, 0, 0.003, 0.39386817)
E = 0.07854502 ; T/2 = 17.808.

The family is stable, with a stability index decreasing from the value +2, but tangent to the horizontal line k = +2 in the stability diagram.

References

Bouckaert, L. 1934, "II. Trajectoires Voisines de l'Equateur; dans: Contributions à la Theorie des Effets de Latitude et d'Asymetrie des Rayons Cosmiques," pp. 174-193, Vol. 54, Ann. Soc. Sc. de Brux. A

Braun, M. 1970, "Structural Stability and the Stormer Problem," Indiana Univ. Math. J. Vol 20, pp. 469-497.

Braun, M. 1970, "Particle Motions in a Magnetic Field," J. Diff. Eq. Vol. 8, pp. 294-232.

Braun, M. 1979, "On the Stability of the Van Allen Radiation Belt," Siam J. Appl. Math. Vol. 37, pp. 464-468.

Braun, M. 1981, "Mathematical Remarks on the Van Allen Belt: A Survey of old and New Results," SIAM Review, Vol. 23, No 1, pp. 61-93.

Contopoulos, G. 1970, "Orbits in Highly Perturbed Dynamical Systems. Part 2: Stability of Periodic Orbits," Astron. Journal, Vol. 75, pp. 108-130.

Deprit, A. and J. Henrard. 1967, "Natural Families of Periodic Orbits," Astron. Journal, Vol. 72, Nr2, pp. 158-172.

Deprit A. and J. Price. 1965, "The Computation of Characteristic Exponents in the Planar Restricted Problem of Three Bodies," Astron. Journal, Vol. 70, pp. 836-846.

deVogelaere, R. 1958, "On the Structure of Symmetric Periodic Solutions of Conservative Systems," In: S. Lefschetz, Contributions to the Theory of Non-Linear Oscillations, Princeton, Vol. 4, pp. 53-84.

Dragt, A.J. 1965, "Trapped Orbits in a Magnetic Dipole Field," Rev. Geophys. Vol. 3, pp. 255-298.

Goudas, C.L., A.A. Halioulias, V.V. Markellos and G. Macris. 1976, "Periodic Oscillations of the First Generation in the Störmer Problem," In: Szebehely and Tapley: Long-Time Pred. in Dynamics, pp. 276, Reidel (Cortina Nato Inst. 1975).

Godart, O. 1938, "Determination des Exposants Characteristiques des Trajectoires Periodiques," dans: Contributions a la Theorie des Effets de Latitude et d'Asymetrie des Rayons Cosmiques, Ann. de la Soc. de Br. Vol. 58, pp. 27-41.

Hénon, H. 1965, "Exploration Numerique du Problème Restreint," Part 2, Annales d'Astrophysique, Vol. 28, No. 6, pp. 992-1007.

Lemaitre, G. 1934, "Trajectoires Infiniment Voisines de l'Equateur," Dans: Contributions à la Theorie, Ann. Soc. Sc. de Brux. A, Vol. 54, pp. 162-174.

Störmer, B. 1907, Arch. Sc. Phys. Nat. Vol. 24, pp.113.

A MULTIMODE INVESTIGATION OF GRANULAR AND SUPERGRANULAR MOTIONS I: BOUSSINESQ MODEL

R. Van der Borght and P. Fox

Monash University, Clayton, Victoria, 3168, Australia

Two types of large-scale convective motions, granulation and supergranulation, are observed in the outer layers of the sun and their observed characteristics can be summarized as follows:

Characteristic	Granulation	Supergranulation
Average Diameter	2000 km	30,000 km
Horizontal and Vertical Velocities	1 km/sec	0.3 - 0.5 km/sec
Average Life-Time	20 minutes	20 hours
Intensity Fluctuation	15%	not observed

These convective motions take place in a highly turbulent medium, and in the outermost layers of the sun the eddy thermal diffusivity and eddy kinematic viscosity are well in excess of their radiative and molecular values. The thermal diffusivity is evaluated usually by the mixing-length theory or its extensions, whereas there does not seem to exist an adequate theory to determine, a priori, the variations with depth of the eddy viscosity.

The accuracy of the methods used to derive the eddy thermal diffusivity and kinematic viscosity could be tested if it were possible to compare the observed characteristics of granulation and supergranulation with the theoretical ones derived from a model of deep convection in a compressible medium. Some progress has been made in this direction over recent years [1,2] with the help of the one-mode approximation.

A. Berger (ed.), The Big Bang and Georges Lemaitre, 269–276.

So far, most attempts have concentrated on granulation or supergranulation in isolation although a detailed theory should take into account the co-existence of these two types of convective motion. Such an approach requires the use of multi-mode expansions and in a highly compressible medium, such as the sun, one should consider multi-mode expansions with as few simplifying assumptions as possible. The full equations have in fact been derived but, in this preliminary investigation, we shall confine ourselves to a Boussinesq model to gain an insight into the behaviour of interacting convective motions.

The appropriate equations, within the multimode-approximation are given below and have already been integrated [3] for certain combinations of modes. Depending on the number of modes and the value of the Rayleigh number the motions have been found to be steady, periodic or aperiodic. These computations were carried out for a Rayleigh number which is not depth dependent, i.e. for the case of constant buoyancy across the layer. In the case of the sun however, the uppermost layers have a temperature gradient which is highly superadiabatic, whereas in the deeper regions the temperature gradient is only slightly superadiabatic.

In what follows we shall assume that the Rayleigh number R varies with depth and is well in excess of its linear value R_ℓ in the top 10% of the convective layer but is almost equal to its linear value in the remaining 90%.

Within the framework of the Boussinesq approximation, the basic hydrodynamic equations to be solved are as follows

$$\operatorname{div} \underset{\sim}{u} = 0 \tag{1}$$

$$\frac{\partial \underset{\sim}{u}}{\partial t} + \underset{\sim}{u}.\operatorname{grad} \underset{\sim}{u} + \frac{1}{\rho_0} \operatorname{grad} p + \underset{\sim}{g} - \nu\nabla^2 \underset{\sim}{u} = 0 \tag{2}$$

$$\frac{\partial T}{\partial t} + \underset{\sim}{u}.\operatorname{grad} T - \kappa\nabla^2 T = 0 \tag{3}$$

where we have assumed that the kinematic viscosity ν and thermal diffusivity κ are constant and the effects of viscous dissipation can be neglected.

The continuity equation (1) will be automatically satisfied if we assume that the velocity $\underset{\sim}{u}$ can be expressed in the following form

$$\underset{\sim}{u} = \left(\sum_i \frac{DW_i}{a_i^2} \frac{\partial f_i}{\partial x}, \sum_i \frac{DW_i}{a_i^2} \frac{\partial f_i}{\partial y}, \sum_i W_i f_i \right) \tag{4}$$

where the $W_i(z,t)$ are functions to be determined, the a_i are horizontal wave numbers, the functions $f_i(x,y)$ satisfy the following differential equations

$$\frac{\partial^2 f_i}{\partial x^2} + \frac{\partial^2 f_i}{\partial y^2} = - a_i{}^2 f_i \tag{5}$$

and $D \equiv \frac{\partial}{\partial z}$.

For instance, in the case of convective cells with hexagonal planform the functions $f_i(x,y)$ take the following form

$$f_i(x,y) = \left(\frac{2}{3}\right)^{\frac{1}{2}}\left[\cos a_i y + \cos a_i\left(\frac{\sqrt{3}}{2}x + \frac{y}{2}\right) + \cos a_i\left(\frac{\sqrt{3}}{2}x - \frac{y}{2}\right)\right] \tag{6}$$

Similarly, for the temperature profile we adopt the following expression

$$T = T_0(z,t) + \sum_i F_i(z,t) f_i(x,y) \tag{7}$$

A process of horizontal averaging yields the following modal equations

$$\left< \frac{\partial T}{\partial t} + \underset{\sim}{u}.\text{grad } T - \kappa\nabla^2 T \right> = 0 \tag{8}$$

$$\left< f_i\left[\frac{\partial T}{\partial t} + \underset{\sim}{u}.\text{grad } T - \kappa\nabla^2 T\right]\right> = 0 \tag{9}$$

$$- \frac{1}{a_i{}^2} D \left< \frac{\partial f_i}{\partial x} M_x + \frac{\partial f_i}{\partial y} M_y \right> + \left< f_i M_z \right> = 0 \tag{10}$$

where M_x, M_y and M_z are the components of the equation of motion,

$$\left< \quad \right> = A \iint \left[\quad \right] dxdy \tag{11}$$

and the constant A is chosen in such a way that

$$\left< f^2 \right> = 1 \tag{12}$$

A lengthy, but otherwise straightforward averaging procedure leads to the following system of partial differential equations in one space variable [3].

$$\frac{1}{\sigma}(D^2 - a_i{}^2)\left(\frac{\partial W_i}{\partial t}\right) + \frac{1}{\sigma}\sum_{j,k}\frac{C_{ijk}}{(a_j a_k)^2}\Big\{a_{kij} W_k (D^2 - a_j{}^2) DW_j + (a_{kij} + a_{ijk}) DW_k (D^2 - a_j{}^2) W_j\Big\} = - Ra_i{}^2 F_i + (D^2 - a_i{}^2)^2 W_i \tag{13}$$

$$\frac{\partial F_i}{\partial t} + \sum_{j,k} \frac{C_{ijk}}{(a_i a_k)^2} \left\{ a_{ijk} F_j DW_k + 2a_i{}^2 a_k{}^2 W_k DF_j \right\} = - W_i DT_0 + (D^2 - a_i{}^2) F_i \quad (14)$$

$$\frac{\partial T_0}{\partial t} + \sum_j D(W_j F_j) = D^2 T_0 \quad (15)$$

where

$$a_{ijk} = a_i{}^2(a_j{}^2 + a_k{}^2 - a_i{}^2) \quad (16)$$

and

$$C_{ijk} = \tfrac{1}{2} \langle f_i f_j f_k \rangle \quad (17)$$

Assuming that granular and supergranular motions extend over the same depth we require at least a two-mode approximation, one with wave number a_1 of order unity to represent supergranulation and another with a much larger wave number a_2 to represent granulation which has a much smaller horizontal extent and would be represented by much more elongated convection cells.

Other investigations [3] show that the most appropriate model is in fact a three-mode one with horizontal wave numbers a_1, $a_2 = na_1$ and $a_3 = (n + 1)a_1$ for which it can be shown that the constants C_{ijk} vanish unless i, j and k are either all equal, for which $C_{ijk} = C = 1/\sqrt{6}$, or all different, in which case $C_{ijk} = C/2$.

Since we are trying to model both granulation and supergranulation we shall assume that the horizontal wave number a_1, corresponding to supergranulation, has a value of $\pi/\sqrt{2}$ as suggested by the linear theory for maximum instability. This corresponds to an aspect ratio of 3.77 for supergranular motion, i.e. to a depth of 7958 km of the convective layer.

Granules on the other hand have an average horizontal extent of 2000 km as compared to 30,000 km for supergranules. The corresponding value a_2 of the wave number is therefore much larger and $n = a_2/a_1 = 15$. This is the value adopted for the numerical integrations.

The eddy thermal diffusivity κ varies from a value of $1.66\ 10^{14}$ at the top of the solar convective region to a value of $8.03\ 10^{10}$ at a depth of 8000 km. Since we are interested mainly in the behaviour of the granules we shall adopt here a representative value of $\kappa = 5 \times 10^{13}$ for the uppermost layers. We have also adopted for the Prandtl number σ a value of 0.1 as indicated by earlier investigations [2].

The buoyancy in the uppermost layers of the sun's convective region is several orders of magnitudes larger than in the deeper regions. For the purpose of the present numerical calculations we have adopted the following values of R

$$R = 5.0\ 10^6 \quad \text{when} \quad 0.9 \leqslant z \leqslant 1$$

$$R = 650 \quad \text{when} \quad 0 \leqslant z \leqslant 0.9$$

In addition, the following boundary conditions have been adopted:

z = 0	z = 1	
$T_0 = 0$	$T_0 = -1$	uniform temperature at the boundaries
$F_i = 0$	$F_i = 0$	
$W_i = 0$	$W_i = 0$	no overshooting
$D^2W_i = 0$	$D^2W_i = 0$	free boundaries

The results of the numerical integrations are given in the following figures and the main characteristics of the flow can be summarized as follows:

In the initial stages the vertical velocity W and the root-mean-square horizontal velocity $V = DW/a$ are considerably larger for modes a_2 and a_3, associated with granulation, than for mode a_1 which represents supergranulation. Ultimately the granular modes die out and the energy is transported mainly by super-granulation motions (Figures 1a and 1b). When numerical integrations are carried out even further oscillatory motions set in.

In the sun, a particular granule has a short life-span of the order of 20 minutes on average. In the present model the granular motion reaches vertical and horizontal velocities of the order of 1 km/sec after an elapsed time-span of 27 minutes.

It is seen, in Figure 1c, that the flux modulation $I = 3.2\ DF/DT_0$ reaches a value of 10% after a time-span of 27 minutes. The observed intensity modulation in granules has an average value of 12 to 15%.

At the same epoch the flux modulation in the supergranules is only of the order of 1% and may not be large enough to be observed.

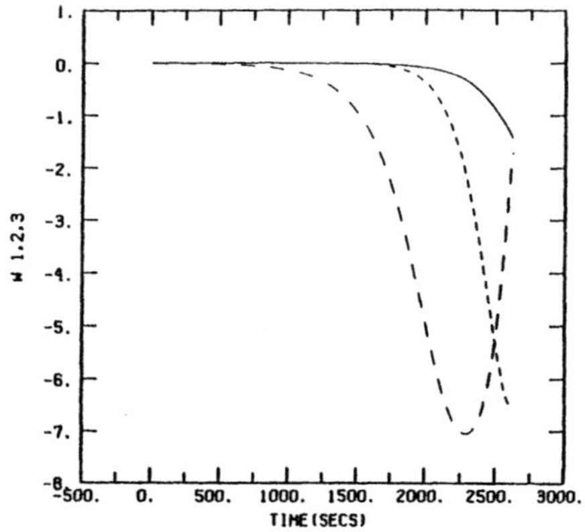

Fig. 1a

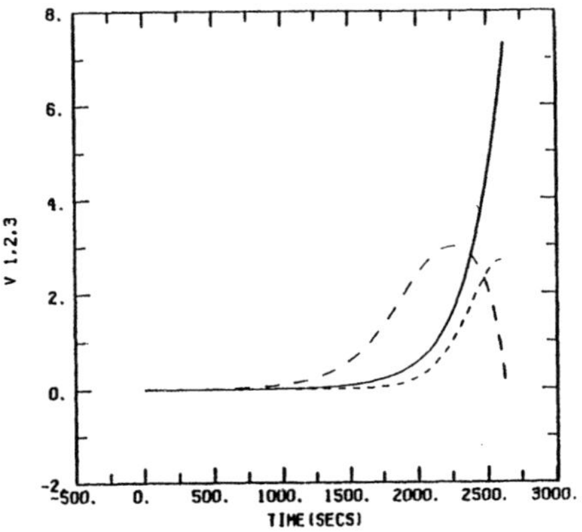

Fig. 1b

Time variation of the vertical velocity W_i and the root-mean-square horizontal velocity V_i, in km/sec, at the top of the convective layer for modes 1 (————), 2(— — — —) and 3 (- - - - - -).

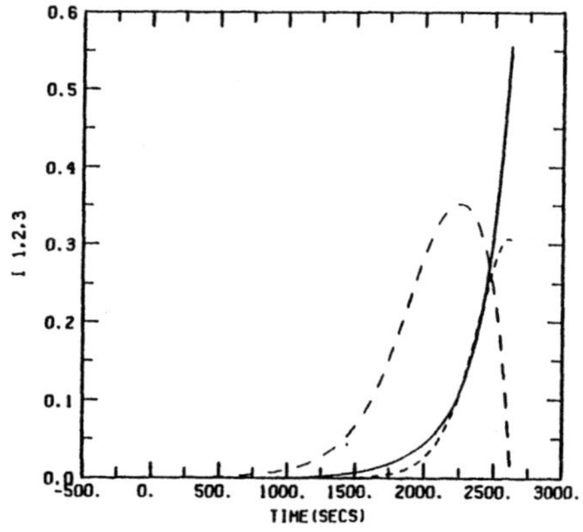

Fig. 1c

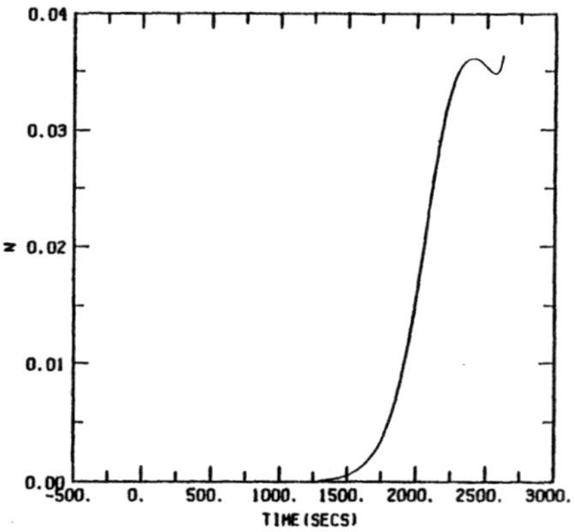

Fig. 1d

Time variation of the flux modulation at the surface I_i, for modes 1 (————), 2 (— — — —) and 3 (- - - - - -), and the fraction of energy N carried by the convective motions.

In addition (Figure 1d) the flux carried by the convective motions is only a fraction of one percent. One-mode compressible investigations point to a value of 4% [2]. In any case it appears that large-scale convective motions do not modify significantly the transport equation and the energy transport in the outer layers of the sun seems to be influenced mainly by turbulent motions.

The distribution of vertical velocities with depth as a function of time is illustrated in Figures 2a and 2b. We see that the granular motions overshoot significantly into the marginally stable layer and occupy, at maximum intensity, 20% of the entire layer as compared to 10% for the unstable layer.

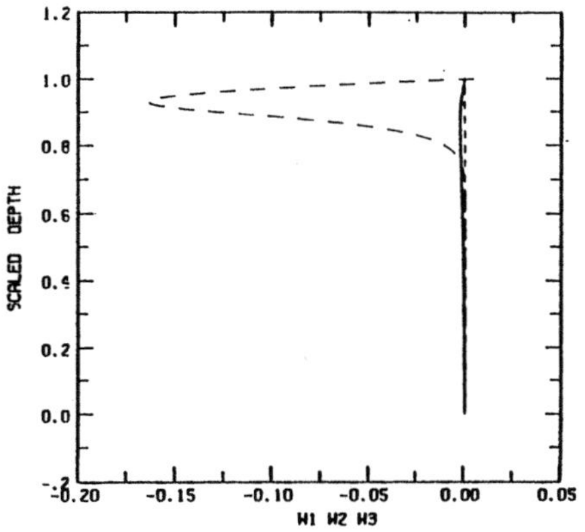

Fig. 2a

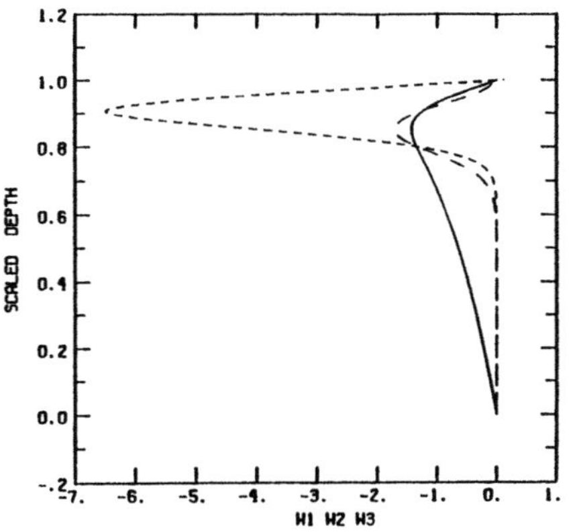

Fig. 2b

Depth variation of the maximum vertical velocity W_i, in km/sec, for modes 1 (———), 2 (— — — —) and 3(- - - - - -) at two different times; t = 1021 secs (Fig. 2a) and t = 2620 secs (Fig. 2b). Note that level z = 1 represents the surface.

This preliminary investigation has shown how the multi-mode expansions can be used to study convective motions in a medium with depth dependent buoyancy. The model used to illustrate this technique is based on the Boussinesq approximation and it is shown that it is possible to obtain numerical results close to the observed characteristic of granulation and supergranulation for values of the parameters, such as eddy conductivity, Prandtl number and buoyancy which lie within the range of generally accepted values.

In order to use calculations of this type as a diagnostic test to decide on the validity of proposed models of the solar convection zone it will be necessary to extend the present work to a full compressible model which takes into account the depth dependence of the various parameters. Such work is now in progress and it is hoped that a report on this work will be available in the near future.

References

[1] Van der Borght, R. 1977, Proc. Astr. Soc. Aust., 3, 91.
[2] Van der Borght, R. and Fox, P. 1983, Proc. Astr. Soc. Aust., 5 (to appear).
[3] Toomre, J., Gough, D.O. and Spiegel, E.A. 1982, J. Fluid Mech., 125, 99.

Part III
Structure of the Universe and Cosmic Rays

GALAXY FORMATION REVISITED

Joseph Silk

Department of Astronomy
University of California, Berkeley
California 94720, U.S.A.

The statistical properties of galaxies are used to infer the distribution of specific binding energy with surface density for a range of Hubble types. It is argued that the inferred locus characterizes physical conditions at the end of the dissipative phase of protogalactic evolution. Cosmological fluctuations in density provide the initial conditions at the onset of galaxy formation, and the critical surface density for radiative cooling to have occurred in protogalactic gas clouds provides a necessary lower bound for star formation. If gas is supported in, and eventually driven from, galactic potential wells by energy input from forming and dying stars, many of the statistical correlations found for galaxies can be understood, including those involving luminosity, effective radius, velocity dispersion or maximum rotation velocity, and metallicity.

I. INTRODUCTION

While the universe is statistically homogeneous and isotropic, the presence of the galaxies is essential if the big bang theory, based on the cosmological principle, is to truly confront the real universe. Georges Lemaître was very much aware of this lack in the theory that he and Alexander Friedmann independently discovered, and he developed the first inhomogeneous model of the expanding universe. It is therefore appropriate in this symposium to return to the challenge of galaxy formation, and ask how theory has fared in the past half-century.

In this undertaking, I will take a narrow view of galaxy

A. Berger (ed.), The Big Bang and Georges Lemaitre, 279–297.

formation, and consider the formation of the galaxies themselves but not the evolution of large-scale structure, insofar as this involves the clustering of pre-existing galaxies. Beautiful computer simulations have been made of the development of large-scale clustering, where only gravitational forces are relevant for describing the motion of discrete points that are identified with galaxies. Unfortunately, galaxy formation necessarily involves complex and inadequately understood physical processes involving hydrodynamic dissipation and star formation. It is for this reason that our understanding of galaxy formation is far from complete.

My approach will be twofold. First I will describe what big bang cosmology predicts for the conditions at the onset of galaxy formation. There are constraints, in that one does not have total freedom, but the available options are numerous. One would be hard-pressed to predict that galaxies had to form if one did not already observe them. This is a well-trodden path, and my review will be a cursory one. The reader is directed to a recent comprehensive review (Efstathiou and Silk 1983) for full references and further details.

It is the second approach that will be emphasized here. Galaxies are slowly evolving systems, and one can study them as one would a fossil record to unveil their properties immediately after the epoch of formation. Cosmology provides the initial conditions, while observations of the galaxies around us demarcate the endpoint of formation. Can we infer the evolutionary pathway that links these two phases? That is the goal of the first section of this article.

II. COSMOLOGICAL ORIGINS

Particle physicists have hastened to plough the fertile field of big bang theory, as the very early universe provides a unique environment where particle energies attain high enough energies to probe schemes of grand unification. The immense extrapolation back in time from the earliest epoch even indirectly observable, namely the epoch of primordial nucleosynthesis, has seemed a glorious fairytale to hard-headed astronomers, and inevitable to many cosmologists. Lemaître himself devoted much effort to concocting a scheme that ultimately failed to avoid the past singularity. We know now that the singularity was inevitable under rather general conditions, and this has set the scene for the extremes of energy that occur toward the Planck time at 10^{-43} sec, when

$$kT \sim 10^{19} \text{ GeV}.$$

Of prime interest for galaxy formation has been the prediction of the amplitude and spectrum of density fluctuations which arise at the epoch of grand unification. Two complementary schemes have emerged. One involves a slow roll-over at symmetry breaking from the false vacuum of grand unification to the true, asymmetric vacuum state. The slow roll-over allows a phase transition in which while the energy of the false vacuum dominates, the universe enters a de Sitter phase. Once the transition is completed, the Friedmann-Lemaître phase resumes. This novel beginning solves a number of puzzles in the standard big bang, possibly including that of the origin of fluctuations. Quantum fluctuations in the de Sitter phase are amplified to the same level on all scales when they enter the particle horizon of the Friedmann-Lemaître universe--as the phase transition is completed. While a scale-invariant fluctuation spectrum is predicted, the amplitude is uncertain. Excessively large amplitudes ($\delta\rho/\rho \gg 1$) are predicted unless the potential that determines the phase transition is carefully fine-tuned; alternatively, supersymmetry may come to rescue the situation. This latter approach introduces a second scale, m_{pl}, into the supersymmetric potential in addition to m_{GUT}, and predicts $\delta\rho/\rho \sim (m_{GUT}/m_{pl})^2 \sim 10^{-4}$, just what is required to form large-scale structure.

A second scheme invokes strings. These topological survivors of grand unification symmetry are expected to be generic to unification groups larger than SU(5), and proton decay limits have now effectively ruled out SU(5). Inflation would dilute the string density to an uninteresting level, but if inflation did not occur, one would be left with a universe now containing many strings. Other considerations suggest that string-like effects may also develop much later, at kT ~ 1GeV. Simple estimates of string production rates (essentially one per horizon volume) and decay rates (of order 10^5-10^8 expansion time-scales due to gravitational radiation) enable the spectrum of baryonic density fluctuations that are generated by the strings to be computed.

In either scheme, baryosynthesis follows grand unification symmetry breaking, and one therefore ends up with isentropic or adiabatic fluctuations, predicted to have a scale-invariant spectrum, where the density fluctuation is measured when a given scale first enters the particle horizon. Observational constraints from the large-scale isotropy of the cosmic microwave background, combined with the need to form galaxies, fix $\delta\rho/\rho \sim 10^{-4}$. From the particle physicist's perspective, this input allows him to specify m_{GUT}. Little if any further consequence occurs until the universe first becomes dominated by non-relativistic matter at redshift z_{eq}. The scale-invariance now breaks: unless massive neutrinos dominate the universe, fluctuations of scale $\gtrsim ct_{eq}$ undergo growth within the horizon, while smaller-scale fluctuations do not. Actually, the uninterrupted growth guarantees constant curvature. For the smaller-scale fluctuations, the associated curvature perturbations are decreased, essentially because they were radiation-dominated at horizon crossing, and the radiation redshifted away. The asymptotic limits for a scale invariant spectrum are curvature fluctuation δK=constant, $\delta\rho/\rho \propto L^{-2}$ for comoving wavelength $L \gg ct_{eq}$ and $\delta K \propto L^2$, $\delta\rho/\rho$=constant for $L \ll ct_{eq}$, provided that damping of baryons by radiation diffusion or of collisionless particles by free streaming is unimportant on small scales in affecting $\delta\rho/\rho$. These latter effects simply impose a short-scale cut-off to the fluctuation spectrum. If massive neutrinos dominate the present universe and Ω=1, say, the horizon scale at the relevant epoch z_{nr} when the neutrinos first become non-relativistic also fixes the maximum scale over which free streaming erases density fluctuations. The only surviving fluctuations have scale $\gtrsim ct_{nr}$, and reflect the initial fluctuation spectrum, with an exponential cut-off on smaller scales.

The only scale-invariant spectrum of fluctuations in accord with observational constraints from the galaxy distribution, galaxy peculiar velocities, and the microwave background isotropy is one with a coherence length substantially smaller than ct_{eq}. In fact, it must be smaller than or comparable to the galaxy clustering scale, $\sim 5h^{-1}$ Mpc. The limiting case of a scale-invariant spectrum with zero initial coherence length is especially simple, and arises if cold collisionless particles, which decoupled non-relativistically, dominate the universe. It is also produced by very massive collisionless particles, such as GeV photinos, or by primordial black holes. Such a spectrum leads to a hierarchical model of galaxy formation: small scales corresponding to the Jeans mass after decoupling ($\lesssim 10^6$ M$\odot$) form first, and aggregate into larger and larger systems.

On the other hand, if one drops the assumption of a scale-invariant spectrum, massive neutrinos satisfy most, if not all, constraints, and lead to a pancake theory of galaxy formation in which large scales ($M_\nu \gtrsim 10^{15}$ M$\odot$, the horizon mass at t_{nr}) collapse and fragment to form galaxies.

We shall make use of these two alternative scenarios for the emerging fluctuation spectrum in a later section. No details of how anything resembling the observed galaxies can be directly inferred from these considerations. Hence in order to ascertain how the fluctuation spectrum, once it becomes nonlinear, transforms itself into galaxies, we turn now to consider the observational constraints on galaxy formation.

III. GALACTIC ORIGINS

Three pathways to observationally constrain galaxy formation are possible. Direct observation of a protogalaxy would be the ideal probe, but this approach has hitherto been unsuccessful. Large-scale structure, specifically the galaxy correlation function, has provided useful limits, but these do little more than normalize the various theories of density fluctuation spectra described in the previous section. Some theories survive the normalization and meet the other large-scale constraints, while others fail, but the amount of freedom is depressingly large. We cannot say, for example, whether the first nonlinear structures are much larger or much smaller than a characteristic galaxy scale. Galaxies, as we now observe them, evolve remarkably slowly. Their characteristic properties can therefore shed considerable light on the formation process. It is this line of reasoning that will be pursued here in some detail.

The two-body relaxation time-scale in a system containing N stars is $t_R \sim (N/\log N)$ crossing times, and generally far exceeds a Hubble time for galaxies. However, the galaxy bulge could have formed from amalgamations of stellar subunits each containing $\sim 10^{6-7}$ stars, say: the evolution would initially have been sufficiently rapid for morphology to develop. This type of inhomogeneous collapse is the basis of van Albada's (1983) simulations of density profiles in elliptical galaxies. In the absence of dissipation, the final binding energy changes by at most a factor of two (if no mass is lost). In this case, the observed dynamical state of a galaxy tells us its mean state at formation. Dissipation by gas cloud interactions provides another way to accelerate the evolutionary time-scale. However, for significant evolution of the observed galaxy to have occurred, it must have been predominantly gaseous. Then, the dynamical state one now observes specifies the mean state of the galaxy at the end of the dissipative phase.

For the spheroidal component of a galaxy, if an isotropic velocity distribution is assumed and rotation is neglected, any two of the following quantities, mass M, half-mass radius R, mean velocity dispersion σ, suffice to determine its equilibrium state. In practice, it is the luminosity L and half-light radius that are directly measured. Adoption of a universal initial stellar mass function allows us to calculate the mass-to-light ratio for a system containing any specified mix of populations I and II, as long as we are only concerned with the luminous mass of the galaxy. Since it is the spheroidal component that probes the formation phase, we define an equivalent spheroid for disk galaxies by using the maximum rotational velocity to infer the velocity dispersion of the spheroid. The light from the old disk stars and any bulge contribution can then be used to specify the luminous mass initially in the spheroid at the formation epoch.

From the condition of virial equilibrium $\sigma^2 \sim GM/R$, we deduce the relation

$$\sigma^4 \sim L(L/R^2)\,(M/L)^2. \qquad (1)$$

From this, we see that rather than use individual determinations of σ, L or R, we can make use of the statistical correlations found between L and σ or L and R. Adoption of a mean M/L means that these two correlations are equivalent: if both are available for a given Hubble type, one can infer the appropriate M/L value for the luminous component. The parameters that one chooses are velocity dispersion σ and surface mass density $\Sigma \equiv M/R^2$. Plotting Σ against σ then provides a way of displaying the various Hubble types, which have a dispersion over both parameters. The spheroid surface density is computed, according to (1), from the (L,σ) or, equivalently, the (L,v_{max}) correlations, for ellipticals and spirals respectively. These relations distinguish between Hubble types, which correlate with v_{max}, with surface brightness, and with L (at a given value of v_{max}). For ellipticals, the wide range in luminosity (and by inference, mass) is reflected in a possibly significant flattening in the (L,σ) relation ($L \underset{\sim}{\propto} \sigma^3$ rather than $L \underset{\sim}{\propto} \sigma^4$) at low luminosities, but first-ranked cluster ellipticals are indistinguishable within the considerable scatter from the typical elliptical. Moreover, there is no data on the (L,σ) correlation at $L < 10^{10} L_\odot$. However, the (L,R) correlation spans a much wider range, incorporating dwarf spheroidals, and clearly reveals that at both extremes of luminosity, ellipticals decrease in surface brightness (Figure 1). Hence for the dwarf ellipticals and for the brightest cluster members, we have utilized the (L,R) correlation with equation (1) to infer the (Σ,σ) distribution. In fact, there is no need, where there is adequate data on L,σ, and R, to assume a mean M/L: it can be evaluated statistically from combining the two correlations.

The resulting (Σ,σ) correlations are displayed in Figure 2. The two-dimensional parametrization of Hubble type clearly separates early-type from late-type galaxies. It is especially interesting that the dwarf ellipticals and spheroidals continue the Hubble sequence to low surface density and velocity dispersion. At the other extreme, after early-type spirals, S0's and E's, the cD galaxies are found. In general, all galaxies are found to fall on a statistically well-defined region in the (Σ,σ) plane. A similar track is also found for groups and clusters of galaxies. Again, a ratio of mass to light appropriate only to the luminous matter has been adopted. The region occupied by groups and clusters in the (Σ,σ) plane lies parallel to that of the galaxies, but substantially displaced towards lower surface density. These tracks reflect the distribution of binding energy in systems of different mass, and we turn now to an explanation of them.

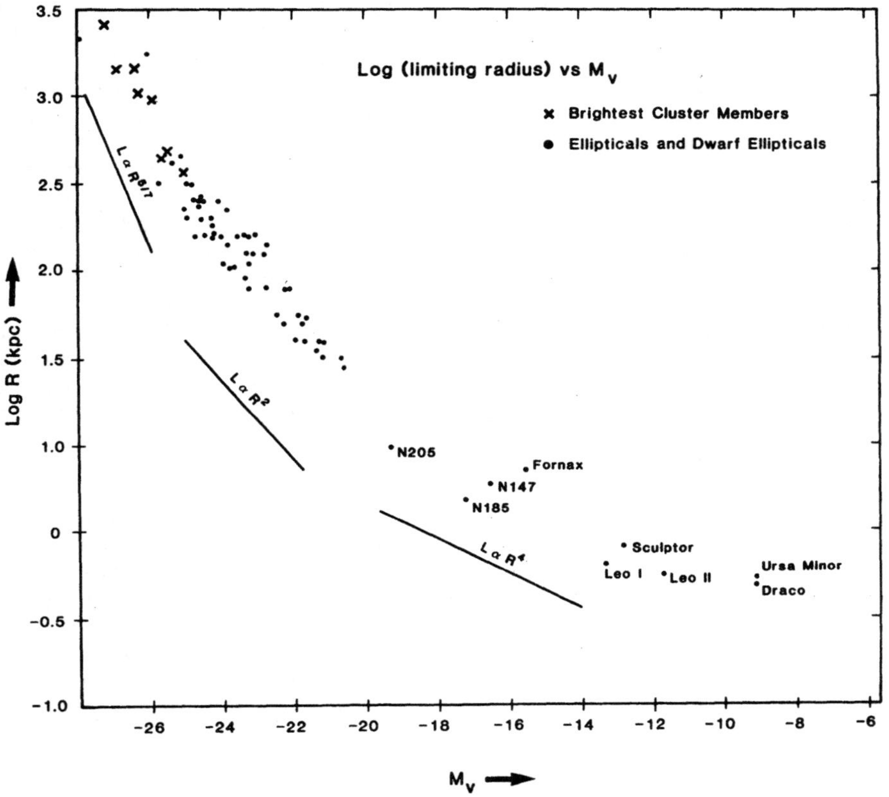

Figure 1. Luminosity - limiting radius relation for elliptical galaxies and dwarf spheroidals. Data is from Oemler (1976), Kormendy (1977), and Malamuth and Kirsher (1981).

Consider the evolution of gas that fills the potential well of a collapsing protogalactic cloud. It is irrelevant for the following argument whether this gas is initially a coherent cloud of galactic mass, or consists, more plausibly, of a number of smaller clouds responding to the protogalactic gravitational field. Furthermore, we assume only that the initial gas distribution has mass equal to that of the old stellar population in the galaxy: it is of little import, for the moment, whether or not the gas itself is self-gravitating. Gas motions in the protogalactic potential well will initially be highly supersonic, until shock fronts develop. Under most plausible cosmological initial conditions, the gas is initially neutral. Now a necessary, although not sufficient, condition for star formation is that radiative shocks develop. The ensuing density increase does not guarantee gravitational instability of the gas: whether or not this occurs depends partly on the shock geometry. For example, a three-dimensional compression will be destabilizing, whereas a one-dimensional compression will not be, in general, until enough mass is swept up for the compressed layer to become gravitationally unstable to modes normal to direction of shock propagation.

The condition for a shock to be radiative in the least favorable, namely plane shock, geometry can be expressed as a lower bound on the cloud surface density that depends only on the post-shock temperature. This, in turn, depends only on the specific binding energy σ^2 of the potential well, since the relative velocity $2^{1/2}\sigma$ between colliding clouds will yield a similar post-shock temperature to the virial temperature of gas filling the potential well. This minimal surface density for a planar radiative shock is

$$\Sigma_{cool} = \rho\sigma t_{cool}, \qquad (2)$$

where the post-shock cooling time-scale

$$t_{cool} = 3/2kT\left(\Lambda(T)\rho\right)^{-1} \qquad (3)$$

and $\Lambda(T)$ is the atomic cooling function (in erg $cm^3 s^{-1}$). The cooling column density Σ_{cool} is shown in Figure 2 for two cases, corresponding to a primordial abundance mixture of H and ten percent He, and to a gas of solar abundance. Heavy element cooling enhances the cooling rate over $10^5 \lesssim T \lesssim 10^7$K and at $T < 10^4$K, consequently reducing Σ_{cool}.

Comparison of Σ_{cool} with the locus of galaxies in the (Σ, σ) plane reveals the striking fact that dissipation must have occurred during galaxy formation. For the luminous matter to have attained its column density in excess of Σ_{cool}, there must have been a preceding phase of gaseous dissipation and radiative cooling in the protogalactic era. While dynamical relaxation during collapse could also have enhanced Σ, there would be no reason for the

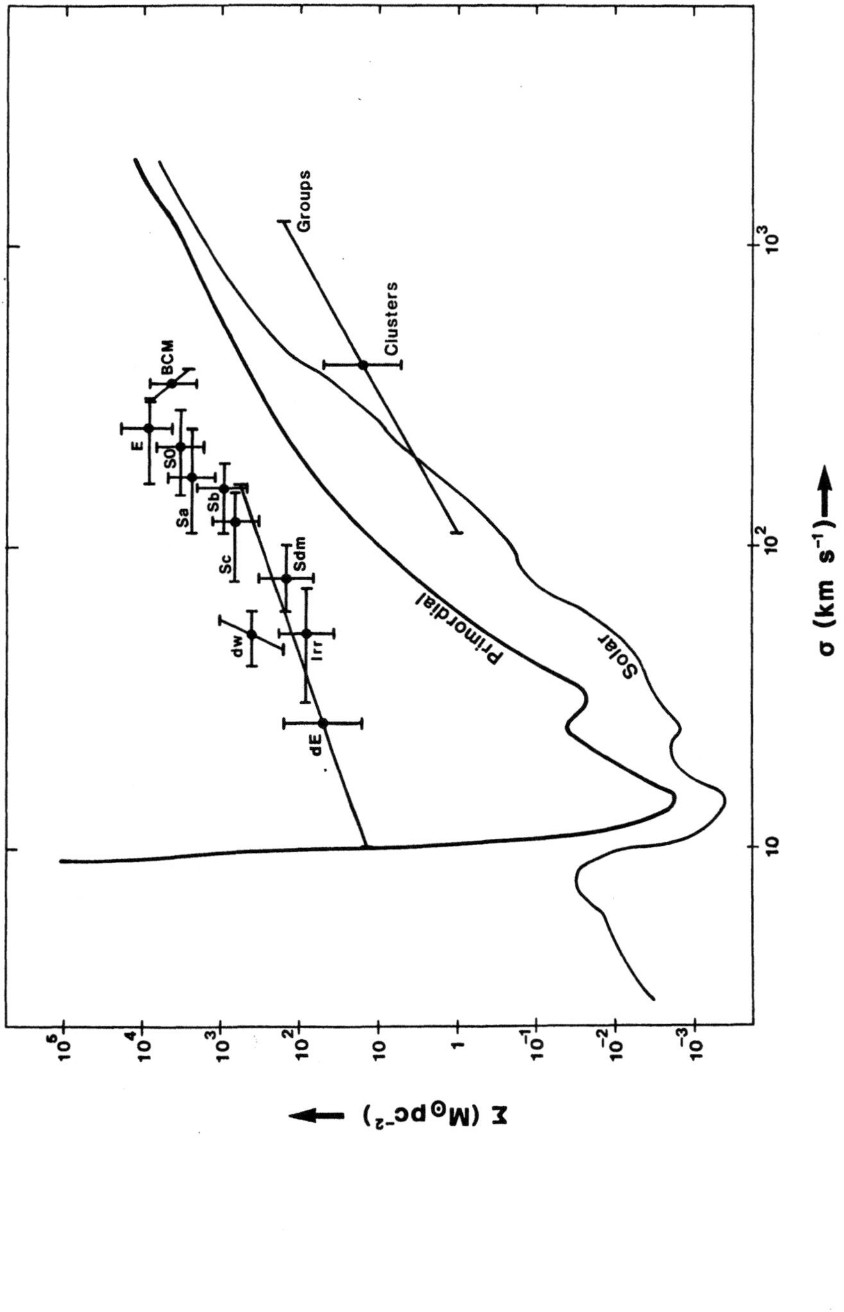

Figure 2. Surface density Σ versus velocity dispersion σ. Data sources are summarized in Silk (1983), and updated to include more recent work. Adopted mass-to-light ratios are taken from Faber and Gallagher (1979). Curves represent condition for gas cloud at virial temperature to cool for either primordial or solar abundance mixture.

observed distribution of Σ to hug the critical surface density curve for there to have been substantial cooling. On the other hand, the surface density of luminous matter in galaxy groups and clusters is less than Σ_{cool} over the appropriate range in σ. Evidently dissipation was unimportant over large scales. This distinction between galaxies and clusters in the dissipation diagram was first realized by Faber (1982) and Gunn (1982), and further extended by Silk (1983).

The dissipation argument was originally applied to protogalaxies by Rees and Ostriker (1977) and Silk (1977), who inferred an upper limit of $\sim 10^{12}$ M$\odot$ to the mass of a self-gravitating gas cloud that is capable of undergoing radiative cooling. It is also apparent (Silk 1983; also Figure 4 below) that there is a lower bound to the mass of such a cloud amounting to $\sim 10^{6}$ M$\odot$ if there has been no heavy element enrichment. However, one can do far more than simply evaluate critical mass-scales, as will now be described.

IV. STAR FORMATION IN PROTOGALAXIES

We would like to understand the underlying correlations of galactic dynamical parameters responsible for the locus of galaxies in Figure 2, namely the (L,σ) and (L,R) relations. I will argue here that star formation provides the key, and enables one to also infer the chemical evolution of galaxies.

While we are far from a full understanding of how stars form, the significance of at least one aspect of star formation is being realized. This is that interstellar clouds do not fragment and form stars over a free-fall time-scale, but are relatively long-lived for ~ 100 free-fall time scales. The most likely means of support against collapse is energy input from ongoing star formation. Clearly, this is likely to be a stabilizing, if not a disruptive process, and numerous observations of outflows from premain sequence stars testify to the role of star formation.

Conditions in a protogalaxy are sufficiently different that one should exercise caution before drawing too close an analogy with conventional star formation. However, the nucleosynthetic evidence from our halo suggests that the initial mass function of the first stars did not differ too drastically from the present one. Theoretical arguments support this inference. Moreover, there are important observational correlations that can most simply be understood if protogalaxies were predominantly gaseous. At the same time, dynamical relaxation elegantly explains other aspects, most notably the Hubble profiles of spheroids. One can most easily reconcile these demands with the natural assumption that, just as at present, stars formed from gas clouds of mass $\sim 10^{6}$ M$\odot$. Such masses are also more or less expected from cosmological considerations (Bond and Szalay 1983). With such masses, the overriding consideration is to avoid premature collapse and star formation on a cloud free-fall time which is much less than the dynamical time scale over which a galaxy can be built up. Energy input associated with ongoing star formation again provides the likely resolution of this problem.

In fact, cloud collisions in a galactic potential well will be supersonic and cause considerable disruption. Silk and Norman (1982) argued that cloud collisions will be inelastic and trigger star formation. However, simulations of supersonic cloud collisions (Chieze and Lazareff 1981; Hausman 1982) suggest that only the overlapping portions of clouds colliding at a typically oblique angle can radiate bulk energy of motion and coalesce: the bulk of the gas is disrupted and fills the potential well of the protogalaxy. This effect increases the gas lifetime against dissipation by up to an order of magnitude (Scalo and Pumphrey 1982).

It is crucial to decide whether the gas remains bound and provides a potential source of new clouds, or is ejected from the galaxy. The relevant criterion is whether or not energy input from forming and dying stars is sufficient to drive the gas out in a steady wind from the protogalactic potential well. If this occurs, one depletes the gas supply and rapidly cuts off star formation within a few crossing times. Renewed infall could then occur. Hence it seems logical to hypothesize that the rate of star formation will tend to regulate itself so as to maintain a gas reservoir, or at least, to deplete the gas very slowly. This certainly is the situation at present in spirals, when the gas fraction is 1-10 percent of the stellar mass. Moreover, following Larson (1974), the simplest way to understand why lower mass galaxies are systematically metal poor is to assume that the gas enriched by early star formation was somehow driven out before it could form stars. A galactic wind is one possibility, and stripping by interaction with other galaxies or intergalactic gas is an alternative possibility, although this would seem to create possible differences in metallicity between cluster and field galaxies that are not observed.

Suppose then that stellar energy sources, which we illustrate with the example of supernovae, keep the primordial gas from prematurely collapsing. It is unlikely that a wind can be driven from a predominantly gaseous massive protogalaxy, because the high gas density will guarantee that cooling occurs. Indeed, the relevant criterion for a radiatively unstable wind is equivalent to $\Sigma > \Sigma_{cool}$. Likewise, the presence of a massive halo inhibits any possible wind. The condition that star formation via stellar energy input to the gas is self-regulating is

$$E_S \dot{M}_* \sigma/v_s = 1/2\, M_{gas}\sigma^2/t_{ff}, \qquad (4)$$

where $\dot{M}_*$ is the rate of star formation,

$$\xi_{SN} = (10^{51}\ \mathrm{erg}/100\mathrm{yr})(1M_\odot/\mathrm{yr})^{-1} \qquad (5)$$

is the specific energy input per unit star formation rate, presumed to be (but this is not essential) from supernovae, $v_s \sim 300$ km/s is the velocity below which a spherical shock wave enters the momentum-conserving phase and sweeps up a dense shell, and the free-fall time-scale

$$t_{ff} \sim GM/\sigma^3. \qquad (6)$$

Numerical calculations show that V_s is insensitive to density, increasing by a factor 2 if the density increases by a factor $\sim 10^3$, but V_s may be expected to depend more on metallicity.

Since $\dot{M}_* = -\dot{M}_{gas}$, we can integrate (4) to yield

$$\dot{M}_* = \sigma^4 \left(\frac{V_s}{2G\xi_{SN}}\right) \exp\left(\frac{-1}{2}\frac{\sigma V_s}{\xi_{SN}}\frac{t}{t_{ff}}\right), \qquad (7)$$

where $\sigma V_s/\xi_{SN} \approx 0.06\,\sigma_{100}$ and $\sigma_{100} = \sigma/100$ km s^{-1}. Hence

$$\dot{M}_*/M = 7.1 \times 10^{-10}\mathrm{yr}^{-1}(\sigma_{100}^4/M_{10})\exp(-0.03\sigma_{100}t/t_{ff}). \qquad (8)$$

In our galaxy at present $\dot{M}_*/M \approx (3M_\odot/\mathrm{yr})(10^{11}M_\odot)^{-1} = 3 \times 10^{-11}\mathrm{yr}^{-1}$. Also $M_{10} \equiv (M/10^{10}M_\odot) \approx \sigma_{100}$ is the approximate normalization for the (L,σ) relation. Thus equation (8) implies that the star formation rate per unit mass is about 20 times larger in young gas-rich systems than at the current epoch. Moreover, the star formation rate declines exponentially over $\sim 30\,\sigma_{100}^{-1}$ dynamical times. Since M_* during the protogalactic phase measures the luminosity in the old stars, we finally infer that

$$L \underset{\sim}{\propto} \sigma^4 \qquad (9)$$

The protogalactic star formation computed in this manner also yields the enrichment. If y is the yield (net heavy element production per unit mass of forming stars), then the fractional heavy element abundance

$$Z = yM_*/M = y\left(1-\exp(-1/2 \frac{\sigma V_s t}{\xi_{SN} t_{ff}})\right). \quad (10)$$

The enriched gas must be lost before the exponential term is appreciable, that is to say, at $t \lesssim 30\ t_{ff}/\sigma_{100}$. In this case, the final heavy element abundance is

$$Z \approx 1/2\ y\ V_s \sigma/\xi_{SN} \approx 0.15\ y\ \sigma_{100}. \quad (11)$$

Some recent metallicity determinations for ellipitcals are plotted versus luminosity in Figure 3. If M_* is linearly proportional to L, equation (10) predicts $Z \propto L^{-1/4}$, a slightly weaker dependence than is observed. The same trend is found for irregular galaxies with strong emission lines, and continues down to Z~0.03 of the solar value (Comte and Stasinska 1983).

Perhaps the most significant inference from Figure 2 is that galaxies, apart from the most luminous systems, lie along a locus equivalent to

$$L \underset{\sim}{\propto} R^4. \quad (12)$$

A simple explanation can be given for this result in terms of a threshold density for star formation following suggestions by Mathews (1972) and Tayler (1976). Consider a self-gravitating mixture of inert dark matter, taken to be universally present at the same density in all systems, and gas. In order for the gas to be unstable to perturbations of wave number k,

$$\rho_{gas}/\sigma^2_{gas} + \rho_{dark}/\sigma^2 > k^2/4\pi G, \quad (13)$$

where ρ_{gas}, ρ_{dark} are the densities of gas and dark matter, and σ_{gas}, σ are the velocity dispersions of these two components. For a universal halo density, ρ_{dark} = constant implies

$$M \propto \sigma^3. \quad (14)$$

If $\sigma_{gas}=\sigma$, the gas self-gravity dominates and allows star formation once $\rho_{gas}>\rho_{dark}$; more generally, we require

$$\rho_{gas}>(\sigma_{gas}/\sigma)^2\ \rho_{dark}. \quad (15)$$

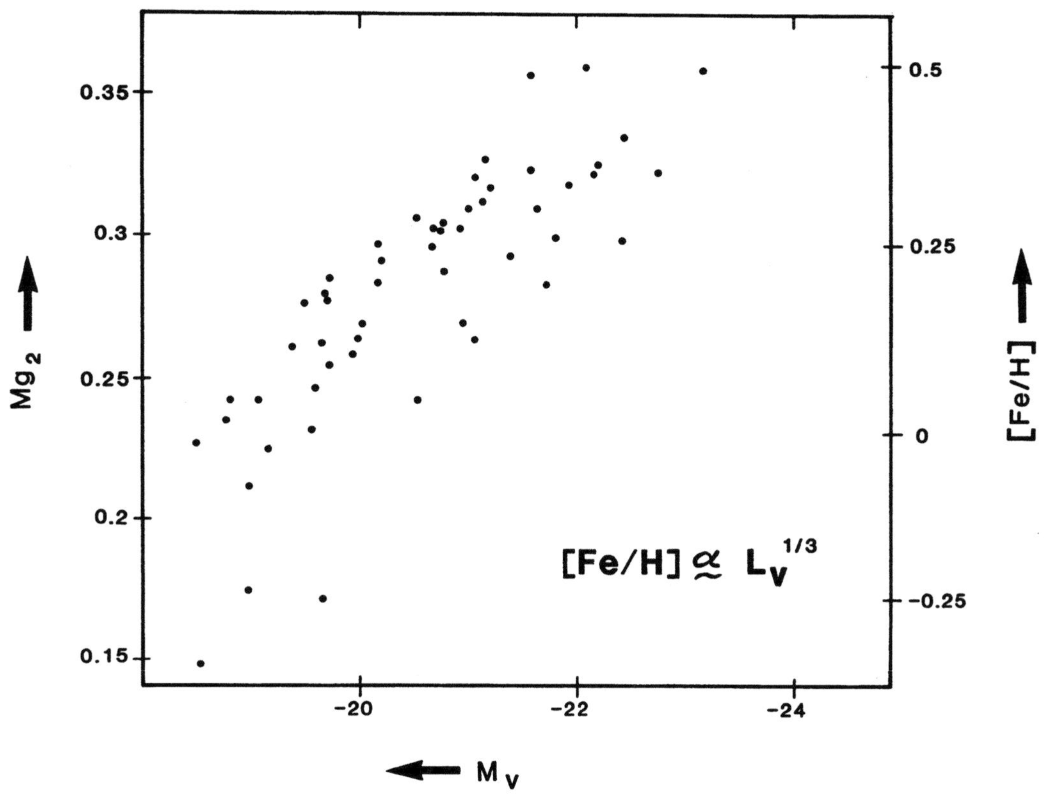

Figure 3. Luminosity versus metallicity for elliptical galaxies. Data from Dressler (1983).

Combining equations (9) and (14) yields

$$M/L \propto \sigma^{-1} \propto L^{-1/4} \quad \text{and} \quad L \propto R^4 \quad . \qquad (16)$$

At fixed ρ_{dark}, one can tolerate proportionally less luminous, star forming gas in the weaker potential wells of the low σ galaxies. This effect accounts for the decrease in surface brightness at low luminosities of Figure 1 (and for the galaxy locus in Figure 2). There is tentative evidence in favour of an enhanced M/L ratio in at least one nearby dwarf spheroidal: the inferred value ($M/L\sim 30$, $L\sim 10^5$ $L_\odot$ for Draco) (Aaronson 1983) is consistent with (16). This latter relation evidently saturates (ρ_{dark} becoming negligible), and leads to $M/L \sim$ constant at $L > 10^8$ $L_\odot$.

V. COSMOLOGICAL CONFRONTATION

Since the gas density must locally exceed that of the dark matter, or a considerable fraction of it, according to (15), this immediately explains how one can avoid the fractionation problem. That is to say, despite the fact that one may have, say, a universe with $\Omega=1$ in which the average baryonic mass fraction is only of order a percent, the luminous stars will necessarily contain a baryonic component comparable in mass to that of the dark matter in order for stars to have formed via gravitational instability.

However, our explanation of relations (16) does rely on the assumption that ρ_{dark} has a universal value for all galaxies. There is one attractive cosmological scenario that leads to precisely this conclusion. This is when cold collisionless relics dominate the mass density of the universe. In this case, $\delta\rho_{dark}/\rho_{dark} \approx$ constant on scales well below M_{eq}, or the comoving scale $L_{eq}\approx 30$ Mpc. Hence for both galaxies and galaxy clusters alike, we infer a scale invariant density at formation ($\delta\rho_{dark}/\rho_{dark}\approx 1$), provided that appropriate epochs of formation are selected. If ρ_{gf} denotes the dark matter density at galaxy formation, then we infer that

$$\Sigma \approx \sigma(\rho_{gf}/G)^{1/2} \qquad (17)$$

over $L<L_{eq}$. The hierarchical clustering locus for cold relics is shown in Figure 4, with ρ_{gf} set equal to the present density. This is only appropriate for large-scale structure that is still in the linear or weakly nonlinear regime. Fitting (17) to the region occupied by galaxies yields an upper bound on the redshift of galaxy formation: $z_{gf}<40$. This is only an upper bound, because dissipation acts to raise the galaxy locus to higher Σ. For groups and clusters, dissipation is unimportant: the epoch of cluster formation is inferred to be at $z\sim 2$. Since the microwave background

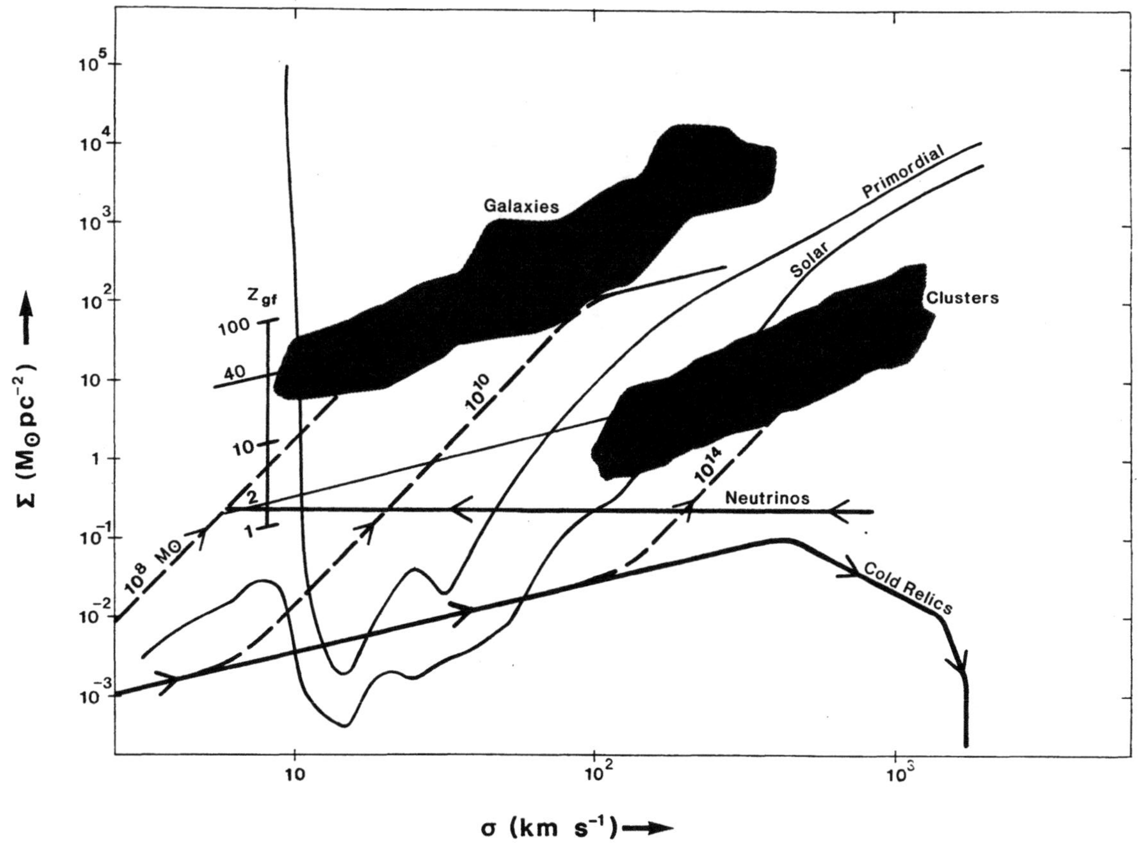

Figure 4. Surface density versus velocity dispersion, as Figure 2, but with evolutionary tracks indicated for galaxy formation either via pancake fragmentation or by hierarchical clustering of cold collisionless particle-dominated fluctuations. Upper bounds to the redshift of galaxy and of cluster formation are shown for the latter case.

temperature $\lesssim 100$ K at galaxy formation if $z_{gf}<40$, Compton cooling can be neglected.

Utilization of the mean density of luminous matter to infer an upper bound on z_{gf} is a well known argument. What is novel about the present result is that the binding energy distribution, or equivalently the approximate $L \propto R^4$ locus in Figure 2 or 4 for luminous matter in galaxies and in clusters, can be explained in terms of a unique cosmological origin.

Pancake theory does not fare so well in this regard. A schematic track for pancake collapse at z=3 is shown in Figure 4. Fragmentation leads to smaller and smaller mass scales. Evolution along locii of constant mass eventually reaches the regime in Figure 4 occupied by galaxies. However, there is no compelling reason for $\delta\rho/\rho$, and hence ρ_{dark}, to be scale invariant, and no simple explanation for equation (16).

Either pancake fragmentation or hierarchical clustering of cold relics ensures a prolific gas supply at the galaxy formation epoch. This is because galaxies form from smaller gas clouds that are produced at precisely the epoch of galaxy formation. Of course, galaxies form very recently in the pancake theory, since one has to wait for the large scales to go nonlinear, but this in itself is not necessarily a fatal objection. Dissipation on galactic scales is therefore a natural implication in either scenario.

Whether galaxies are still substantially gaseous when clusters form is more problematical. This would be expected in the pancake model, and be likely to occur in the cold relic picture if the dissipational collapse of a protogalaxy is sufficiently slow. The various arguments of the previous section suggest that it will be: note that duration of collapse for

$$\sim 10\tau_{ff} \sim 10^9 \text{ yr}$$

for a typical galaxy yields a timescale comparable to the typical cluster crossing time. Moreover, supersonic cloud collisions naturally yield such a time-scale for dissipation of bulk kinetic energy (Scalo and Pumphrey 1982) even if energy input from forming or dying stars plays a lesser role than envisaged in §IV.

If indeed gas-rich protogalaxies are present when clusters form, then other outstanding problems may be resolvable. Silk and Norman (1982) accounted for the dependence of morphological type on local galaxy density found by Dressler (1980) with a model involving protogalactic mergers, a prerequisite for which was the dominant gas content. Predominantly steller systems do not merge to yield sufficiently deep potential wells (Ostriker 1981), as confirmed by Merritt (1983) in his study of cD galaxy formation. Another interesting effect recently addressed by Wang and Scheurle (1983) is that of tidal torque generation between neighboring protogalaxies. This can be enhanced during a prolonged gas-rich phase as clusters form, and may explain why spirals acquire more specific angular momentum than ellipticals, provided that the latter are stripped of gas more rapidly.

In summary, dissipation may be the key process that enables one to make a connection between the density fluctuations emerging from the early universe and the gaLaxies around us. Many of the fossilized properties of galaxies arose long ago when galaxies were forming out of gas-rich clouds. I suspect that Lemaître would have appreciated this result, for he spent many years fruitlessly trying to establish that cosmic rays were a relic of the big bang. One week before his death, he was delighted to learn about the discovery of the cosmic microwave background radiation, the faded remnant of the primordial fireball. Galaxies, stared at in awe by astronomers for over a century, might yet provide an equally impressive relic of, as well as evidence for, the big bang theory.

REFERENCES

Aaronson, M. 1983, Ap.J. Letters, 266, L11.
Bond, J. R. and Szalay, A. S. 1983, Ap.J. (in press).
Chieze, J. and Lazareff, B. 1980, Astr. and Ap., 91, 290.
Comte, G. and Stazinska, G. 1983 (private communication).
Dressler, A. 1983, preprint.
Dressler, A. 1980, Ap.J. 236, 351.
Efstathiou, G. and Silk, J. 1983, Fund. Cosm. Physics (in press).
Faber, S. 1982 in Astrophysical Cosmology, eds. H.A. Bruck, G. V. Coyne, and M. S. Longair (Rome: Pontificia Academia Scientarum).
Faber, S. and Gallagher, J. 1979, Ann.Rev.Astr.Ap., 17, 135.
Gunn, J. 1982 in Astrophysical Cosmology, eds. H.A. Bruck, G.V. Coyne, and M. S. Longair (Rome: Pontificia Academia Scientarum).
Hausman, M. 1981, Ap.J., 245, 72.
Kormendy, J. 1977, Ap.J., 218, 333.
Larson, R. 1974, M.N.R.A.S., 169, 229.
Malamuth, R. and Kirshner, R. 1981, Ap.J., 251, 508.
Mathews, W. 1972, Ap.J., 174, 108.
Merritt, D. 1983, in preparation.
Oemler, A. 1976, Ap.J., 209, 693.
Ostriker, J. 1980, Comm. Ap., 8, 177.
Rees, M. and Ostriker, J. 1977.

Scalo, J. and Pumphrey, W. 1982, Ap.J. Letters, 258, L29.
Silk, J. 1977, Ap.J., 211, 638.
Silk, J. 1983, Nature, 301, 574.
Silk, J. and Norman, C. 1981, Ap.J., 267, 59.
Tayler, R. J. 1976, M.N.R.A.S., 177, 39.
van Albada, T. 1982, M.N.R.A.S., 201, 939.
Wang, Y. and Scheurle, H. 1983, Astron. and Ap. (in press).

CLUSTERS AND SUPERCLUSTERS

J.H. Oort
Sterrewacht, Leiden, The Netherlands

The problem of the formation of galaxies and clusters of galaxies has intensely occupied Lemaître's thoughts. It is therefore of interest to compare his ideas with what is at present thought about the birth of galaxies and about their distribution in space.

Twenty-five years ago, on June 9, 1958, Lemaître described his theory at the 11e Conseil de Physique Solvay in Brussels, where some 40 physicists and astronomers had gathered to discuss the structure and the evolution of the universe. The attendance included the physicists Bragg, Klein, Oppenheimer, Pauli, Perrin and Wheeler, while among the astronomers were Ambartsumian, Baade, Bondi, Gold, Heckmann, Hoyle, McCrea, Shapley, van de Hulst, Ledoux and Schatzman. Ledoux, McCrea and myself are the only scientists at the *present* meeting who where at that Solvay Conference. Lemaître gave the opening lecture. He held the opinion that the "cosmological constant" played an essential role in the evolution. This formed an important aspect of the discussions at the meeting, as did also Hoyle's theory of the steady state Universe and the continuous creation of matter.

The introduction of a cosmical force of repulsion proportional to the radius R of the Universe led Lemaître to consider in particular the phase in which the repulsive force was equal, but opposite in sign, to the gravitational attraction, and the Universe was therefore temporarily in equilibrium. The equilibrium is unstable and the Universe re-assumes its accelerated expansion after a certain time. During the equilibrium phase, which he assumed to have taken place when R was about 1/10th of its present value, conditions would, as Dr. Peebles has discussed, have been favourable

A. Berger (ed.), The Big Bang and Georges Lemaitre, 299–314.

for the formation of gas clouds, for their gathering into galaxies, and finally for the formation of galaxy clusters. Lemaître pictured the latter as a sort of waves, populated by galaxies moving through. In this picture the virial theorem cannot be used directly to estimate masses.

The duration of the equilibrium phase cannot be determined from observations. Lemaître estimated that it might have been of the same order or even longer than the Hubble time, and that therefore the age of the Universe might be sensibly higher than had been previously thought.

During the last 25 years important developments have occurred in our knowledge of clusters of galaxies. These concern principally two aspects:

(a) The structure of clusters and the motions of cluster galaxies; and in particular the discovery that many clusters contain a large mass of hot gas, of which the temperature as well as the distribution has been measured by X-ray observations. The new observations have shown rather definitely that most of the galaxies we see in a cluster are essentially permanent members, and not just passing through.

(b) The continually growing evidence that the clusters are part of much larger structures, "superclusters", which may pervade the entire Universe as a kind of "network".

(a) STURCTURE, DYNAMICS AND EVOLUTION OF GALAXY CLUSTERS

Examples of the structure are shown in Figures 1a, b taken from an investigation by Geller and Beers (1982), and Figure 2 from an article by Dressler (1980). From the former it is evident that several clusters have highly irregular structures. Geller and Beers estimate that approximately 40% of their clusters show significant substructures, while many of the remaining clusters are elongated. Dressler's picture illustrates the great differences in central concentration which occur.

Observations with the Einstein X-ray Observatory have shown that the hot gas in clusters has generally a similar distribution as the galaxies (cf. Forman & Jones 1982; Jones 1983).

In the regular, centrally concentrated clusters the X-ray data permit an independent determination of the gravitational potential. This appears to agree approximately with that derived from the distribution and motions of the galaxies. Both lead to total masses between 10^{15} and 10^{16} $M_\odot$ for the richer clusters, and even higher for the richest ones. These values are an order of magnitude higher than what had been expected from the masses of the member galaxies as estimated from their luminosities, and indicate the presence

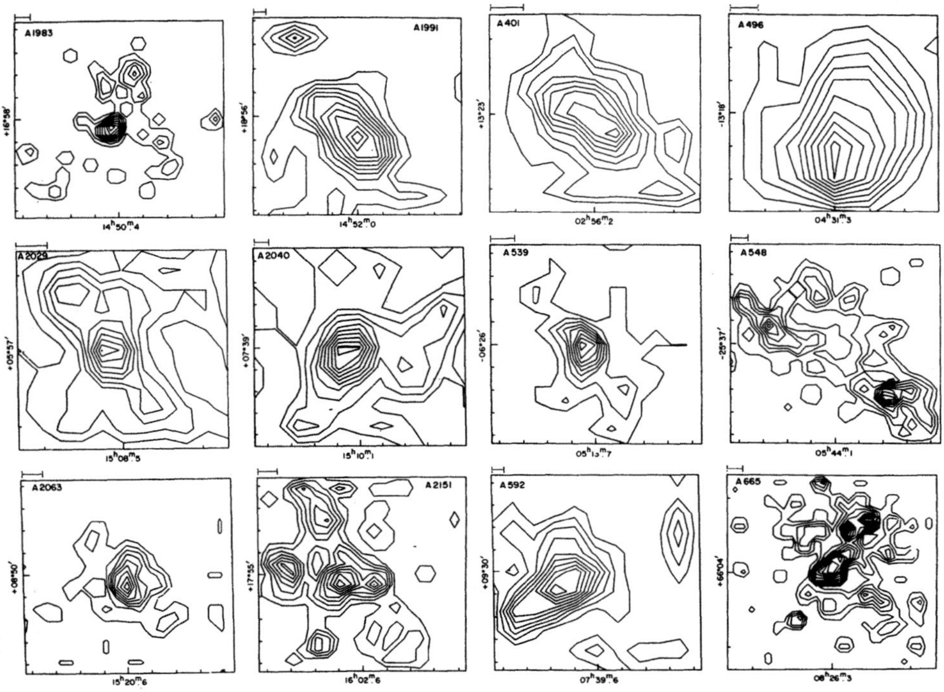

Figure 1. Contours of galaxy distribution in rich clusters (Geller and Beers 1982).

of large non-luminous masses, and unexpected large values for the ratio of mass to light (M/L). Not more than a small fraction of the "dark" mass can be provided by the intracluster gas. Most of it may reside in the galaxies. In fact, the rotation curves of spirals, as well as the velocity dispersions in giant ellipticals show that matter with a very high M/L ratio dominates in the outer regions of galaxies. The nature of this matter has been the subject of much discussion. Probably it is largely non-baryonic, because if it consisted of baryons the abundance of deuterium produced in the early Universe should be lower than what is observed.

From the evident irregularities in their structures we infer that we live in an era where the galaxy clusters are being formed. This does not imply that they are very young. From the few cases of irregular clusters where sufficient velocity data are available we infer that they might well have been in existence for a major fraction of the Universe's age.

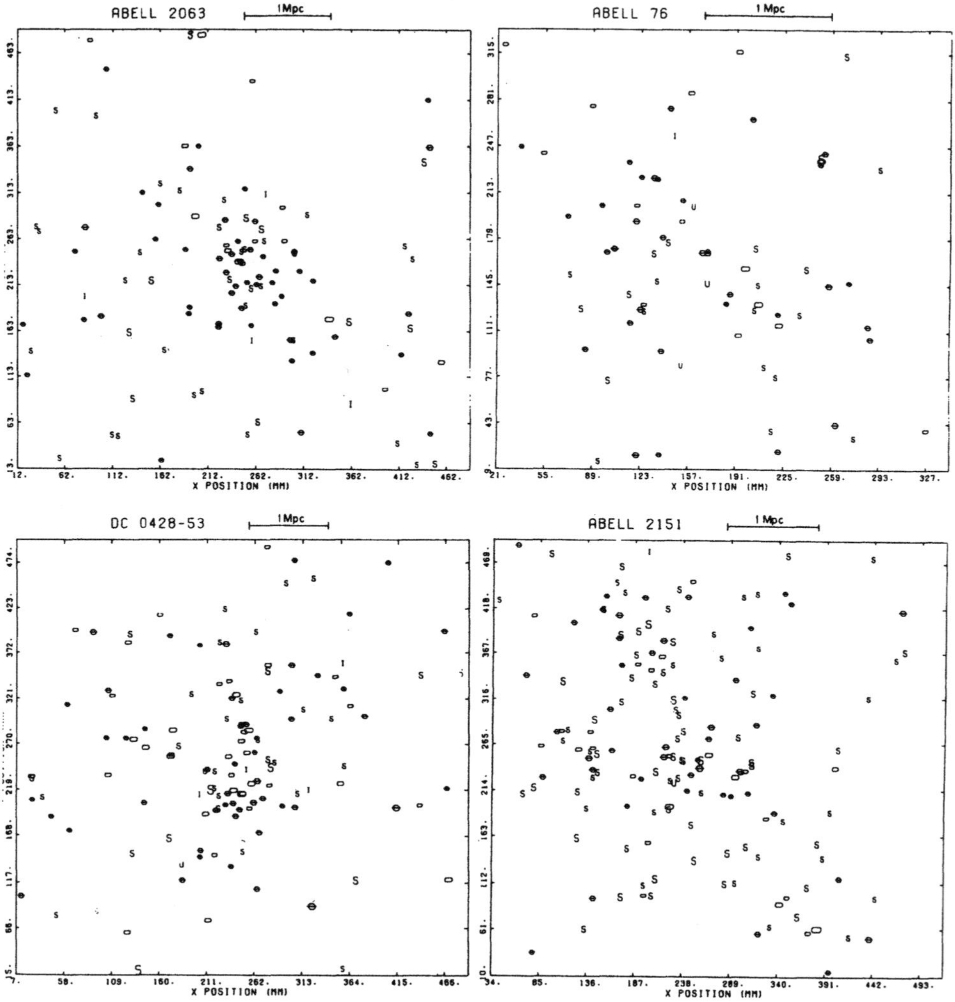

Figure 2. Galaxy distribution in clusters. Left: high-concentration, regular clusters; right: low-concentration, irregular clusters (Dressler 1980).

(b) SUPERCLUSTERS

That structures exist which are much larger than the rich clusters has long been evident. It is shown most clearly in the plots made 50 years ago by Harlow Shapley and Adelaide Ames

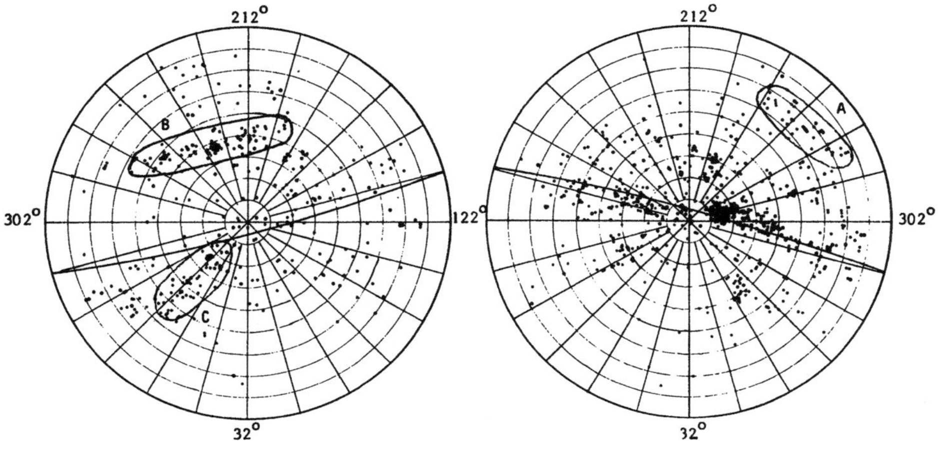

Figure 3. The distribution of galaxies brighter than the 13th photographic magnitude. The right-hand panel shows the north galactic hemisphere, the left-hand the south galactic hemisphere. The galactic poles are at the centers, the circles are at intervals of 10° latitude; ℓ^{II} is shown at the circumference. A, B and C are probably small separate superclusters. (Adapted from Shapley and Ames 1932).

(Figure 3). The picture, which I used also at the Solvay Conseil of 1958, illustrates all the various characteristics of the distribution of galaxies, from the smallest-scale clumpiness to their concentration in a large cluster, the Virgo cluster, and finally the very uneven distribution on a still larger scale, with particularly a "filamentary" structure stretching from the Virgo cluster through the north galactic pole in the direction of roughly 140° galactic longitude and a similar extension from the cluster in the opposite direction, towards $\ell \sim 310°$. The cluster has a diameter of about 10°, or 2.5 Mpc, while the extended structure has a length of $\sim$ 30 Mpc. This is called the Local Supercluster. It has been studied extensively, by de Vaucouleurs (1956, 1978, 1983), Tully (1982) and many others. It has a complicated structure, part of which is concentrated towards a plane, which de Vaucouleurs has termed the supergalactic plane. The intricate structure is illustrated in Figures 4 and 5 which show the distribution projected on planes perpendicular to this equator. The plane in Figure 4 passes through the Virgo cluster and its appendages, the plane in Figure 5 is perpendicular to this. The space co-ordinates were determined by assuming distances roughly proportional to the radial velocities. The Sun is at the centre, and is the origin of the rectangular co-ordinates. Figure 5

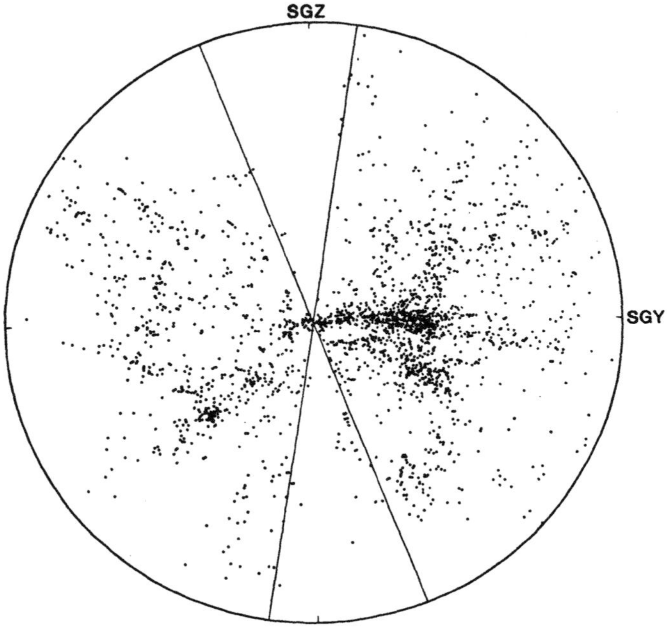

Figure 4. All 2175 galaxies in the Nearby Galaxy Catalog (Fisher and Tully 1981) projected onto the SGY-SGZ plane. The SGY-axis is directed toward supergalactic longitude 90°, supergalactic latitude 0° (ℓ^{II} = 227°, b^{II} = +83°.7), the SGZ-axis toward supergalactic latitude 90° (ℓ^{II} = 47°.4, b^{II} = + 6°.3). The radius of the outer boundary is 60 Mpc. The galactic zone of avoidance (b < 15°) is contained within the opposed wedges tilted by 6° with respect to the SGZ-axis. There is a zone of incompletion (δ < -45°) which is projected across most of the southern supergalactic hemisphere. Reproduced by courtesy of R.B. Tully.

contains only the galaxies in the north galactic hemisphere. The concentration toward the supergalactic plane is further illustrated in Figure 6. This contains all galaxies brighter than $M_B = -19.5$ in a cylinder perpendicular to the supergalactic plane with a radius of about 20 Mpc around the Virgo cluster. The galaxies in the Virgo cluster itself have been omitted.

If we penetrate to larger distances it becomes increasingly difficult to outline the superstructure, even if approximate distances are known from radial velocities. As an example I show two Figures taken from a large radial-velocity survey made at the Harvard Center for Astrophysics (Huchra *et al* 1983). It extends

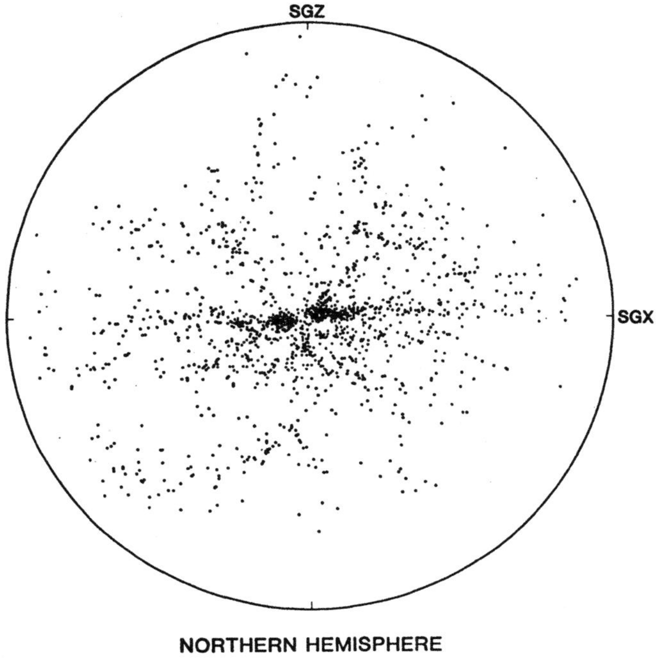

Figure 5. Projection on the SGZ-SGX plane for the galaxies in the north galactic hemisphere.

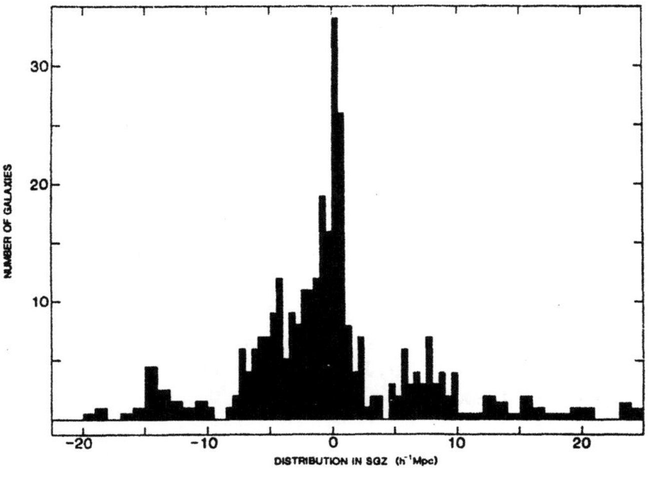

Figure 6. The distribution of Tully and Fisher's NBG galaxies in the surroundings of the Virgo cluster normal to the plane of the Local supercluster.

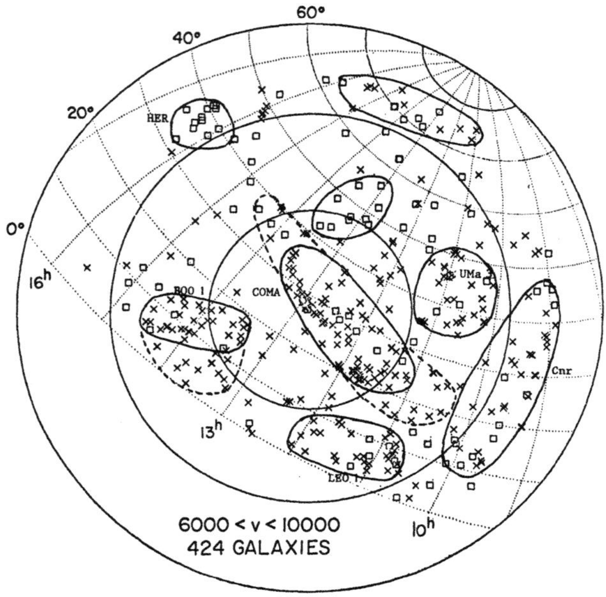

Figure 7. An equal-area plot in galactic co-ordinates of galaxies brighter than m_B = 14.5 above b = +40°. The galactic pole is at the centre, the circles are at b = 30°, 50°, 70°. Right-ascensions and declinations are indicated by dotted curves. Galaxies whose absolute magnitudes are fainter than −20.0 (on the H_0 = 50 distance scale) have been omitted. The various symbols denote the following velocity bins: x (6000–8000), □ (8000–10 000 km s^{-1}). Reproduced by courtesy of Davis *et al* (1981) and the *Astrophysical Journal*, except for the contours.

to approximately twice the distance to which the Shapley-Ames Catalogue extended, but covers only part of the sky. Figure 7 shows the distribution of galaxies with velocities between 6000 and 10 000 km s^{-1} in the north galactic hemisphere above 45° latitude; Figure 8 shows a specimen of a plot of radial velocity against right-ascension for galaxies in the declination zone 20° to 30°. I have made an attempt to outline the principal superstructures that can be distinguished. This evidently involves considerable arbitrariness. Several may well be just chance configurations. The plots fail moreover to show the interconnections which probably exist.

The superclusters indicated have major diameters ranging roughly from 10 to 100 Mpc. The larger ones contain one or a few dense clusters like, for instance, the Coma cluster and Abell 1367 in the Coma supercluster. The structures seem to have a tendency

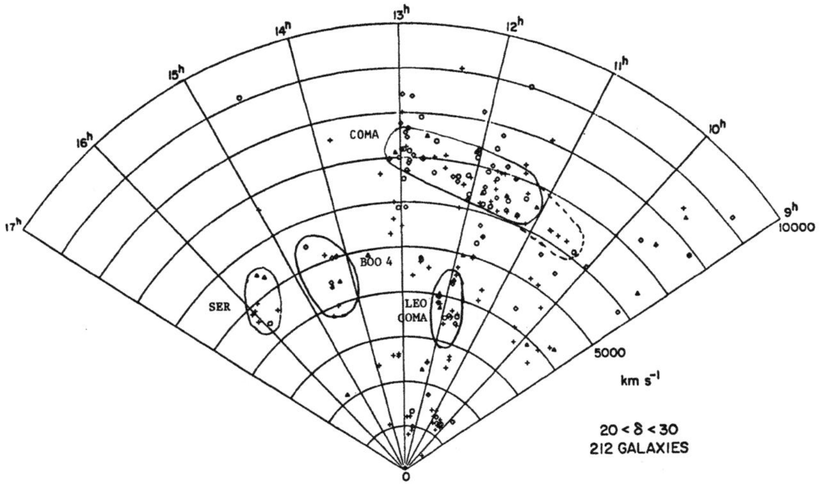

Figure 8. Distribution of galaxies in right-ascension and velocity in the declination zone 20° - 30° (cf. Figure 7).

for elongated shapes. There are indications that whenever a supercluster contains rich clusters these appear to be elongated in the direction of the supercluster branch in which they are situated; a striking example is the Coma cluster, which is strongly elongated in the direction of Abell 1367, the other rich cluster in the Coma supercluster. A similar phenomenon is observed in the Perseus supercluster, where the three dense clusters Abell 426, 347 and 262 are elongated along the direction of the principal branch of the supercluster. That such preferential orientations are probably a general phenomenon has been shown in an investigation by Binggeli (1982) who has indicated that cluster major axes in general have a tendency to point in the direction towards their nearest neighbour cluster.

The difficulty of outlining large structural features in the distribution of galaxies can be significantly diminished by looking at the distribution of *clusters* instead of individual galaxies.

The two principal lists of clusters are those by Abell (1958) with 2712 clusters, and the *Catalog of Galaxies and Clusters of Galaxies* by Zwicky *et al* (1961-1968). Both were made from the Palomar Sky Survey plates, but they differ greatly in character: Abell's criteria for defining a cluster were much stricter than those applied by Zwicky and his co-workers whose catalogue contains an order of magnitude more entries, but is less homogeneous.

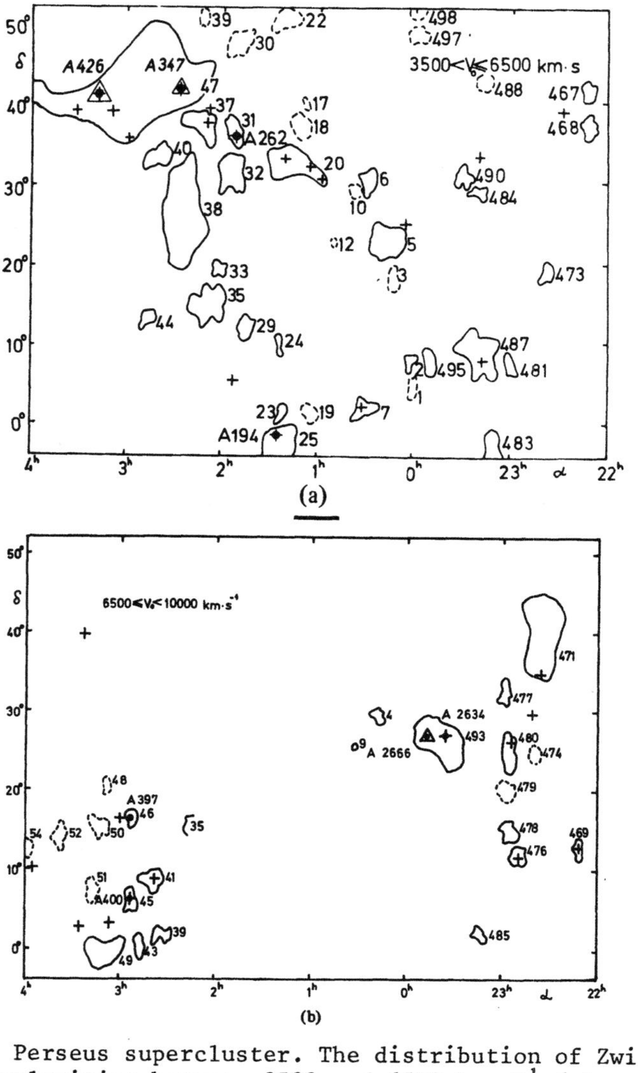

Figure 9. Perseus supercluster. The distribution of Zwicky clusters with velocities between 3500 and 6500 km s^{-1} (a), and between 6500 and 10 000 km s^{-1} (b) in the south galactic hemisphere. Solid contours show the clusters with measured redshifts, dotted contours indicate those with distances estimated from magnitudes and cluster diameters. The numbering is from Nilson (1973).
Abell clusters are indicated by solid circles and by their numbers in Abell's catalog (1958); A 426 is the Perseus cluster.
Δ: (X-ray sources), +: (radio sources). Reprinted by courtesy of Einasto *et al* (1980).

The Zwicky data have been used extensively by astronomers at the Tartu Observatory to investigate the supercluster structure of the Universe. As an example Figure 9 shows two plots for a region of about 90° x 60° in the south galactic hemisphere (Einasto *et al* 1980); the upper is for velocities between 3500 and 6500 km s^{-1}, the lower for those between 6500 and 10 000 km s^{-1}. The Zwicky clusters are indicated by contours; the three Abell clusters mentioned earlier are in the upper left-hand part. The supercluster is defined by the slanting row of Zwicky clusters in the upper part of (a), an almost vertical branch extending from A 347 to A 194 at the bottom, and a third branch extending from A 194 towards lower right-ascensions.

The distribution of individual galaxies having radial velocities between 3000 and 7500 km s^{-1} is shown for approximately the same region in three panels of Figure 10, due to Giovanelli *et al* (1983). Probably most of the galaxies plotted belong to the Perseus supercluster. They show much the same distribution as the Zwicky clusters. They indicate an interesting phenomenon, viz., that the earlier types appear to define the structure of the supercluster more sharply than the Sc, Sd and Irr types. This is confirmed in the upper left panel where E and SO systems have been plotted; this panel contains a number of galaxies with unknown radial velocities.

An extensive analysis of Zwicky clusters has likewise been made by the Tartu astronomers for the region surrounding the Coma supercluster (Tago *et al* 1983). In this article the authors elaborate on the interconnections between the Coma complex and other surrounding large agglomerations: the Local Supercluster, and the Abell 779 and Hercules superclusters; they conclude that the superclusters "merge to a single connected network".

Structures of very large scale have been studied with the aid of Abell's clusters. Unfortunately radial velocities are presently available for only a small fraction of these clusters. N. Bahcall and R. Soneira (1983) have analysed correlations in a complete sample of 104 clusters with known velocities up to a distance of $\sim$ 500 Mpc, in which they listed all agglomerations with a minimum space density enhancement factor of 20 (Figure 11). They found some extremely large superclusters; the largest (No. 12 of their catalogue) has a diameter of 360 Mpc, and contains 15 rich Abell clusters. Whether this ia a unique case, and whether it represents a limiting size can only be decided when radial velocities of many more rich clusters will have been determined. There are indications that No. 12 may indeed be not far from the largest structures existing.

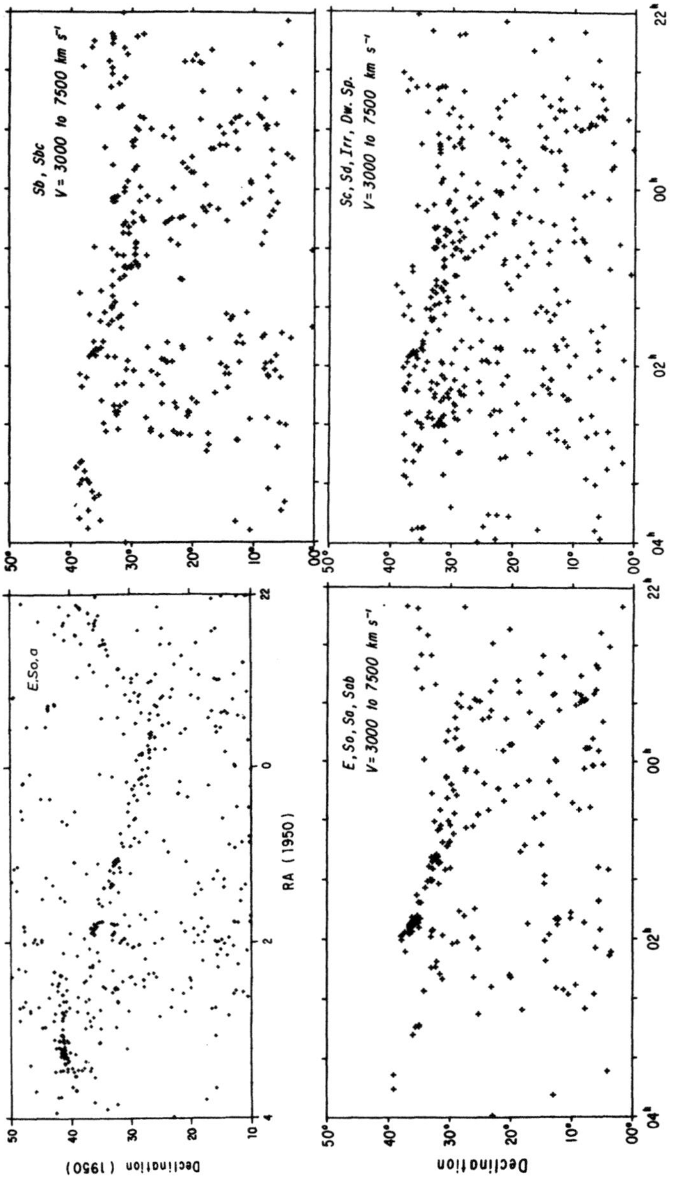

Figure 10. Distribution of galaxies from the catalogue by Zwicky *et al* (1961-68) in the region of the Perseus supercluster. Top left: E-SO,a for all velocities; the other panels are limited to galaxies with velocities between 3000 and 7500 km s^{-1} (Giovanelli *et al.*, in preparation). I am greatly indebted to the authors for putting their most recent material at my disposal in advance of publication.

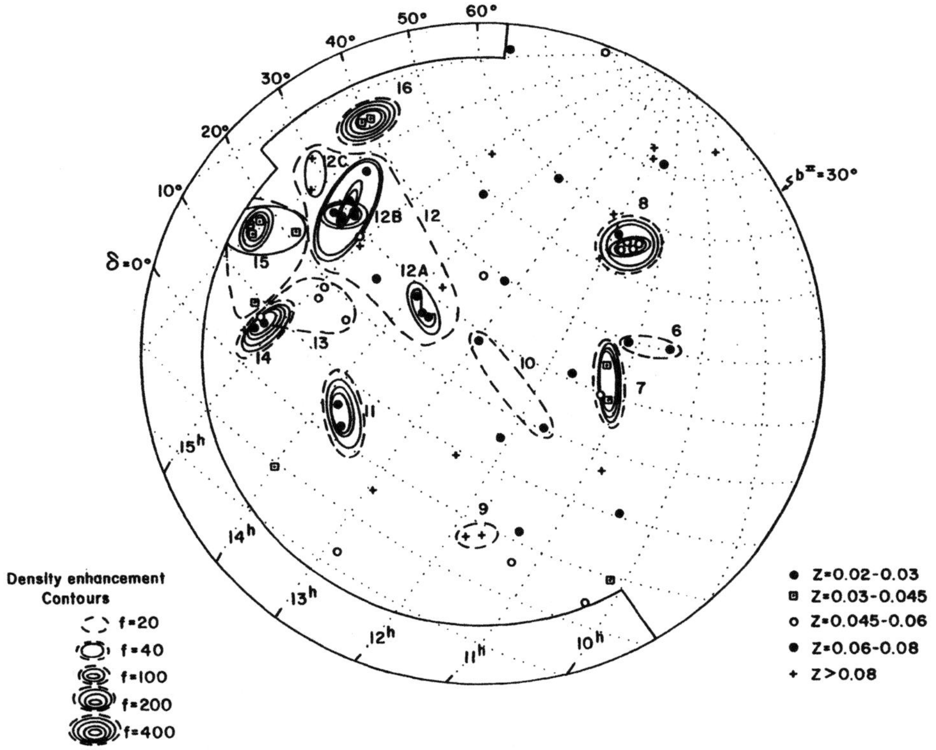

Figure 11. Giant superclusters in the north galactic hemisphere above 30° latitude. The outermost contour is b = 30°; the inner contour is the completeness limit of the sample. Small contours show the space density enhancement factor over the mean space density at the distance of the supercluster. Numbers refer to the authors' list of superclusters. The elongated contour number 10 is the Coma supercluster. Reproduced by courtesy of N. Bahcall and R. Soneira (1983).

We have observed that many galaxy clusters are still in the period of their formation. *Super*clusters are *all* still in a rudimentary stage of development. Except for their collapse they are unlikely to have changed their structure since their origin. In a sense they reflect the original "waves" in the Universe.

SPECULATIONS ON THE ORIGIN OF SUPERCLUSTERS

What can superclusters teach us about the history of their formation? The problem is intimately connected with that of the origin of the galaxies. As is well-known two different scenarios have been proposed for this origin (Zeldovich *et al* 1982). The one, which has been first discussed by Zeldovich and the Moscow school, is that the density fluctuations in the Universe which gave rise to its structure were "adiabatic", matter and radiation density fluctuating together. In this case the only fluctuations which can have survived the radiation period are those with masses in excess of 10^{14} to 10^{15} $M_{\odot}$, which corresponds with the mass range of superclusters. After decoupling these fluctuations separated out from their surroundings, and finally collapsed. The collapsed regions will generally be flat (Zeldovich's "pancakes"), or filamentary. It is tempting to identify the strongly aspherical superclusters with such features. In this scenario galaxies would have formed as a consequence of the large-scale collapse. Because galaxies are old (most quasars were born between $z = 2.5$ and 3.5) the collapse should have occurred not much later than $z \sim 5$, when the Universe was roughly 1 milliard years old if $\Omega = 1$, or about 2 milliard years if $\Omega = 0.1$.

The alternative scenario is one where there would have been isothermal fluctuations, presumably most numerous for small masses. In this scenario the first objects to condense after decoupling would have masses of the order of 10^6 $M_{\odot}$, the Jeans mass at the time of decoupling. Gravitational interaction between these earliest condensations would lead to their agglomeration into larger and larger masses, finally into galaxies. Clustering could continue among the galaxies, and lead to the clumpy Universe we see around us. The hierarchical clustering process *might* by the present time even have formed the large clusters and perhaps superclusters.

The fundamental, as yet unanswered, question is: Did galaxies form first and superclusters afterwards, or were superclusters the first structures to form, and did galaxies form as a consequence of their collapse?

Apart from the question of whether isothermal fluctuations have existed, the hierarchical clustering hypothesis is faced with two specific difficulties: Could it in the time evolved since the formation of protogalaxies have led to structures as large as the largest known superclusters, and, secondly, could it have formed the strongly flattened and filamentary shapes indicated in some superclusters? Dekel, West and Aarseth (1983) have shown by a numerical many-body simulation that the orientation of the long axes of clusters along supercluster chains cannot be understood in the purely hierarchical scenario, but requires that the clusters have been formed in conjunction with these chains. But the evidence for this preferential orientation is not yet entirely compelling. Moreover, the galaxies might very well have existed previously, formed from isothermal fluctuations. The only requirement is that superimposed on the isothermal fluctuation spectrum there would have existed large waves leading to the formation of the superstructures.

It should be stressed that even if the purely adiabatic scenario is adopted hierarchical clustering must have played a decisive role in later stages, and must be responsible for the smaller-scale clumpiness of the Universe and for the form of the covariance function as observed today.

Some insight into whether superclusters formed in the gaseous phase of the Universe might ultimately be obtained from phenomena such as the segregation of galaxy types which was mentioned in connection with the Perseus supercluster.

The extreme smoothness of the microwave background radiation, with an upper limit $\delta T/T < 10^{-4}$ for fluctuations on the scale of superclusters and large clusters, puts interesting constraints on their origin. If density fluctuations corresponding to the above limit started to separate out at the time of decoupling they could not have collapsed in time for the formation of quasars and galaxies. Other phenomena, such as the large virial masses of clusters, the large rotation velocities in the outer region of spirals, the abundance of deuterium, have suggested that most of the mass of the Universe consists of invisible, non-interacting particles (for instance, heavy neutrinos). These same particles might also explain the discrepancy between the smallness of the background fluctuations and the epoch of supercluster formation.

For more detailed information on superclusters, see Oort (1983).

REFERENCES

Abell, G.O. 1958, Ap.J. Suppl. 3, p. 211.
Bahcall, N.A., Soneira, R.M. 1983, Ap.J. Submitted for publication.
Dekel, A., West, M.J., Aarseth, S.J. 1983, Ap.J., in press.
de Vaucouleurs, G.H. 1956, Vistas Astron. 2, p. 1584, and in The Large-Scale Structure of the Universe, IAU Symp. No. 79, ed. M.S. Longair, J. Einasto, p. 205. Cf. also: Meeting on Clusters and Groups of Galaxies (Trieste 1983), eds. F. Mardirossian, G. Giurcin and M. Mezzetti.
Dressler, A. 1980, Ap.J. 236, p. 351.
Einasto, J., Joêveer, M., Saar, E. 1980, MNRAS 193, p. 353.
Forman, W., Jones, C. 1982, Ann.Rev.Astron.Astrophys. 20, p. 547.
Geller, M.J., Meer, T.C. 1982, Publ.Astr. Soc. Pacific 94, p. 421.

Giovanelli, R., Haynes, M.P., Chincarini, G.L., 1983. In preparation.
Jones, Christine 1984, Meeting on Clusters and Groups of Galaxies (Trieste 1983) eds. F. Mardirossian, G. Giurcin and M. Mezzetti.
Nilson, P. 1973. Uppsala General Catalogue of Galaxies, Uppsala Obs. Ann. 6.
Oort, J.H. 1983, Ann.Rev.Astron.Astrophys. 21, p. 273.
Shapley, J., Ames A. 1932, Harvard Obs. Ann. 88, p. 43.
Tago, E., Einasto, J., Saar, E. 1983, MNRAS 206, p. 559.
Tully, R.B. 1982, Ap.J. 257, p. 389.
Zeldovich, Ya.B., Einasto, J., Shandarin, S.F. 1982, Nature 300, p. 407.
Zwicky, F., Wild, P., Herzog, E., Karpowicz, M., Kowal, C.T. 1961-1968. Catalogue of Galaxies and Clusters of Galaxies, Vols. 1-6, Pasadena, Calif. Inst. Technology.

COSMIC RAY SOURCES AND CONFINEMENT IN THE GALAXY

L.Koch-Miramond

Service d'Astrophysique, Centre d'Etudes Nucléaires, Saclay

ABSTRACT

Recently the Saclay-Copenhagen spectrometer on board the satellite HEAO 3 has provided extremely accurate data on the elemental composition and energy spectra of cosmic ray nuclei from Be to Zn in the energy range 0.7 to 25 GeV/n. These data have been interpreted in the framework of galactic diffusion model, either homogeneous or with a halo. The data tend to favor a rather flat halo, ~ 1 kpc thick. The mean escape pathlength from the Galaxy is found to decrease with the magnetic rigidity of the particles R as $R^{-0.6}$. The momentum spectra at the source for the main primary elements between C and Fe are well fitted by a single power law with an exponent $\gamma = -2.41 \pm 0.05$. Implications for cosmic ray sources and confinement of recent results on high energy cosmic nuclei, electrons and gamma rays are reviewed.

INTRODUCTION

Most of the present results in cosmic ray astrophysics agree with a galactic origin for cosmic rays. It is generally assumed that cosmic ray sources (CRS) are uniformly distributed in the galactic disk and that particles diffuse rapidly in a confinement volume with relatively slow leakage accross the boundary: this is the so-called leaky-box model (Davis, 1959; Cowsik et al, 1967). But wether cosmic rays are (i) free to wander around in the Universe, (ii) confined in a galactic halo, in the galactic disk or (iii) compelled to stay close to sources, are still open questions.

A. Berger (ed.), The Big Bang and Georges Lemaitre, 315–325.

The relevant observations are the cosmic ray composition and momentum distribution, the anisotropy in arrival directions at Earth, the gamma ray diffuse galactic emission and the radio synchroton diffuse galactic emission.

COSMIC RAY COMPOSITION AND MOMENTUM SPECTRA FROM HEAO 3

The experiment results from a collaboration initiated in 1968 between the Danish Space Research Institute headed by B.Peters and the Centre d'Etudes Nucléaires de Saclay in France. A detailed description of the instrument can be found in Bouffard et al (1982), and the main observational results in the Proceedings of the 17th and 18th International Cosmic Ray Conference held in Paris (1981) and Bangalore (1983).

The main results are the following:

1) Cosmic ray composition

Very accurate measurements of the relative abundances of 27 elements between Be and Zn as a function of energy in the range 0.7 to 25 GeV/n have allowed us to study the propagation of cosmic rays in the framework of galactic diffusion models, either homogeneous or with a halo (Koch-Miramond, 1981). The mean escape length λ_e of cosmic rays from the confinement volume has been derived from the measurement of the secondary over primary ratios B/C and (Sc + Ti + V + Cr)/Fe; the best fit to these data is given by:

$$\lambda_e = (22 \pm 2)\ R^{-0.60\pm0.04}\ \mathrm{g/cm^2} \text{ for } R > 5.5 \text{ GV}$$

$$\lambda_e = 7.9 \pm 0.7\ \mathrm{g/cm^2} \text{ for } R < 5.5 \text{ GV}$$

where R is the magnetic rigidity of the particle in GV and the medium traversed assumed to be pure hydrogen. Surprisingly enough the simple leaky-box formalism appears to fit adequately all the available data. In terms of diffusion models these results can be interpreted as implying that either the diffusion coefficient K is $\alpha\ vR^{0.6}$(v is the cosmic ray velocity) or that the size H of the confinement volume decreases with R : H α $R^{-0.6}$(H is the height of the halo in one-dimensional models; Cesarsky,1980). The radioactive cosmic ray clock ^{54}Mn has been used to derive the size of the galactic halo as seen by cosmic rays in the GeV/n range. The data tend to favour a rather flat halo ~ 1 kpc thick, i.e. only ~ 10 times thicker than the galactic gas disk.

The abundances of 16 elements have been derived at the cosmic ray source with an accuracy only limited by our knowledge on formation and destruction cross sections of cosmic ray nuclei in the interstellar medium. There are large differences between the source composition and the local galactic and solar system composition. However the overabundances in the cosmic ray source seem to be correlated with the first ionization potential of the elements (Cassé and Goret,1978). It is significant that the same correlation holds for the nuclei accelerated during solar flares (Meyer,1981). All nuclei between C and Ni in cosmic ray sources and in solar energetic particles have identical overabundances as compared to solar system abundances, with the remarkable exception of C which is twice as abundant in cosmic ray sources.

The data are compatible with the suggestion of an injection of cosmic ray particles by stellar flares in a two-stage acceleration process (Meyer 1983). The acceleration to high energy has to be prompt i.e must take place on a timescale short with respect to the cosmic ray mean escape time from the galaxy which is ~ 8 million years at 1 GeV/n (Wiedenbeck et al, 1981). The inferred mean density seen by cosmic rays is $n_H \sim 0.3\ cm^{-3}$, similar to the interstellar density within 1 kpc around the sun. The overabundance of C mentioned above may be related to the anomalous neon isotopic composition of cosmic rays (^{22}Ne being found 3 to 4 times more abundant in cosmic ray sources), if a minor component of cosmic rays is accelerated from material having undergone a specific nucleosynthetic process, such as quiescent helium burning in massive stars (Meyer 1983, Cassé, 1984).

But the main acceleration stage requires much more energy than can be provided by the flaring of ordinary stars. Since supernova explosions provide most of the energy into the interstellar medium, the resulting shock fronts have recently been suggested to be the sites of acceleration of suprathermal stellar flare particles (Axford et al 1977, Krimsky 1977, Ellison and Eichler 1984).

2. Energy spectra

The distribution in energy of He nuclei appears to follow the same power law from 20 to 10^6 GeV/n. The spectral index is $\gamma \simeq -2.83 \pm 0.20$ between 5.10^3 and 5.10^5 GeV/n (Burnett et al, 1983) and $\gamma \simeq -2.77 \pm 0.05$ between 15 and 500 GeV/n (Ryan et al, 1972). The H differential spectrum exhibits no drastic change of slope from 50 to 10^6GeV: $\gamma \simeq -2.7 \pm 0.2$ over the whole range.

The HEAO 3 data have been used to deduce the source spectra of C,N,O,Ne,Mg,Si,Ca and Fe at source after appropriate corrections for solar modulation, energy losses and nuclear interactions in the interstellar medium, and escape from the confinement region. The observed spectra of B,O,Si and Fe are shown in figure 1. Figure 2 shows the observed spectrum of O between 0.5 and 300 GeV/n using all the available data. The curve is the propagated source spectrum $dJ/dE \propto \gamma P^{-2.4}$ (P = momentum of

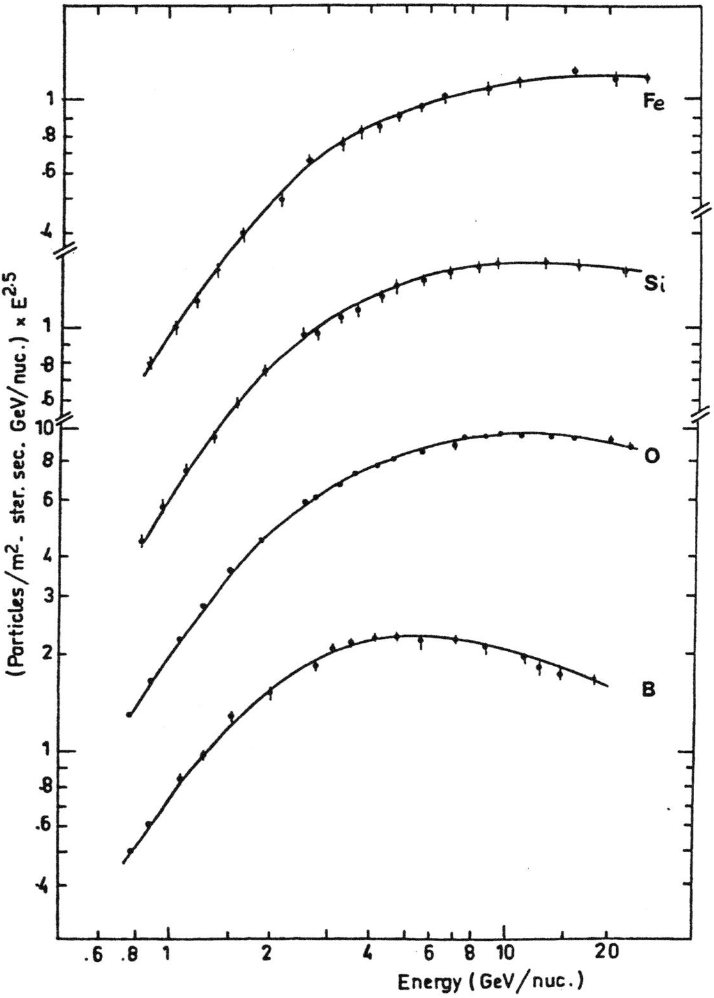

Figure 1.: Examples of energy spectra observed at Earth by the Saclay-Copenhagen experiment on-board HEAO 3 for mostly primary (Iron, Silicon and Oxygen) and for secondary (Boron) nuclei. Differential spectra have been multiplied by $E^{2.5}$. The curves are drawn only to guide the eye.

the particle), using a deceleration parameter in the solar cavity ϕ = 600 MV. The energy spectra at the source for the main primary elements from O to Fe are well fitted between 1 and 25 GeV/n by a single power law in momentum with an exponent $\gamma = -2.41 \pm 0.05$ (Engelmann et al,1984). At higher energies the spread in experimental data for Z > 2 particles is too large to draw any conclusion about the behaviour of their spectral shape. Only the He and H spectra are sufficiently well known between 50 and 5 10^5 GeV/n. They show a constant slope within the experimental errors. In the leaky-box model, the mean confinement time of particles τ_e is proportional to the mean escape length λ_e. Assuming that the confinement time of all nuclear species is the same, $\tau_e(R)$ must be a single power law at least up to $\sim 10^6$ GV (which seems likely considering the constancy of the slope of the observed spectra in the energy region where the escape losses are dominating). It follows that the He spectrum at the source has

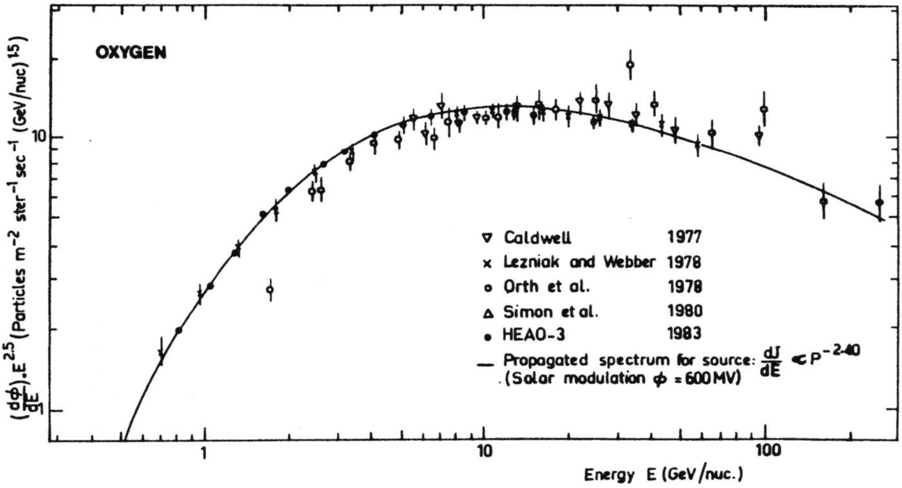

Figure 2: Comparison of the Oxygen intensities versus energy measured by different experimenters. The varius data sets have been normalized at 10 GeV/n. The differential spectra are multiplied by $E^{2.5}$. Continuous curves are propagated spectra for source dJ/dE α $P^{-2.4}$ and a solar modulation parameter ϕ=600 MV.

an index $\gamma \simeq -2.45 \pm 0.1$ between 10 and 30 GeV/n and $\gamma \simeq -2.25 \pm 0.1$ between 50 and 5.10^5 GeV/n. Hence the He source spectrum seems to become flatter when the energy increases. This is probably true also for the heavier nuclei since, as shown by extensive air shower measurements the average primary mass of cosmic rays does not vary appreciably up to 10^5 GeV (Lindsley, 1983).

This variation of the shape of the distribution of cosmic rays with energy could be explained in the framework of the acceleration by shock fronts if the Mach numbers are larger than 3.5 (Ellison and Eichler,1984), but this solution is not unique.

COSMIC RAY ELECTRON SPECTRA AND GALACTIC RADIO SYNCHROTON DIFFUSE EMISSION

1. Electron Spectra

The best recent results on the intensity of cosmic ray electrons versus energy (see Webber, 1982 for a review) show that the slope of the electron spectrum at earth is ~ -3.0 at 10 GeV and becomes steeper ~ -3.3, at higher energies. Below 10 GeV the solar modulation effects obscure the true spectrum. It can be estimated from the galactic radio emission which is mainly due to the synchroton losses of a few hundred MeV electrons moving in the galactic magnetic fields. Thus a measurement of the spectrum and spatial distribution of this radio emission provides a direct measure of the electron spectral shape in interstellar space.

Γ being the exponent of the electron spectrum power law, the radio spectrum spectral index $\alpha = \frac{1}{2}(\Gamma+1)$. The radio spectrum from our galaxy is well known from 1 MHz to $\sim$ 10000 MHz corresponding to electron energies of $\sim$ 150 MeV to $\sim$ 15 GeV. At 100 MHz, corresponding to electron energy of 1.5 GeV, one has $\alpha = -0.62 \pm 0.04$ giving $\Gamma = -2.2 \pm 0.1$ for the electron spectrum. To deduce the electron spectrum at source i.e after acceleration it is necessary to correct for all the losses. Since synchroton losses dominate at $E > 30$ GeV, from the exponent of the electron spectrum observed at high energy, $\Gamma \sim -3.3$, one deduces the exponent at source $\Gamma \sim -2.3$. It is quite similar to the exponent of the nuclear component of cosmic rays at source. But many problems remain in the interpretation of radio continuum data and in the evaluation of energy losses.

2) Electron Distribution in the Galaxy

In our Galaxy, the synchroton radio emission is partially correlated with spiral arms. It gives us some indications on the distribution of cosmic ray electrons in the galactic plane, but their interpretation is heavily hindered by our lack of knowledge of the magnetic field distribution. Although difficult to ascertain the presence of a radio halo in our Galaxy is suggested by the data and would imply that the half-thickness of the electron distribution perpendicular to the disk is $\sim$ 1 kpc. In the disk itself both radio continuum and gamma-ray diffuse emission maps (Fichtel et al,1978) suggest a moderate intensity contrast (~ a factor of 2) between the spiral arms and the interarm regions.

COSMIC RAY NUCLEI AND ELECTRON GRADIENTS AND GAMMA RAY DIFFUSE EMISSION

Detailed observations of the galactic gamma radiation have now been obtained by instruments on board of two satellites, SAS 2 and COS B, in the energy ranges 35 to 100 MeV and 50 to 5000 MeV respectively, (Fichtel et al, 1978, Mayer-Hasselwander et al, 1982). Two main processes contribute to the diffuse galactic gamma ray emission in this energy range: i) the decay of neutral pions generated by collisions of nuclear cosmic rays of energy greater than ~ 700 MeV with interstellar medium particles, mainly atomic and molecular hydrogen, ii) bremsstralhung emission which involves electrons of energy comparable to that of the gamma rays.

Lebrun et al (1983) have shown that the gamma ray emission at $E > 300$ MeV follows the interstellar matter density as given by the Berkeley 21 cm line HI survey (Heiles and Habing, 1974) and the CO line survey (Dame and Thaddeus,1983), CO being the tracer of H_2 . Gamma rays at $E > 300$ MeV originate mainly from cosmic ray proton interaction and can be explained if uniformly distributed cosmic ray proton interact with the interstellar gas. No cosmic ray proton gradient (within a factor 2) was found within the disk from the galactic center to 20 kpc from the center, in the anti-center direction (Bloemen et al 1984).

From the gamma ray emissivity at $E < 100$ MeV (an energy range where gamma rays originate mainly from electron bremstralhung), the same authors found that the relativistic electron density in the disk falls rapidly for galactocentric distances greater than 10 kpc (the solar distance), being ~ 0 at 18 kpc. Hence a strong cosmic ray electron gradient seems present outside the solar circle within the galactic equatorial plane. Significant variations in the intensity of cosmic ray electrons over the galaxy are also found with the SAS 2 data by Issa et al, 1980: a fall of at galactocentric distances greater than 10 kpc and a reduction by at least an order of magnitude, compared with the local intensity, for galactocentric distances smaller than 2 kpc.

ANISOTROPIES AT HIGH ENERGY AND CONFINEMENT

The COS B result argue in favor of a very big halo of cosmic ray protons of a few GeV in our galaxy. Is it still true at very high energy ? The recent results from anisotropy measurements have been put together by Hillas in Figure 3 and show very small anisotropies $\leqslant 0.2\%$ up to $E \sim 10^6$ GeV. This result gives confidence in the diffusion models used to interpret the observations. At higher energy a significant anisotropy appears, increasing as $E^{0.5}$ at least to ~ 10^{10} GeV. In the range 2.10^8 - 10^{10} GeV there is an intensity gradient in galactic latitude (Astley et al, 1981; Efimov et al, 1983), amounting to 0.2% per degree of latitude at 4 10^9 GV and corresponding to a deficit of flux from the north. The origin of the knee in the anisotropy and

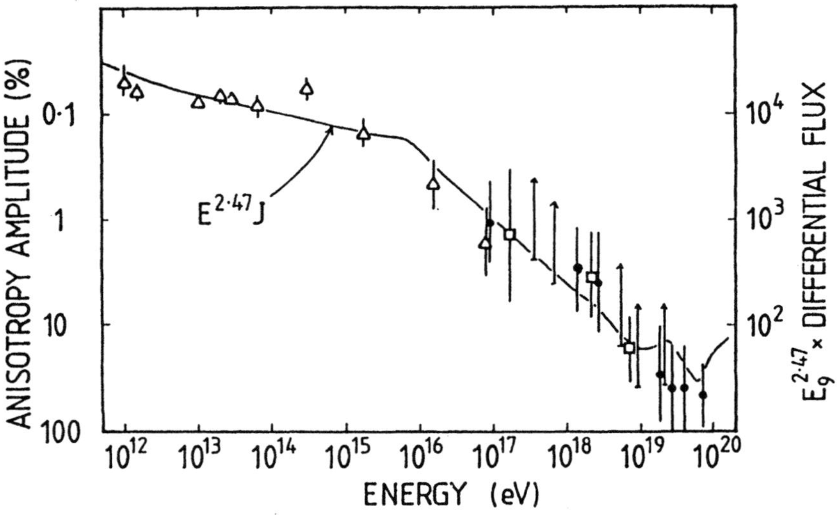

Figure 3: According to Hillas (1984) the observed values (corrected for solar motion below 10^{15} eV) of the amplitude A of the first Fourier component (24 hour sidereal period) of the cosmic ray intensity variation as the sky passes overhead (the simplest measure of anisotropy) versus energy. The inverse of A, as a measure of the mean residence time of cosmic rays in the confinement volume, is compared with variation in differential flux J(E) versus energy. As pointed out by Hillas the continuus curve $E^{2.47}$. J(E), which is in good agreement with the observed points, might suggest a single power law source spectrum $E^{-2.47}$ extending to 10^{19} eV.

differential flux vs energy around a few 10^6 GeV is unclear. Is it due to a leackage from the galaxy becoming more rapid for the more energetic particles or to a specific physical process arising in the sources (e.g. photonuclear reactions as particles from a pulsar escape through the radiation field of a very young supernova, Hillas, 1983) ? As shown by Berezinsky and Mikhailov (1983), the amplitude of the anisotropy is in reasonable agreement with cosmic rays originating in the galaxy up to 10^{10} GeV if one assumes a small component of the galactic magnetic field normal to the disk, even without requiring very much of a halo.

CONCLUSIONS

1) Cosmic ray nuclei abundances and momentum spectra for 30 species in the range 0.7 to 25 GeV/n can be consistently interpreted in the framework of the simplest homogeneous diffusion model in the Galaxy with escape mean free path $\lambda_e \propto R^{-0.6\pm0.1}$ for $5.5 \leq R \leq 50$ GV and λ_e constant below, and sources uniformly distributed in the galactic disk.

2) Identical source momentum spectra are found for all ($Z > 2$) primary nuclei, the spectral index at source being -2.41 ± 0.05 from 1 to 25 GeV/n.In this energy range, the spectral index obtained for He is in good agreement with that of heavier nuclei if the same escape law is assumed. This escape law holds probably at least till 10^6 GeV/n ; hence spectrum at source should have the same exponent, -2.2 ± 0.1, for all nuclei at energy greater than ~ 300 GeV/n (including H at energy greater than 50 GeV).Thus the source spectrum might become flatter when the energy increases.

3) This variation in spectral index at source versus energy could be explained in the framework of shock acceleration, with the efficiency of acceleration increasing with energy for strong shock fronts.

4) The cosmic ray electron spectrum at source appears similar (exponent -2.3 at 30 GeV) but the difficult evaluation of losses in the interstellar medium weakens this conclusion.

5) Gamma ray astronomy results from COSB tend to argue for a very big halo of relativistic cosmic ray protons in our galaxy with no apparent cosmic ray gradient from the galactic center to 20 kpc away. In contrast a very sharp relativistic electron gradient is seen outside the solar circle.

A galactic origin for cosmic rays even at the highest energies is favored by the present data although the smoothness of the energy spectrum from 10^7 to near 10^{10} GeV presents a great challenge to physicists working on cosmic accelerators. Hence the question "where do cosmic ray originate"? is still open to a lively debate.

....

Finally I would like to point out that our cosmic ray instrument on board the NASA Observatory HEAO3 was heavily relying-in order to derive the isotopic composition of cosmic rays from flux measurements-on the pioneering work done in the early 1930's by Georges Lemaître on the interaction of cosmic rays with the geomagnetic field (Lemaître and Vallarta, 1936).

REFERENCES

Astley S.M., Cunningham G., Lloyd-Evans, J., Reid R.J., Watson A.A. 1981, 17th Int.Cosmic Ray Conf.,Paris,2,156.
Axford W.I., Leer E., Skadron G.,1977, 15th Int.Cosmic Ray Conf.,Plovdiv, 11,132.
Berezinsky V.S., Mikhailov A.A., 1983, 18th Int.Cosmic Ray Conf., Bangalore, 2, 174.
Bloemen J.B.et al, the Caravane Collaboration,1984, Astron.Astrophys.,in press.
Bouffard M., Engelmann J.J., Koch-Miramond L., Lund N., Rasmussen I.L. Peters B., Soutoul A., and the Saclay - Copenhage Collaboration, 1982, Astr.and Sp.Sci. 84,3.
Burnett T.H. et al, the JACEE Collaboration,1983, 18th Int.Cosmic Ray Conf., Bangalore, 2,17.
Caldwell J.H.,1977,Ap.J., 218, 269.
Cassé M., Goret P., 1978, Ap.J.,221, 703.
Cassé M., 1984, Stellar Nucleosynthesis, D.Reidel, pp.55-75.
Cesarsky C., 1980, Ann.Rev.Astron.Astrophys., 18, 289.
Cowsik R., Yash Pal, Tandon S.N., Verma R.P., 1967, Phys.Rev.,140,B1147.
Dame T.M., Thaddeus P. 1983, in preparation.
Davis Jr L., 1959, 6th Int.Cosmic Ray Conf., Moscow, 3, 220.
Efimov N.M., Mikhailov A.A., Pravdin M.I., 1983, 18th Int.Cosmic Ray Conf., Bangalore, 2,149.
Ellisson D.C., Eichler D.,1984, Ap.J.,submitted
Engelmann J.J. et al, the Saclay-Copenhagen Collaboration, 1983, 18th Int.Cosmic Ray Conf.,Bangalore,OG2-4.
Engelmann J.J. et al the Saclay -Copenhagen Collaboration 1984,in preparation.
Fichtel C.E., Simpson G.A., Thompson D.J.,1978, Ap.J.,222,833
Goret P.et al., the Saclay-Copenhagen Collaboration,1983, 18th Int.Cosmic Ray Conf., Bangalore, 2,17.
Greisen K., 1966, Phys.Rev.Lett.,16,748.
Heiles C., Habing H.J.,1974, Astron.Astrophys.Suppl.14, 1.
Hillas A.M., 1975, 14th Int.Cosmic Ray Conf., Munich, 14,717.

Hillas A.M., 1983, Composition and Origin of Cosmic Rays ed.M.Shapiro, Reidel, pp.125-148.
Hillas A.M., 1984, Annual Rev.of Astron. and Astrophys.,22, in press.
Issa M.R.,Riley P.A., Strong A.W., Wolfendale A.W.,1980,Nature, 287,810.
Koch-Miramond L. et al, the Saclay-Copenhagen Collaboration 1981, Astron. and Astrophys.102, L9.
Koch-Miramond L.,1981,17th Int.Cosmic Ray Conf.,Paris,12,21.
Koch-Miramond L., et al, the Saclay-Copenhagen Collaboration, 1983,18th Int.Cosmic Ray Conf., Bangalore, OG5 2.12.
Lebrun F. et al., the Caravane Collaboration, and Dame T.H., Cohen R.S., Thaddeus P., 1983, Ap.J., submitted.
Lemaître G., Vallarta S.M., 1936, Phys.Rev.49,719.
Lezniak J.A., Webber N.R.,1978,Ap.J.,223,676.
Lindsley J., 1983, 18th Int.Cosmic Ray Conf., Bangalore, Rapporteur paper.
Lloyd-Evans, J., et al., 1983, Nature, 305,784
Mayer-Hasselwander H.A. et al, the Caravane Collaboration, 1982,Astron. Astrophys., 105, 164.
Meyer J.P., 1981, 17th Int. Cosmic Ray Conf.,Paris, 2, 281.
Meyer J.P., 1984, Ap.J. Suppl., in press.
Orth C.D., et al., 1978, Ap.J.,226,1147.
Ryan M.J., Ormes J.F., Balasubrahmanyan V.K., 1972, Phys.Rev.Lett.28,985.
Samorski M., Stamm W., 1983, Ap.J.Lett.268,L17.
Simon M. et al., 1980,Ap.J.,239,712.
Webber W.R., 1983, Composition and Origin of Cosmic Rays, ed.M.Shapiro, Reidel, pp.25-45.
Wiedenbeck M.E., Greiner D.E., 1980, Ap.J.Lett.,239, L 139.
Wiedenbeck M.E., Greiner D.E., 1981, Phys.Rev.Lett. 46, 682.

STRONG EVIDENCE FOR METAGALACTIC SHOCK WAVES AT REDSHIFTS $z \sim 2-3$

L.M. Ozernoy and V.V. Chernomordik

Department of Theoretical Physics
Lebedev Physical Institute
Moscow 117924, USSR

Explosions of quasars and young galaxies are believed to proceed at large redshifts - an assumption which is examined in the present work. When occured these explosions should create blast waves which are propagated in the metagalactic medium. The shock waves formed can produce, during radiative cooling stages, dense cold spherical shells around the epicentres of explosions. But even before that, at the stage of adiabatic expansion , each spherical shock wave front if it lies on the line of sight with a more distant quasar , can imprint into the quasar spectrum a specific absorption "doublet" with a distance between the components (in the rest frame) $\Delta\lambda_0 \lesssim 3$ Å. The components of each doublet have a small but the same equivalent width $W_0 \lesssim 0.3$ Å , the ratio $W_0/\Delta\lambda_0$ weakly depending on W_0. We demonstrate here that such doublets of Lyα lines are really present among the 'Lyα forest' in the absorption spectra of distant ($z= 2-3$) quasars which are now commonly believed to be of mostly intervening (and not intrinsic) origin.

Further accumulation of data on absorption doublets , which can serve as direct indicators of metagalactic shock waves , may provide valuable information about physical conditions in the intergalactic gas at large redshifts.

A full account of this work is published in Astrophys. Space Sci. 97, 19 (1983).

A. Berger (ed.), The Big Bang and Georges Lemaitre, 327.

ALGEBRAIC PROGRAMMING IN GENERAL RELATIVITY AND COSMOLOGY.

Y. De Rop, A. Moussiaux, P. Tombal*, A. Ronveaux

Département de Physique
Facultés Universitaires N.D. de la Paix
61, rue de Bruxelles, 5000 NAMUR - Belgium.

J. Demaret, J.L. Hanquin*

Institut d'Astrophysique
Université de Liège
4200 COINTE-OUGREE, Liège - Belgium.

* Boursier IRSIA.

ABSTRACT.

Some algebraic programs, developed recently in the framework of a collaboration between two teams of physicists from Liège and Namur, are briefly described.

They deal with topics in general relativity and cosmology such as : the field equations for Bianchi cosmological models, the general relativistic Hamiltonian formalism and its application to some vacuum inhomogeneous space-times and the search for metrics of stationary axisymmetric space-times.

These programs are written in the algebraic languages REDUCE 2 and 3 as well as in LISP (in the framework of the general program SHEEP developed in Stockholm and London).

INTRODUCTION.

The recent development of algebraic programming methods on computer has made possible the direct calculation of the explicit form of many analytic expressions and equations which are syste-

A. Berger (ed.), The Big Bang and Georges Lemaitre, 329–340.

matically present in miscellaneous topics of mathematical physics.

A collaboration between two teams of physicists, one from the "Département de Physique" of Namur, the second, from the "Institut d'Astrophysique" of Liège, is recently born; its aim is, on the one hand, to apply some existing algebraic programs to theoretical problems in general relativity, group theory and gauge theories and, on the other hand, to develop with the help of available algebraic languages (REDUCE 2 and 3 [1] , LISP, SHEEP [2] and, coming soon, MACSYMA) new original programs in mathematical physics.

More particularly, in the fields of general relativity and cosmology, we have recently implemented in Namur and Liège, the program SHEEP (written in LISP and developed in Stockholm and London), which constitutes an indispensable tool for the relativist.

We describe here briefly some programs recently developed in the framework of this collaboration : they deal with the field equations for Bianchi cosmological models, the Hamiltonian formalism in general relativity and its application to some vacuum inhomogeneous space-times and with stationary axisymmetric metrics.

Some other applications, in the field of general relativity, are also in progress and, especially, the extension of the Hamiltonian program to non-vacuum space-times (perfect fluids, electromagnetic and Yang-Mills fields).

BIANCHI MODELS.

Investigations of these spatially homogeneous cosmological models by computer were systematically done by one of us and published in a book [3] . Let us summarize briefly the standard approach [4].

The Bianchi-type metrics are written in a synchronous reference frame in the following way :

$$ds^2 = - dt^2 + g_{\alpha\beta}\, dx^\alpha\, dx^\beta \qquad (\alpha,\beta = 1,2,3) \tag{1}$$

The physical hypothesis of space homogeneity leads to the Killing equations, expressing the vanishing of the Lie derivative of the spatial part of the metric, with respect to the generators of the corresponding isometry group :

$$\mathcal{L}_\xi\, g_{\alpha\beta} = 0. \tag{2}$$

The Killing vectors, solutions of the Killing equations, define the vector fields :

$$X_a = \xi_a \frac{\partial}{\partial x^\alpha} \qquad (a = 1,2,3) \qquad (3)$$

with commutation relations written as :

$$[X_a, X_b] = C^c_{ab} X_c \quad . \qquad (4)$$

The systematical search for isometry groups is done by investigation of all real three dimensional Lie algebras which generate the nine Bianchi types. Using these symmetries, the metric for each model is written :

$$ds^2 = -dt^2 + \gamma_{ab}(t)\, \omega^a \omega^b \qquad (5)$$

with $\omega^a = e^a_\alpha dx^\alpha$, where the invariant basis vectors,

$$Z_a = e^\alpha_a \frac{\partial}{\partial x^\alpha}$$

are defined by $[X_b, Z_a] = 0$.

For each of the nine Bianchi models, the structure constants of the corresponding algebra are introduced in the computer (in REDUCE). The next choice refers to the diagonal or non-diagonal character of the γ_{ab}'s.

The program gives directly the form of the Ricci tensor in terms of the γ_{ab}'s and their two first time derivatives :

$$R_\alpha^{\ \beta} = R_\alpha^{\ \beta}(\gamma_{ij}, \dot{\gamma}_{kl}, \ddot{\gamma}_{mn}) \quad . \qquad (6)$$

For instance, for Bianchi IX model (homogeneity group = SO(3)), the input of a diagonal $\gamma_{aa}(t)$ in Misner's parametrization [5] ($\Omega, \beta_\pm$, functions of t) :

$$\begin{aligned} \gamma_{11} &= e^{-2\Omega} e^{2[\beta_+ + \sqrt{3}\,\beta_-]} \\ \gamma_{22} &= e^{-2\Omega} e^{2[\beta_+ - \sqrt{3}\,\beta_-]} \\ \gamma_{33} &= e^{-2\Omega} e^{-4\beta_+} \end{aligned} \qquad (7)$$

gives for the non zero mixed components of the Ricci tensor :

$$R^0_0 = 3(-2\,\dot{\beta}_-^2 - 2\,\dot{\beta}_+^2 - \dot{\Omega}^2 + \ddot{\Omega})$$

$$\begin{aligned} R^1_1 = {} & 3\sqrt{3}\,\dot{\beta}_-\,\dot{\Omega} + 3\,\dot{\beta}_+\,\dot{\Omega} - 3\,\dot{\Omega}^2 - \sqrt{3}\,\ddot{\beta}_- - \ddot{\beta}_+ + \ddot{\Omega} \\ & - \frac{1}{2}\,e^{2\Omega + 4\beta_+ + 4\sqrt{3}\,\beta_-} + \frac{1}{2}\,e^{2\Omega + 4\beta_+ - 4\sqrt{3}\,\beta_-} \\ & - e^{2\Omega - 2\sqrt{3}\,\beta_- - 2\beta_+} + \frac{1}{2}\,e^{2\Omega - 8\beta_+} \end{aligned} \tag{8}$$

$$\begin{aligned} R^2_2 = {} & -3\sqrt{3}\,\dot{\beta}_-\,\dot{\Omega} + 3\,\dot{\beta}_+\,\dot{\Omega} - 3\,\dot{\Omega}^2 + \sqrt{3}\,\ddot{\beta}_- - \ddot{\beta}_+ + \ddot{\Omega} \\ & + \frac{1}{2}\,e^{2\Omega + 4\beta_+ + 4\sqrt{3}\,\beta_-} - \frac{1}{2}\,e^{2\Omega + 4\beta_+ - 4\sqrt{3}\,\beta_-} \\ & - e^{2\Omega + 2\sqrt{3}\,\beta_- - 2\beta_+} + \frac{1}{2}\,e^{2\Omega - 8\beta_+} \end{aligned}$$

$$\begin{aligned} R^3_3 = {} & -6\,\dot{\beta}_+\,\dot{\Omega} - 3\,\dot{\Omega}^2 + 2\,\ddot{\beta}_+ + \ddot{\Omega} + \frac{1}{2}\,e^{2\Omega + 4\beta_+ + 4\sqrt{3}\,\beta_-} \\ & - e^{2\Omega + 4\beta_+} + \frac{1}{2}\,e^{2\Omega + 4\beta_+ - 4\sqrt{3}\,\beta_-} - e^{2\Omega - 8\beta_+} \,. \end{aligned}$$

The field equations in a vacuum or non-vacuum case can therefore easily be constructed.

The 200 pages Moussiaux's tables [3] give the Ricci tensors in almost all situations : a fully non-diagonal γ_{ab} matrix for some Bianchi models and a non-diagonal γ_{ab} matrix with at least one off-diagonal element for all Bianchi models.

The extension of this problem to spatially homothetic metrics [6] is also investigated and is in progress.

HAMILTONIAN FORMALISM.

The Hamiltonian formulation of general relativity initiated by Dirac [7] and Arnowitt, Deser and Misner (ADM) [8] appears as the privileged tool for the study of the dynamics of relativistic space-times of cosmological and astrophysical interest.

Hamiltonian cosmology, i.e. the study of cosmological models (essentially, spatially homogeneous and isotropic Friedman-

Lemaître-Robertson-Walker models and spatially homogeneous but anisotropic Bianchi and Kantowski-Sachs models) by this method, has attracted considerable interest in the last years ([4] ,[9]).

The Hamiltonian formalism, beyond its intrinsic interest in classical cosmology, constitutes also the first step of the important method of canonical quantization of classical space-times and, especially, cosmological models (quantum cosmology [9], [10]).

Progress in the study of the Hamiltonian formulation and canonical quantization of more complex space-times as, for example, inhomogeneous cosmological models or axisymmetric space-times, requires one to have at his disposal the expressions of the super-Hamiltonian and supermomenta constraints as well as of Hamilton's canonical equations in terms of the canonical variables.

The super-Hamiltonian, $\mathcal{H}$, and the super-momenta, $\mathcal{H}^i$, characeristic of the Hamiltonian formulation of vacuum space-times, are expressed as functions of the canonical variables, g_{ij} (i,j, = 1,2,3) and π^{ij} (the momenta canonically conjugate to the g_{ij}'s, cfr. [8] and [11] , Chapter 21), as follows :

$$\mathcal{H} = - \sqrt{g}\ \{R + g^{-1}\ [\tfrac{1}{2}\ (\pi^i_i)^2 - \pi^{ij}\ \pi_{ij}\ ]\} \tag{9-a}$$

$$\mathcal{H}^i = - 2\pi^{ij}{}_{|j} \tag{9-b}$$

g and R are respectively the determinant and the scalar curvature of the spatial metric tensor and the vertical stroke in (9-b) denotes covariant derivation with respect to the spatial metric. The constraint equations : $\mathcal{H} = 0$ are equivalent to the G_{oo} and G_{oi} vacuum field equations. The canonical equations, equivalent to the remaining vacuum field equations are given by

$$\dot{g}_{ij} = 2Ng^{-1/2}\ [\pi_{ij} - \tfrac{1}{2}\ g_{ij}(\pi^\ell_\ell)\ ] + N_{i|j} + N_{j|i} \tag{10-a}$$

and

$$\dot{\pi}^{ij} = - Ng^{1/2}\ (R^{ij} - \tfrac{1}{2}g^{ij}R) + \tfrac{1}{2}\ Ng^{-1/2}\ g^{ij}\ [\pi^{k\ell}\pi_{k\ell} - \tfrac{1}{2}(\pi^\ell_\ell)^2\ ]$$

$$- 2Ng^{-1/2}\ [\pi^{im}\pi^j_m - \tfrac{1}{2}\ \pi^{ij}(\pi^\ell_\ell)\ ] + g^{1/2}(N^{|ij} - g^{ij}N^{|m}{}_{|m}$$

$$+ (\pi^{ij}N^m)_{|m} - N^i_{|m}\pi^{mj} - N^j_{|m}\pi^{mi} \tag{10-b}$$

The dot denotes the time derivative and the components of the reciprocal spatial metric tensor, g^{ij}, are given by :

$$g_{ij}g^{jk} = \delta_i^{\ k} \tag{11}$$

N and N^i are, respectively, the lapse and shift functions given by

$$N = (-{}^{(4)}g^{oo})^{-1/2} \tag{12-a}$$

$$N^i = {}^{(4)}g^{oi}(N)^2 \ , \ N_i = {}^{(4)}g_{oi} \tag{12-b}$$

where the indes "o" corresponds to the time variable and the superscript (4) denotes space-time quantities.

We have developed a series of subroutines written in the algebraic languages REDUCE 2 and SHEEP, described respectively in [12] and [13], which allow one to obtain the explicit form of the constraints (9-a,b) and of the canonical equations (10-a, b), in a spatial Cartan basis (independent of time), characterized by its structure coefficients, or in a natural basis.

These programs use as input the components of the basis one-forms in a natural basis as well as the components of the metric tensor in the corresponding Cartan basis and automatically yield the complete set of constraint and canonical equations.

The REDUCE 2 program has been applied to Bianchi models as well as to a series of cosmological and non-cosmological inhomogeneous space-times. The results are described in detail in [12] : we give here, as an example, the results obtained for one model, the Kompaneets cylindrically symmetric model [14], for which the metric tensor in a natural basis is non-diagonal and is given by

$$g_{ij} = \begin{bmatrix} e^{2(\gamma-\psi)} & 0 & 0 \\ 0 & e^{2\psi} & \omega e^{2\psi} \\ 0 & \omega e^{2\psi} & \omega^2 e^{2\psi} + \alpha^2 e^{-2\psi} \end{bmatrix} \tag{13}$$

where the canonical variables ψ, γ, ω and α are functions of $x^1 = r$ and t. The canonically conjugate momenta, π^{ij}, automatically calculated by the program are given by :

$$\pi^{ij} = \begin{bmatrix} \frac{1}{2}\pi\gamma e^{-2(\gamma-\psi)} & 0 & 0 \\ 0 & \frac{e^{-2\psi}}{2}\{\pi\psi+\pi\gamma+\frac{\omega\ \pi_\alpha}{\alpha}\ e^{4\psi}+\alpha\pi_\alpha-2\omega\pi_\omega\} & \frac{1}{2}\{\ e^{-2\psi}\ -\ \frac{\omega\pi_\alpha}{\alpha}e^{2\psi}\} \\ 0 & \frac{1}{2}\ \{e^{-2\psi}\pi_\omega\ -\ \frac{\omega\pi_\alpha}{\alpha}\ e^{2\psi}\} & \frac{\pi_\alpha}{2\alpha}\ e^{2\psi} \end{bmatrix}$$

where π_ψ, π_γ, π_ω and π_α are the momenta canonically conjugated to ψ, γ, ω and α, respectively. (14)

The super-Hamiltonian and supermomenta have the following form :

$$\mathcal{H} = \frac{1}{2\alpha}\ e^{-\gamma+\psi}\ \{4\alpha''\alpha+4\psi'^2\alpha^2\ -\ 4\alpha\alpha'\gamma'\ +\ e^{4\psi}\omega'^2 -\ \pi_\alpha\pi_\gamma\alpha+\frac{1}{4}\pi_\psi^2\ +\ e^{-4\psi}\pi_\omega^2\alpha^2\}$$

$$\mathcal{H}^1 = e^{2(\psi-\gamma)}\{\pi_\omega\ \omega'\ +\ \pi_\gamma\gamma'\ +\ \pi_\alpha\alpha'\ -\ \pi_\gamma'\}$$

$$\mathcal{H}^2 = 0$$

$$\mathcal{H}^3 = 0 \qquad (15)$$

where the prime denotes the derivative with respect to r. Finally, the canonical equations are given by :

$$\dot{\psi} = \frac{e^{\psi-\gamma}}{4\alpha}\ N\pi_\psi\ +\ N^1\psi'$$

$$\dot{\gamma} = -\ \frac{e^{\psi\ \gamma}}{2}\ N\pi_\alpha\ +\ N^1\gamma'\ +\ N^{1\prime}$$

$$\dot{\omega} = \alpha e^{-\gamma-3\psi}N\pi_\omega\ +\ N^1\omega'$$

$$\dot{\alpha} = -\ \frac{e^{\psi-\gamma}}{2}N\pi_\gamma\ +\ N^1\alpha'$$

$$\dot{\pi}_\psi = \frac{e^{-\gamma+\psi}}{2\alpha}\ \{N\ [\ 8\alpha^2\psi''\ -\ 8\alpha^2\psi'\gamma'\ +\ 8\alpha\psi'\alpha'\ +\ 4\alpha^2\psi'^2\ +\ 4\alpha\gamma'\alpha'\ -\ 4\alpha\alpha''$$

$$-\ 10\ e^{4\psi}\omega'^2\ +\ \alpha\pi_\alpha\pi_\gamma\ +\ 3e^{-4\psi}\alpha^2\pi_\omega^2\ -\ \frac{1}{4}\ \pi_\psi^2\ ]\ +\ 8N'\alpha^2\psi'\}$$

$$+\ N^1\pi_\psi'\ +\ N^{1\prime}\ \pi_\psi$$

$$\dot{\pi}_\gamma = \frac{e^{-\gamma+\psi}}{2\alpha}\{N[4\alpha^2\psi'^2 - 4\alpha\psi'\alpha' + e^{4\psi}\omega'^2 - \alpha\pi_\alpha\pi_\gamma + e^{-4\psi}\alpha^2\pi_\omega^2 + \frac{1}{4}\pi_\psi^2] - 4N'\alpha\alpha'\} + N^1\pi_\gamma' + N^{1'}\pi_\gamma$$

$$\dot{\pi}_\omega = \frac{e^{5\psi-\gamma}}{\alpha^2}\{N[\alpha\omega'' + 5\alpha\psi'\omega' - \alpha\gamma'\omega' - \alpha'\omega'] + N'\alpha\omega'\} + N^1\pi_\omega' + N^{1'}\pi_\omega$$

$$\dot{\pi}_\alpha = \frac{e^{-\gamma+\psi}}{2\alpha^2}\{N[-8\alpha^2\psi'^2 + 4\alpha^2\psi'\gamma' + e^{4\psi}\omega'^2 - e^{-4\psi}\alpha^2\pi_\omega^2 + \frac{1}{4}\pi_\psi^2 - 4\psi''\alpha^2] + N'[4\alpha^2\gamma' - 8\alpha^2\psi'] - 4N''\alpha^2\} + N^1\pi_\alpha' + N^{1'}\pi_\alpha \quad (16)$$

The other inhomogeneous space-times to which this program has been applied include some inhomogeneous generalizations of Bianchi and Kantowski-Sachs models, plane-symmetric and Gowdy T^3 universes, spherically symmetric and axisymmetric space-times and, more recently, spatially homothetic space-times, Gowdy $S^1 \otimes S^2$ and S^3 cosmological models [15] and generalized Einstein-Rosen metrics [16] .

The first examples treated with the SHEEP program show that this program is of the order of three times more rapid than the original REDUCE program.

The original program dealt only with the gravitational (or "vacuum") part of the Hamiltonian formalism. Recent work to extend it to space-times with perfect fluids [17] as well as electromagnetic and Yang-Mills fields as sources of the gravitational field, is in progress.

The possibility of providing the theoretical tools essential to the study of the dynamics and of the canonical quantization of space-times as complex as the axisymmetrical cases, clearly demonstrates the interest of the algebraic programming in the Hamiltonian formalism.

STATIONARY AXISYMMETRIC METRICS AND ERNST'S EQUATION.

Algebraic programming is particularly useful to construct axisymmetric metrics, solutions of Einstein's field equations. It is well known, since Ernst [18] that the search for solutions of the vacuum field equations for stationary axisymmetric space-times reduces to the resolution of one scalar (non linear) complex equation and that knowing any solution of this equation, all independent g_{ij}'s (three) can then be obtained by algebraic manipulation or by integration only.

Let us describe the mathematical recipe in a few steps. The stationary axisymmetrical metric is written in the Weyl form :

$$ds^2 = f^{-1} [e^{2\gamma}(d\rho^2 + dz^2) + \rho^2 d\phi^2] - f(dt - \omega \, d\phi)^2 \tag{17}$$

where f, γ and ω are functions of ρ and z (the usual cylindrical coordinates) only.

The Ernst (complex) equation is the following :

$$(\xi\xi^\star - 1)\Delta\xi = 2 \, \xi^\star \, \vec{\nabla}\xi \, . \, \vec{\nabla}\xi \tag{18}$$

where $\Delta\xi$ and $\vec{\nabla}\xi$ are respectively the cylindrical Laplacian and gradient of the complex solutions $\xi(\rho,z)$. ($\xi^\star$ is the complex conjugate of ξ). Now the function $f(\rho,z)$ is simply given by :

$$f = \mathrm{Re}\left(\frac{\xi - 1}{\xi + 1}\right) \tag{19}$$

and the functions $\gamma(\rho,z)$ and $\omega(\rho,z)$ are given by a path integral from the two integrable systems :

$$\frac{\partial\gamma}{\partial\rho} = \frac{\rho}{(\xi\xi^\star - 1)^2} \left\{ \frac{\partial\xi}{\partial\rho}\frac{\partial\xi^\star}{\partial\rho} - \frac{\partial\xi}{\partial z}\frac{\partial\xi^\star}{\partial z} \right\}$$

$$\frac{\partial\gamma}{\partial z} = \frac{2\rho}{(\xi\xi^\star - 1)^2} \mathrm{Re}\left(\frac{\partial\xi}{\partial\rho}\frac{\partial\xi^\star}{\partial z}\right) \tag{20}$$

$$\frac{\partial\omega}{\partial\rho} = \frac{2\rho}{(\xi\xi^\star - 1)^2} \mathrm{Im} \, (\xi^\star + 1)^2 \frac{\partial\xi}{\partial z}$$

$$\frac{\partial\omega}{\partial z} = - \frac{2\rho}{(\xi\xi^\star - 1)^2} \mathrm{Im} \, (\xi^\star + 1)^2 \frac{\partial\xi}{\partial\rho} \tag{21}$$

The path integration for these two functions is usually performed in prolate coordinates, x and y, related to ρ and z by :

$$\rho = k(x^2 - 1)^{1/2} (1 - y^2)^{1/2} \tag{22}$$

$$z = k\,x\,y \qquad (k = \text{positive constant})$$

From any solution $\xi(x,y)$ of the transformed Ernst equation, the right-hand sides of the f-equation (19) and of the γ- and ω-systems ([20] and [21]) are known; the algebraic program wirtten in REDUCE 3 [19] gives the f-solution and the γ- and ω-integrals (from the point x_o,y_o to the point x,y by integration with respect to x first, followed then by a y-integration).

Let us illustrate this by two simple examples :

1. The following elementary function $\xi = px - iqy$ (p and q are real constants) is injected in the Ernst operator by a subroutine TESKS which gives :

$$2(xp^3 + xpq^2 - xp + iyp^2q + iyq^3 - iyq) = 0 \quad . \tag{23}$$

   Now a factorization procedure called VTESKS gives :

$$2(px + iqy)\,(p^2 + q^2 - 1) = 0 \tag{24}$$

   The condition $p = (1 - q^2)^{1/2}$ is then introduced in the γ- and ω-integrals and the computer gives automatically the corresponding solutions, leading to the Kerr metric [20] .

2. The same procedure is used for a more sophisticated metric (Tomimatsu-Sato type [21]) :

$$\xi(x,y) = \frac{p^2(x^4 - 1) - 2\,ipqxy(x^2 - y^2) - q^2(1 - y^4)}{2px(x^2 - 1) - 2\,iqy\,(1 - y^2)} \tag{25}$$

   VTESKS factorizes now in a non-trivial way the resulting Ernst operator, generating again the condition $p^2 + q^2 = 1$, and the complete metric is then reconstructed by more complicated integrations.

ACKNOWLEDGMENT.

We are grateful to the Belgian "Fonds National de la Recherche Scientifique" for financial assistance in the development of this research project connected with algebraic programming in mathematical physics.

REFERENCES.

1. Hearn, A.C., "Reduce 2, User's Manual", Univ. of Utah, Salt Lake City, Utah 84112, U.S.A.
Hearn, A.C., "Reduce, User's Manual", version 3.0. The Rand Corporation, Santa Monica, Ca. 90406, U.S.A.
2. Frick, I., "SHEEP Manual", Inst. of Theoretical Physics, Univ. of Stockholm, Vanadisvagen 9, 11346 Stockholm, Sweden.
3. Moussiaux, A., General Relativity Table for Homogeneous Cosmological Models, ed. A. Wesmael-Charlier, Namur, Belgium, 1982.
4. Ryan, M.P. and Shepley,L.C., Homogeneous Relativistic Cosmologies, Princeton Univ. Press, Princeton, New Jersey, 1975. MacCallum, M.A.H., Physics of the Expanding Universe, Lecture Notes in Physics, 109, ed. Demianski, M., 1, Springer-Verlag, Berlin, 1979.
5. Misner, C.W., Phys. Rev. 186, 1319, 1969.
6. Eardley, D.M., Commun. Math. Phys. 37, 287, 1974.
7. Dirac, P.A.M., Lectures in Quantum Mechanics, Belfer Graduate School of Science Monograph Series n° 2, Yeshiva University, New York, 1964.
8. Arnowitt, R., Deser, S. and Misner, C.W.,"Gravitation, an introduction to current research",ed. Witten, L., Wiley, New York, p. 227, 1962.
9. Ryan, M., Hamiltonian Cosmology, Lecture Notes in Physics, vol. 13, ed. Ehlers, J., Hepp, K. and Weidenmüller, H.A., Springer Berlin, 1972.
10. MacCallum, M.A.H., Quantum Gravity : an Oxford Symposium, ed. Isham, C.J., Penrose, R., Sciama, D.W., Clarendon, Oxford, p. 174, 1975.
11. Misner, C.W., Thorne, K.S. and Wheeler, J.A., Gravitation, Freeman, San Francisco, 1973.
12. Moussiaux, A., Tombal, P. and Demaret, J., Gen. Rel. Grav. 15, 209, 1983.
13. Moussiaux, A., Tombal, P. and De Rop, Y., Hamil.SHP : "SHEEP" programming of Hamiltonian formalism in general relativity, Fac. Univ. N.D. de la Paix, Namur, Département de Physique, publication interne, 1983.
14. Kompaneets, A.S., Sov. Phys. JETP, 7, 659, 1958.
Chitre, D.W., Güven, R. and Nutku, Y., J. Math. Phys. 16, 457, 1975.
15. Hanquin, J.L. and Demaret, J., J. Phys. A 16, L5, 1983.
16. Hanquin, J.L., in preparation, 1984.
17. Moussiaux, A., Tombal P. and De Rop, Y., Hamir.SHP, Introduction de la matière dans le formalisme hamiltonien, Fac. Univ. N.D. de la Paix, Namur, Département de Physique, publication interne, 1983.
18. Ernst, F.J., Phys. Rev. 167, 1175, 1968.
19. Moussiaux, A. and Tombal, P., "Algebraic computation of the line element from a solution of Ernst equation", preprint,

Namur, 1983.
20. Kerr, R.P., Phys. Rev. Lett. 11, 237, 1963.
21. Tomimatsu, A. and Sato, H., Phys. Rev. Lett. 29, 1344, 1972; Prog. Theor. Phys. 50, 95, 1973.

FINITE EUCLIDEAN AND NON-EUCLIDEAN GEOMETRY WITH APPLICATIONS TO THE FINITE PENDULUM AND THE POLYGONAL HARMONIC MOTION. A FIRST STEP TO FINITE COSMOLOGY.

René De Vogelaere

Department of Mathematics
University of California, Berkeley

ABSTRACT

Because the Universe is both finite and atomic it is desirable to have a geometry which satisfies most of the properties of Euclidean or non-Euclidean geometry for which the number of points on each line is finite. This paper presents essential features of such geometries. The beginning of finite mechanics is present through the application to the finite circular pendulum and to the harmonic polygonal motion, for which time is also discrete.

0. INTRODUCTION

This international symposium is to honor the 50th anniversary of the major contribution of Monseigneur Georges Lemaitre, the theory of the expanding universe. As a former student, I would like to present a summary of research carried on during these last three years which might very well lead to contributions in cosmology many years from now and is in the spirit of Lemaitre's teachings at Louvain.

For many years, I found it puzzling that to study a Universe which is both finite and atomic, we use a model which is both infinite and continuous. The reason goes back 2500 years ago. Before that, the Pythagorian school, using results they had obtained, as well as former results of the Babylonians (e.g. the tablet Plimpton 322 (1)), were convinced that integers (through their ratio) were sufficient to describe all points in geometry.

A. Berger (ed.), The Big Bang and Georges Lemaitre, 341–355.

The scandal of the discovery of irrationals that $x^2 = 2y^2$ has no integer solutions, was immediately and irrevocably applied to geometry with the consequences of forcing the discovery of irrational points and real points and imposing the Euclidean line to be continuous. At no time before the 19th century was it realized that the application of this algebraic result to Euclidean geometry implied the assumption that a circle had points in common with every line through its center. If this had been realized by the Pythagorean school, a finite Euclidean geometry might have developed parallel to the infinite case. Instead, results, which form steps along the way, have been discovered independently.

I will mention a few: modular arithmetic (Aryabatha (2)), the existence of primitive roots (Gauss (3)), Galois fields, finite projective geometry (Veblen (4)), p-adic fields (Hensel (5)), finite elliptic functions (Tate (6)).

The results I will describe, were first conjectured using the computer and later proven in the framework of finite projective geometry using tools from number theory and algebra. They concern finite Euclidean geometry, finite trigonometry, finite non-Euclidean geometry, finite elliptic functions with application to the pendulum, in which "time" is also discretized, and the finite harmonic polygonal motion which presents some analogy with the motion on an ellipse under central force.

1. FINITE PROJECTIVE GEOMETRY

It is well known that to any power of a prime, p^k, corresponds a finite projective geometry. I will exclude $p = 2$ to avoid complete quadrangles with collinear diagonal points.

The axioms for these geometries will now be given, for $k = 1$ in the 2-dimensional case, both in synthetic form and in algebraic form. The relation between these two forms can be found in Dembowski (7).

1.0 Axioms (Synthetic Form)

Given the primaries, points, lines and incidence, and the prime p, p different from 2,

1) Two points are incident to one and only one line.
2) Two lines are incident to one and only one point.
3) (Pappus) Given two distinct lines a and b, three distinct points A0,A1,A2 on a, three distinct points on B0,B1,B2 on b, C2 the point incident to the line c01 which is incident to A0 and B1 and to the line c10 which is incident to A1 and B0, and similarly for C1 and C0,

then C0, C1 and C2 are incident to the same line.

4) There exist one line l, with p+1 distinct point on it.
5) There exist two points not on l.

It follows from these axioms that there exist p^2+p+1 points and lines, that each line has p+1 point incident to it and that each point has p+1 lines incident to it.

1.1 Axioms (Algebraic Form)

Given the prime p and the associated field Z[p], in which addition and multiplication are done modulo p,

1) Two triples P0,P1,P2 and Q0,Q1,Q2 of elements P0,P1,P2, Q0,Q1,Q2 in Z[p] are equivalent if there exists a non-zero element k in Z[p] such that $Q0 = k\,P0$, $Q1 = k\,P1$, $Q2 = k\,P2$.
2) The set of equivalent triples with the notation $P = (P0,P1,P2)$ is called a point.
3) The set of equivalent triples with the notation $l = [L0,L1,L2]$ is called a line.
4) The point $P = (P0,P1,P2)$ is incident to the line $l = [L0,L1,L2]$ if and only if $P0.L0 + P1.L1 + P2.L2$ is congruent to 0 modulo p.

It is easy to see that we can choose one of the equivalent set of triples in such a way that the first element different from zero is equal to 1, and that these points and lines satisfy the axioms in the synthetic form.

These definitions can be extended to the Galois field corresponding to p^k, $k > 1$. In particular if $k = 2$, we have the finite complex projective plane in which the integers modulo p are replaced by $a + b\delta$, with a and b integers modulo p and $\delta^2 = d$, a non-quadratic residue modulo p.

2. FINITE EUCLIDEAN GEOMETRY

There are many ways to derive the classical Euclidean geometric plane from the classical projective plane. I will now describe a method which allows the extension of the process to the finite case.

In classical Euclidean geometry, let A0, A1 and A2 be the vertices of a triangle, let M0, M1 and M2 be the midpoints of the sides (M2 on A0 A1,...), let the barycenter M be the point common to A0 M0, A1 M1 and A2 M2, let H0, H1, H2 be the feet of the perpendiculars from the vertices to the opposite sides. (A0 H0 perpendicular to A1 A2, etc.), let the orthocenter H be the point common to A0 H0, A1 H1, A2 H2. Then M1 M2 is parallel to A1 A2 and therefore meets A1 A2 in an ideal point N0 on the line at infinity, similarly for N1 and N2. Let A0 H0 meet the line at infinity at I0, ... It is well known that N0 I0, N1 I1, N2 I2 are pairs of an involution (projectivity of order 2) on the line at infinity, and that the fixed points of this involution are the non real isotropic points whose homogeneous Cartesian coordinates are

$$(1,\ i,\ 0) \quad \text{and} \quad (1,\ -i,\ 0)\ .$$

The isotropic points are common to all circles, indeed these have as an equation in homogeneous coordinates

$$(x-az)^2 + (y-bz)^2 = R^2z^2 \quad ,$$

and $x=1$, $y=i$ or $-i$, $z=0$, satisfy this equation.

The projective plane is obtained from the Euclidean plane by considering that the ideal points or directions in Euclidean geometry are ordinary points in projective geometry and the line at infinity, consisting of all the ideal points, is an ordinary line.

To obtain the Euclidean plane from the projective plane we have to proceed in a reverse order, we choose one line as the line at infinity and we define on that line a fundamental involution. This can be done by starting from a triangle A0,A1,A2 and two points M and H (such that 3 of the 5 points are not incident to the same line). First we obtain M0 on A1 A2 and A0 M,...; we obtain N0 on A1 A2 and M1 M2,...; the line m at infinity is the line N0,N1,N2; then we obtain I0 on m and A0 H,... . This defines pairs (N0,I0), (N1,I1), (N2,I2), of the fundamental involution, whose fixed points are the isotropic points. Any conic passing through these points, real or not, is defined as a circle.

Other pairs of the fundamental involution are defined as perpendicular directions. If the fixed points of the fundamental involution are not real, the geometry obtained is isomorphic to the 2-dimensional Euclidean geometry. Otherwise, we obtain a geometry which partakes in many properties of Euclidean geometry but should be given another name, say, paraeuclidean.

We move now to the finite case. In the finite projective plane associated to p, each line is incident to p+1 points, each point is incident to p+1 lines. We can repeat the construction given above and obtain a distinguished line, called the ideal line or line at infinity and a fundamental involution. Let us assume that the triangle A0,A1,A2 and the points M and H are such that the fundamental involution has no real points. Then none of the p+1 points on the ideal line are distinguished; we call them ideal points, points at infinity, or directions.

Starting with a polarity (8)(9), like in the infinite case, it is well known that a point conic is defined as the set of points which are on their own polar and that each conic contains p+1 points and intersects a line in 0, 1 or 2 points. A conic defines an involution on any line 1, when we associate to a point P on 1 the intersection with 1 of the polar of P. If the involution defined by a conic on the ideal line m coincides with the fundamental involution, the conic is defined as a circle. The center of a circle is the pole of m with respect to it. The conic passing through the points M0,M1,M2 and through the intersection H0 of A0 H and A1 A2, and H1 and H2 is a circle, which, by analogy with the classical case, will be called the circle of Brianchon-Poncelet (10), also called the circle of Euler or of Feuerbach.

In finite Euclidean geometry, each line has on it one ideal point, on m, and p ordinary points. Any pair of points A,C can be chosen to define a unit of length on that line. To relate units on lines having the same direction we use parallelograms. To relate units on lines which have different directions we can use the circle centered at one of the points C, through the other, A. Because only half of the lines through P intersect the circle, something else has to be done on the other lines. If we take a point P on one of these lines, and define the distance between C and P as δ, distances between any two points on that line can be obtained and then by using circles centered at C through P, we obtain all distances. δ is not arbitrary and is chosen in such a way that the theorem of Pythagoras is satisfied. δ is not real and its square is a non-quadratic residue modulo p.

An elegant proof, justifying the notion of distances in finite Euclidean geometry, is obtained by first defining a finite trigonometry, suggested by the above considerations. This finite trigonometry will be defined in the next section.

An angle is associated to a pair of directions. Addition of angles can be defined and it is possible to prove (11), although the proof is far from trivial, that the abelian group associated to the addition of angles is cyclic and therefore has generators. One of these can be chosen as unit, the others are multiples modulo p+1. Because p+1 is even, angles can be considered as being even or odd. An angle which is odd cannot be bissected; an angle which is even can be bissected by 2 lines. This is the first evidence of a principle which recurs often: the principle of compensation. The sum of the angles of a triangle is even, therefore, if two angles are even the third one is also. Therefore, for about one triangle in four, all angles can be bissected and there are four circles tangent to the three sides. About three triangles out of four do not have inscribed circles. Once this fact is taken into account it is possible to generalize the classical results of Euclidean geometry involving bissectrices and inscribed circles. For instance, for even triangles, the Theorem of Feuerbach holds: The four circles tangent to the three sides of the triangle are tangent to the circle of Brianchon-Poncelet.

One of the advantages of the approach given above is that by simply exchanging the role of M and H, and continuing any construction derived from these points and the original triangle, new results can be obtained, which essentially double the number of theorems in finite as well as classical Euclidean geometry.

It would be inappropriate to give here even a partial list of all the results which have been generalized. Suffice to say that the algebraic proofs devised use the algebraic structure field (corps) and therefore are also valid, in most instances, for the infinite case, rational or real. This allows insurance that points defined in very different ways are either always identical or usually distinct. It has also allowed me to prove a very large number of new results in classical Euclidean geometry which were first conjectured on the computer and then proven in the finite and infinite case.

The results of finite 2-dimensional Euclidean geometry have been generalized to n dimensions using the exterior algebra introduced by Grassmann (12),(13). In the finite 3-dimensional case, the geometry of the tetrahedron and of the orthogonal tetrahedron (14) lead to an interesting new perspective on the subject.

3. FINITE TRIGONOMETRY

Let j denote +1 or -1. Given the sets Z of the integers, Z_p of the integers modulo p, Z_{p+j} of the integers modulo $p+j$, let δ be a square root of a non-quadratic residue of p. The problem addressed here is to construct two functions sin and cos with domain Z and range $\{Z_p, \delta Z_p\}$ which satisfy the trigonometric identities

$$\sin^2(x) + \cos^2(x) = 1 \qquad (0.0)$$

$$\sin(x+y) = \sin(x)\cos(y) + \cos(x)\sin(y) \qquad (0.1.0)$$

$$\cos(x+y) = \cos(x)\cos(y) - \sin(x)\sin(y) \qquad (0.1.0)$$

the periodicity property

$$\sin(2p+2j+x) = \sin(x) \;, \quad \cos(2p+2j+x) = \cos(x) \qquad (1.0)$$

the symmetry properties

$$\sin(p+j+x) = -\sin(x) \quad , \quad \cos(p+j+x) = -\cos(x) \tag{2.0}$$

$$\sin(-x) = -\sin(x) \quad , \quad \cos(-x) = \cos(x) \tag{2.1}$$

$$\sin(p+j-x) = \sin(x) \quad , \quad \cos(p+j-x) = -\cos(x) \tag{2.2}$$

and such that

$$\sin(0) = 0 \quad , \quad \cos(0) = 1 \tag{3.0}$$

$$\sin((p+j)/2) = 1 \quad , \quad \cos((p+j)/2) = 0 \tag{3.1}$$

$$\sin(x) \neq +1 \text{ and } \neq -1 \quad \text{for } 0 < x < (p+j)/2 \quad . \tag{3.2}$$

As usual,

$$\tan(x) = \sin(x)/\cos(x) \quad . \tag{4.0}$$

Several examples follow:
For $p = 11$ and $j = \delta^2 = -1$,

x	sin(x)	cos(x)	tan(x)
0	0	1	0
1	-2	5δ	-4δ
2	2δ	4	-5δ
3	4	2δ	-2δ
4	5δ	-2	3δ
5	1	0	∞

For $p = 11$, $j = 1$, $\delta^2 = -1$,

x	sin(x)	cos(x)	tan(x)
1	8δ	1δ	8
2	6	8	9
3	4δ	4δ	1
4	8	6	5
5	1δ	8δ	7
6	1	0	∞

For $p = 13$, $j = -1$, $\delta^2 = 2$,

x	sin(x)	cos(x)	tan(x)
0	0	1	0
1	-2δ	4δ	6
2	-6	-2	3
3	6δ	6δ	1
4	-2	-6	-4
5	4δ	-2δ	-2
6	1	0	∞

For $p = 13$, $j = 1$, δ^2

x	sin(x)	cos(x)	tan(x)
1	3	3δ	7δ
2	5δ	9	2δ
3	5	12δ	4δ
4	12δ	5	5δ

5	9	5δ	10δ
6	3δ	3	δ
7	1	0δ	∞

Elsewhere (11), I will give a proof that such functions exist. For $j=-1$, the proof depends on properties of primitive roots. For $j=+1$, the proof depends on a generalization to some cyclotomic polynomials. These functions are not uniquely defined, but are interrelated. I will also give there an algorithm to compute these functions for large p.

It is now easy to justify that if the ideal is $z=0$ and the fundamental involution is associated to the circle $x^2+y^2=z^2$, if $P=(P0,P1,1)$ and $Q=(Q0,Q1,1)$ then, if $d(P,Q)$ denotes the distance between P and Q,

$$(d(P,Q))^2 = (Q0-P0)^2 + (Q1-P1)^2 \quad .$$

For instance, for $p=13$ and $\delta^2=2$, if $P=(0,0,1)$ and

$$Q = (1,1,1)\ , \quad d(P,Q) = \sqrt{2} = \delta \quad ,$$

$$Q = (1,2,1)\ , \quad d(P,Q) = \sqrt{5} = 3\delta \quad ,$$

$$Q = (1,3,1)\ , \quad d(P,Q) = \sqrt{10} = 6 \quad .$$

4. FINITE NON-EUCLIDEAN GEOMETRY

The classical non-Euclidean geometry are of two types, elliptic of Bolyai (15) and hyperbolic of Lobatchevski (16). In the finite case such distinction does not exist. A conic being chosen as the ideal, a line which is not tangent to the ideal will intersect this conic at either two points or at no points. There are two scales for the distances, one is modulo p-1 and the other is modulo p+1.

Using homogeneous coordinates, $X=(X0,X1,X2)$, and the notation

$$|A| = \sqrt{A.A} \quad , \qquad A.B = A0\ B0 + A1\ B1 + A2\ B2 \qquad (0)$$

the ideal conic is chosen as

$$X.X = 0 \quad \text{or} \quad X0^2 + X1^2 + X2^2 = 0 \quad . \qquad (1)$$

Let j denote +1 or -1, the distance $d(A,B)$ between two points. A and B is then defined by

$$\cos(d(A,B)) = j\ A.B/(|A|\ |B|) \quad . \qquad (2)$$

Given a trigonometric table, to obtain an unambiguous definition, we have to choose $d(A,B)$ in the "first quadrant". If various points are on the same line, the definition can be made more precise and the addition formula can be proven either modulo p-1 or p+1.

If A is fixed, all the points X equidistant from A are such that

$$|A|^2 \; |X|^2 + k(A.X)^2 = 0 \tag{3}$$

or

$$k'(A.X)^2 = -|X|^2 \quad . \tag{4}$$

$A.X = 0$ is the equation of the polar of A with respect to the ideal. (3) are conics for which p is also a polar of A. If p intersects the ideal at C and D, (3) are the conics through C and D. It is appropriate to give (3) as the algebraic definition of circles with center A in finite non-Euclidean geometry.

A large number of results have been generalized (11). An apparently new result will now be described.

If $A = (A0,A1,A2)$ and A is not an ideal point, let us define

$$A' = A / \sqrt{-A.A} \quad . \tag{5}$$

$|A| = \sqrt{-A.A}$ is called the length of A. Either each component of A' is an integer or each component is an integer divided by δ, in this last case we say that A' is pure imaginary, δ is the square root of a non-residue modulo p.

If A is hyperbolic, A' is real; if A is elliptic, A' is pure imaginary. Moreover, $A'.A' = -1$.

Given two points A and B of the same type, M on AB is called a midpoint of [A,B] if the distances MA and MB are equal. It can be shown that the midpoints of [A,B] are $M^+ = A' + B'$ and $M^- = A' - B'$.

I define as mediatrices $m = A' - B'$ which passes through $M^+ = A' + B'$ and $m^- = A' + B'$ which passes through $M^- = A' - B'$. I define as medians n_i and n_i^- the lines joining a vertex to one of the midpoints of the opposite side. The interior medians n_i have a point G_3 in common. Two exterior medians n_{i-1}^- and n_{i+1}^- meet at a point G_i of the interior median n_i. The interior mediatrices m_i have a point O_3 in common. Two exterior mediatrices m_{i-1}^- and m_{i+1}^- meet at a point O_i of the interior mediatrix m_i. The four lines joining the corresponding points G_i and O_i have a point V in common (which, surprisingly, is not the orthocenter). By analogy with Euclidean geometry, the points G_j are called the center of mass of the triangle, the points O_j are called the centers of the circumcircle of the triangle, and the lines joining G_i to O_i are called the lines of Euler of the triangle. I will call V the center of the triangle.

The points and lines M_i, $\bar{M_i}$, n_i, $\bar{n_i}$ are real only if the vertices of the triangle are all of the same type. But V is always real. In fact an alternate construction of V has been obtained, which succeeds even when not all three vertices are of the same type.

5. FINITE REAL JACOBIAN ELLIPTIC FUNCTIONS

In this section j and j' will again denote +1 or -1. Given p and m different from 0 and 1, we will define a set E = E(p,m) and an operation "+".

5.0 Definition of the Set E

Given s, c, d in Z_p, the elements of E are (s,c,d) such that

$$s^2 + c^2 = 1 \quad \text{and} \quad d^2 + ms^2 = 1 \ . \tag{0}$$

If -1 and -m are quadratic residues, we also have to include $(\infty, c\ \infty, d\ \infty)$, where $c^2 = -1$ and $d^2 = -m$.

5.1 Definition of the moduli k and k1 and of addition of elements in E

$$i := \sqrt{-1} \ , \quad m1 := 1-m \ , \quad k := \sqrt{m} \ , \quad k1 := \sqrt{m1} \tag{1}$$

Let
$$D = 1 - m\ s0\ \ s1 \ . \tag{2}$$

If $D \neq 0$,

$$(s0,c0,d0) + (s1,c1,d1) =$$
$$\big((s0c1d1 + s1c0d0)/D,\ (c0c1 - d0s0d1s1)/D,\ (d0d1 - ms0c0s1c1)/D\big) \ . \tag{3}$$

If $D = 0$, $s0c1d1 = s1c0d0$, $c0 \neq 0$ and $c1 \neq 0$,

$$(s0,c0,d0) + (s1,c1,d1) = (\infty, c\ \infty, d,\ \infty) \ , \tag{4}$$

where $c = c1/(s1d0)$ and $d = d1/(s1c0)$.
If $D = 0$, $s0c1d1 = -s1c0d0$, $c0 \neq 0$ and $c1 \neq 0$,

$$(s0,c0,d0) + (s1,c1,d1) =$$
$$\big((s0^2 - s1^2)/(2s0c1d1),\ (c0^2 + c1^2)/(2c0c1),\ (d0^2 + d1^2)/(2d0d1)\big) . \tag{5}$$

If $D = 0$, $s0c1d1 = j\ s1c0d0$, $c0 = 0$ and $c1 \neq 0$,

$$(s0,c0,d0) + (s1,c1,d1) = (\infty, c\ \infty, d\ \infty) \ , \tag{6}$$

where $c = -d0s1/c1$ and $d = (d0/c1)^3/(ms0)$.

If $D = 0$, $s0c1d1 = j\ s1c0d0$, $c0 \neq 0$ and $c1 = 0$,

$$(s0,c0,d0) + (s1,c1,d1) = (\infty, c\ \infty, d\ \infty) \tag{7}$$

where $c = -d1s0/c0$ and $d = (d1/c0)/(ms1)$.
If $s0 \neq 0$,

$$(\infty, c\ \infty, d\ \infty) + (s0,c0,d0) = (s0,c0,d0) + (\infty, c\ \infty, d\ \infty) = (-cd/(ms0),\ dd0/(ms0),\ cc0/s0)\ . \tag{8}$$

If $s0 = 0$,

$$(\infty, c\ \infty, d\ \infty) + (0,c0,d0) = (0,c0,d0) + (\infty, c\ \infty, d\ \infty) = (\infty, d0c\ \infty, c0d\ \infty) \tag{9}$$

$$(\infty, c0\ \infty, d0\ \infty) + (\infty, c1\ \infty, d1\ \infty) = (0, d0\ d1/m, c0c1) \tag{10}$$

It is straightforward, but longwinded, to show that the set E is closed under addition and forms an abelian group. A cyclic subgroup can be obtained in a systematic way using the notion of quotient group, which allows the definition of finite elliptic functions.

5.2 Example for p of the form $4\ell-1$

With $p = 11$, $m = 3$, $(-1/11) = (-3/11) = -1$,

$$E = \{(0,1,1), (0,1,-1), (0,-1,1), (0,-1,-1), (1,0,3), (1,0,-3), (-1,0,3), (-1,0,-3), (5,3,5), (5,3,-5), (5,-3,5), (5,-3,-5), (-5,3,5), (-5,3,-5), (-5,-3,5), (-5,-3,-5)\}\ .$$

If the elements of E in the above order are abbreviated 0,1,2,...,15, the addition table is

+	0	1	2	3	4	5	6	7	8	9	10	11	12	13	14	15
0	0	1	2	3	4	5	6	7	8	9	10	11	12	13	14	15
1	1	0	3	2	7	6	5	4	13	12	15	14	9	8	11	10
2	2	3	0	1	6	7	4	5	14	15	12	13	10	11	8	9
3	3	2	1	0	5	4	7	6	11	10	9	8	15	14	13	12
4	4	7	6	5	2	1	0	3	10	13	14	9	8	15	12	11
5	5	6	7	4	1	2	3	0	9	14	13	10	11	12	15	8
6	6	5	4	7	0	3	2	1	12	11	8	15	14	9	10	13
7	7	4	5	6	3	0	1	2	15	8	11	12	13	10	9	14
8	8	13	14	11	10	9	12	15	4	1	2	5	0	7	6	3
9	9	12	15	10	13	14	11	8	1	6	7	2	5	0	3	4
10	10	15	12	9	14	13	8	11	2	7	6	1	4	3	0	5
11	11	14	13	8	9	10	15	12	5	2	1	4	3	6	7	0
12	12	9	10	15	8	11	14	13	0	5	4	3	6	1	2	7
13	13	8	11	14	15	12	9	10	7	0	3	6	1	4	5	2
14	14	11	8	13	12	15	10	9	6	3	0	7	2	5	4	1
15	15	10	9	12	11	8	13	14	3	4	5	0	7	2	1	6

5.3 Example for p of the form $4\ell+1$

With p = 13, m = 3, (-1/13) = (-3/13) = 1,

$$E = \{(0,1,1),\ (0,1,-1),\ (0,-1,1),\ (0,-1,-1),\ (\ ,5\ ,6\ ),\ (\ ,5\ ,-6\ ),\ (\ ,-5\ ,6\ ),\ (\ ,-5\ ,-6\ ),\ (6,2,6),\ (6,2,-6),\ (6,-2,6),\ (6,-2,-6),\ (-6,2,6),\ (-6,2,-6),\ (-6,-2,6),\ (-6,-2,-6)\}\ .$$

+	0	1	2	3	4	5	6	7	8	9	10	11	12	13	14	15
0	0	1	2	3	4	5	6	7	8	9	10	11	12	13	14	15
1	1	0	3	2	6	7	4	5	13	12	15	14	9	8	11	10
2	2	3	0	1	5	4	7	6	14	15	12	13	10	11	8	9
3	3	2	1	0	7	6	5	4	11	10	9	8	15	14	13	12
4	4	6	5	7	3	1	2	0	12	14	13	15	11	9	10	8
5	5	7	4	6	1	3	0	2	10	8	11	9	13	15	12	14
6	6	4	7	5	2	0	3	1	9	11	8	10	14	12	15	13
7	7	5	6	4	0	2	1	3	15	13	14	12	8	10	9	11
8	8	13	14	11	12	10	9	15	7	1	2	4	0	5	6	3
9	9	12	15	10	14	8	11	13	1	4	7	2	6	0	3	5
10	10	15	12	9	13	11	8	14	2	7	4	1	5	3	0	6
11	11	14	13	8	15	9	10	12	4	2	1	7	3	6	5	0
12	12	9	10	15	11	13	14	8	0	6	5	3	4	1	2	7
13	13	8	11	14	9	15	12	10	5	0	3	6	1	7	4	2
14	14	11	8	13	10	12	15	9	6	3	0	5	2	4	7	1
15	15	10	9	12	8	14	13	11	3	5	6	0	7	2	1	4

If e0 = (s0,c0,d0) and e = (s,c,d) are elements of the cyclic subgroup, we define using the addition formulas

$$e(0) = e0\ , \qquad e(j) = e(j-1)+e\ , \qquad j = 1, \ldots\ . \tag{11}$$

To e(j) = (sj,cj,dj) we can associate a point

$$P(j) = (2sjcj,\ cj^2 - sj^2,\ 1) \tag{12}$$

on the circle $x^2+y^2 = z^2$.

For j any integer, the P(j) are the vertices of a Poncelet polygon, which is also circumscribed to a circle. Moreover, in finite mechanics, we can consider that P(j) is the position of a mass at "time" j of a pendulum moving in a uniform field in the direction of the y axis.

A sketch of the proof will now be given. First using the work of Hensel (5) in the p-adic field associated to p, we define the derivatives of the functions sn, cn and dn and of the function am defined by (o is the symbol for composition of functions) sin o am = sn, cos o am = cn. We obtain as in the classical case, D am = dn, D sn = cn dn, D cn = -sn dn, D dn = -m sn cn, D (2am) = -m sin o (2am), which corresponds to the equation of the circular motion in the uniform field parallel to the y axis.

The connection between the finite Jacobian elliptic functions and those of Tate (6), which correspond to the Weierstrass p functions, has also been established (11).

6. THE HARMONIC POLYGONAL MOTION

Some 100 years ago, Casey (17) introduced the notion of a harmonic polygon, inscribed in a circle, to generalize results due to Lemoine (18) for the triangle. The notion generalizes trivially to conics.

Let P(0) and P(1) be two distinct points on a conic. Let d be a directrix. P(2) is determined as follows: the tangent t(1) at P(1) intersects d at Q(1); Q(1)P(0) meets the conic at a new point P(2). P(3),... are obtained similarly from P(1) and P(2),... . Consider a motion on an ellipse:

$$x(t) = a\cos(E(t)) \quad , \qquad y(t) = b\sin(E(t)) \quad . \tag{0}$$

The tangent at P(t) meets the cord through P(t + h) and P(t - h) on the directrix x = a /c, if

$$\begin{aligned}&\cos\big((E(t+h) + E(t-h))/2\big) + \\ &\quad + \big(e - \cos(E(t))\sin\big((E(t+h) + E(t-h))/2\big)\big)/\sin(E(t)) \\ &\quad = e\cos((E(t-h) - E(t+h))/2) \quad .\end{aligned} \tag{1}$$

If h is small,

$$E(t+h) = E(t) + hDE(t) + h^2D^2E(t)/2 + \ldots \tag{2}$$

substituting in (1) gives

$$\begin{aligned}&\big(-\sin E(t) + (e - \cos E(t)) / \sin E(t)\big)\, D^2\, E(t) \\ &\qquad = -e(DE(t))^2 + O(h^2) \quad .\end{aligned} \tag{3}$$

to the order of h^2. After integrating we get, for some constant k,

$$DE = k(1 - e\cos(E)) \quad . \tag{4}$$

This should be compared with the integral form of Kepler's equation

$$(1 - e\cos(E))DE = k \quad . \tag{5}$$

We can repeat the construction in finite Euclidean geometry: a focus of the ellipse is obtained as the intersection of two tangents to distinct isotropic points and a directrix is the polar of a focus. In fact, (4) is the differential equation of the motion. This can be verified laboriously by checking that the derivative of the equation (1) relating successive points of the motion is identically zero if (4) is satisfied. The results hold for the finite case as follows from the properties of the derivatives of the p-adic functions sine and cosine.

7. CONCLUSION

Many of the results of classical Euclidean geometry and non-Euclidean geometry generalize to the case where there are p^{k+1} points on each line, where p is a prime. This has, most likely, implication in cosmology as well as atomic physics, as is hinted at by results in finite mechanics, namely, the finite pendular motion and the polygonal harmonic motion. I hope that many of the participants of this symposium and other readers will take up the challenge to pursue this new line of inquiry.

8. ACKNOWLEDGMENT

This paper is dedicated to the memory of Monseigneur Lemaitre. For five years, at Louvain, and for many years thereafter I have had the privilege of learning from his inspirational lectures and conversations on mechanics, astronomy, numerical analysis, number theory, relativity and cosmology. His independent spirit has guided me through the exploration of this subject using many of the tools which he possessed at his fingertips: historical readings, computer programming, number theory, group theory.

The theory of the circular pendulum, its connection with elliptic functions, and the proof of properties of elliptic functions using theorems of classical mechanics, was a topic that was particularly dear to Lemaitre in his teaching and had a direct influence on the development of section 5.

9. REFERENCES

(1) Neugebauer,O. and Sachs,A. 1945. Mathematical Cuneiform Texts, American Oriental Series 29 (American Oriental Society, New Haven, Connecticut).

(2) Aryabhata, 1930. The Aryabhatiya of Aryabhata, translated with notes by Walter Eugene Clark (University of Chicago Press, Chicago, Illinois).

(3) Gauss, Carl, 1801. Disquisitiones Arithmeticae, Lipsiae, Fleicher. Translated by Arthur S.J. Clarke, 473 pp. (Yale University Press, New Haven, CT, 1966).
(4) Veblen,O. and Young,J. 1910 and 1918. Projective Geometry, Vols. I and II. (Wesley, Boston).
(5) Hensel, Kurt, 1908. Theorie der algebraischen Zahlen, 349 pp. (Teubner, Berlin);
Hensel, Kurt, 1913. Zahlentheorie, 356 pp. (Goeschen'sche Verslaghandlung).
(6) Tate,J.T. 1972. The arithmetic of elliptic curves, 43 pp. Colloquium Dartmouth College, Hanover, NH, Aug.29 - Sept.1.
(7) Dembowski,P. 1968. Finite Geometries, 375 pp. Ergebnisse der Mathematik und ihrer Grenzgebiete, Band 44 (Springer, New York).
(8) von Staudt,K.G.C., 1847. Geometrie der Lage, Nuremberg;
von Staudt,K.G.C., 1857. Beitrage zur Geometrie der Lage, Nuremberg.
(9) Coxeter,H.S.M., 1949. The Real Projective Plane, 196 pp. (McGraw-Hill, New York).
(10) Brianchon, Charles and Poncelet, Jean, 1820. See Smith,D.E., 1925, History of Mathematics, Vols. I,II.
(11) DeVogelaere, Rene. Finite Euclidean and non-Euclidean Geometry, in preparation.
(12) Grassman, Hermann G., 1896. Ausdehnungslehre, Gesamm. Math und Phys. Werke, Leipzig.
(13) Forder, Henry G., 1960. The Calculus of Extension (Chelsea Publishing Co., New York).
(14) Neuberg, Joseph, 1886. Mémoire sur le tétraèdre, Mémoires couronnés de l'Académie de Belgique, Vol. 37, pp.3-72.
(15) Bolyai, Janos, 1886. The science absolute of space independent of the truth and falsity of Euclid's Axiom, XI. Translated by Dr. George Brus Halstead, Vol. 3, 71 pp. (The Neomon, Austin, Texas).
(16) Lobachevskii, Nikolai I. See Norden, A., 1958. Elementare Einfuhrung in die Lobachewskische Geometrie, 259 pp. (VEB Deuscher Verlag der Wissenschaften, Berlin).
(17) Casey, John, 1889. Geométrie Elementaire Récente. Mathesis 9, pp.5-70.
(18) Lemoine, Emile, 1886. Propriétés relatives à deux points du plan d'un triangle qui se déduisent d'un point K quelconque du plan comme les points de Brocard se déduisent du point de Lemoine. Mathesis, Ser. 1, Vol. 6, Suppl. 1-27.

Part IV
Georges Lemaître: The Man and His Work

Einstein and Lemaître, Le Coq (Belgium), 1933.

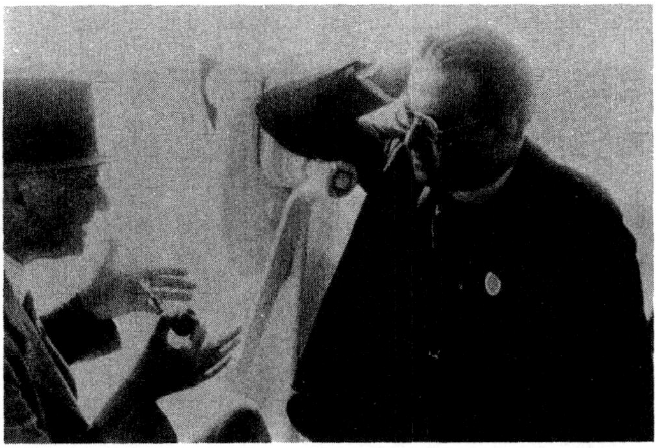

Eddington and Lemaître, IAU Stockholm, August 1938.

His Holiness the Pope Pio XII and Lemaître, Pontificia Academia Scientiarum, Citta del Vaticano, 1939.

Lemaître with some students of the Catholic University of Louvain, Louvain 1950.

From left to right: Lindsey, Oort, Lemaître and Ramberg, Observatory Bloemfontein, South Africa, 1952.

Background: Raphael Boon; sitting in front of the computer: Mrs A. Deprit-Bartholomé and on the right: Monseigneur G. Lemaître; computer room (Burroughs E101) Collège des Prémontrés, Louvain 1959.

MONSIGNOR GEORGES LEMAITRE

Andre Deprit

National Bureau of Standards,
Gaithersburg, MD 20899, U. S. A.

Georges Henri Joseph Edouard Lemaitre was born at Charleroi in Belgium on 17 July 1894, the first child of Joseph Lemaitre and Marguerite Lannoy.

The earliest member of the Lemaitre family of whom there is any certain knowledge is Pierre--Joseph Lemaitre: allegedly unhappy about the re--marriage of his father (an officer in the royal armies?), he migrated with his brother Jean--Joseph (1720--1786) from Bordeaux to Courcelles in the Hainaut where he died in 1774 as a small tenant farmer. The Lemaitre family took root in the Sambre valley around the fortress of Charleroi which the marquess of Castel Rodrigo then Spanish governor-general of the Low Countries had built in 1666, and named after his sovereign, Charles II, the last monarch in the Spanish branch of the Habsburgs. Charleroi endured its condition as a bastion guarding the road to Brussels and Antwerp against the French armies. Besieged, overrun, rebuilt several times, it was given by the Treaty of Utrecht to the Austrian Habsburgs; eighty years later, they lost it to the conscripts of General Dumouriez. With the advent of steam power, Charleroi evolved from a garrison town into a thriving industrial center based on coal mines, iron foundries, glass works, tobacco factories and woolen mills. By the middle of the XIXth century when the Sambre was canalized, the city boomed at a

A. Berger (ed.), The Big Bang and Georges Lemaitre, 363–392.

very fast pace; Vauban's fortifications with their Austrian and Dutch extensions were dismantled between 1868 and 1871 to make room for quays, railways, roads, factories and houses. For the next century, Charleroi prospered as the center of a wide industrial zone which the poet Emile Verhaeren dubbed the Pays Noir ("the Black Country"). The industrial revolution dislocated the social patterns of the rural countryside. In the one--room school of his village, Edouard Severe Joseph Lemaitre (1824--1894) had learned La Fontaine's fables in Walloon; but as a young man he taught himself French grammar and business arithmetic while cleaning and repairing lamps in the store--room of the local coal mine. By cunning and hard work, he rose from the accounting department of the mine to the top position as general manager; now a man of means, he relinquished the position to establish his own business dealing in mining timber. From his marriage with Alexise Catherine Allard (1831--1898) were born six children; Joseph, Msgr Lemaitre's father, was the youngest one.

Having graduated from the Law School at the University of Louvain in 1889, Joseph Lemaitre (1867 -- 1942) received from his father a quarry in the neighborhood of Antwerp and glassworks in Marcinelle, a suburb of Charleroi. At 26, he married Marguerite Lannoy (1869 -- 1956), the daughter of a local brewer. She gave him four sons; Georges was the oldest one. Joseph and Marguerite Lemaitre raised their children to respect and maintain the tenets of the ruling class in the Black Country: personal dignity, professional integrity, fidelity to the Catholic faith, loyalty to the established institutions, civic, social and religious.

1. A good son, a good student

Georges'childhood was a prosaic and comfortable variation on the bourgeois tale of respectability through conformity. Georges received his early education at the grade school of his parish (1899--1904), and proceeded in due course to the Jesuit high school of his native city (1904--1910). Latin and Greek were the basic courses of a curriculum preparing for a university education. As a schoolboy, Georges was conscientious but uninspired. In the senior years, he manifested a disposition for manipulating mathematics; he also exhibited a talent for improvising original solutions to problems in Euclidean geometry. By the time he graduated, Georges had made his decision: he wanted to be both a priest and a research scientist. To his friends and advisers, the mix seemed unusual; they recommended proceeding with caution. Anyway the family circumstances could not permit Georges to go ahead immediately with his personal plans. For years his father had experimented with new processes for stretching molten glass; the research strained the business finances. The rupture point was reached when a simultaneous explosion of the experimental oven and of a furnace wrecked the plant. The mishap convinced the banks to recall their loans. Forced to declare bankruptcy, Joseph Lemaitre made arrangements to repay his creditors, and moved his family from Marcinelle to Brussels. He had found a job as a lawyer with the Societe generale de Belgique, Belgium's main holding bank. For the time being, Georges had to consider earning a professional degree leading to an occupation where he would be in a position to assume his share of the family's burden.

Therefore, in September 1910, Georges entered the College Saint Michel, the Jesuit preparatory school in Brussels, to study intermediate geometry, algebra and trigonometry, thereby preparing the entrance examination at the College of Engineering in Louvain. The major instructor in mathematics was Father Henri Bosmans, reputed for his numerous research notes on the origins of calculus and on Simon Stevin and other mathematicians of the Low Countries during the Renaissance. He infused Georges with his enthusiasm for the history of sciences, of geometry in particular; the student learned from the master the art of reading scientific texts in Latin.

From the year spent in the preparatory class in Saint Michel (1910--1911) dates the long--standing friendship between Georges Lemaitre and Charles Manneback (1894--1974). Together they presented the examination in July 1911, were admitted, and registered the following October at the College of Engineering. But Lemaitre also enrolled for the classes in philosophy offered to laymen by the Institut superieur de Philosophie, formerly the Institut de Philosophie thomiste founded by Desire Mercier. At the time Lemaitre matriculated, Msgr Mercier had resigned his professorship to occupy the seat of archbishop of Malines. Yet it must be noted that Msgr Mercier has had a profound, although indirect, influence on Georges Lemaitre; Mercier's writings on priesthood firmed Lemaitre in his dual project. Having earned his B.A. cum laude in engineering (July 1913), Lemaitre began his professional training as a mining engineer. On the face of his academic record, one may safely conclude that Lemaitre was not interested in pursuing a professional career. From his first three years at the university of Louvain dates his infatuation with Hamiltonian dynamics, an area to which he was introduced by Ernest Pasquier (1849--1926). A protege of Philippe Gilbert (1832--1892), Pasquier had made a name for himself by publishing in Paris a translation of the two volumes of Oppolzer's Lehrbuch zum Bahnbestimmung der Cometen und Planeten.

2. First intermission : a good soldier

The invasion of Belgium by Germany (4 August 1914) had been the turning point in Lemaitre's youth. On 9 August, he and his brother Jacques joined the Fifth Corps of Volunteers in Charleroi. While Falkenheyn's army invested Antwerp, Belgium's main harbor and major fortress, Lemaitre's unit was put on the job of digging trenches between the forts. In October of that year, the brothers

Lemaitre escaped from Antwerp with the Belgian field army led by King Albert I, retreating with them to the coastal depression in the northwestern corner of Belgium on the left bank of the Yzer. At which time, the corps of volunteers was disbanded, and Lemaitre was detailed to the 9th regiment of line infantry (10 October 1914); the assignment entangled him for several days in bloody fightings house to house in the village of Lombartzyde. After the flood plain had been inundated to stop the German infantry, the 9th regiment was placed on the right flank of the Belgian army. On 22 April 1915, Lemaitre watched there the awesome debacle caused by the first attack with chlorine gas; the madness of it would never fade in his memory.

The Belgian army being short of gunners, Lemaitre was transferred to the 3rd regiment of artillery (3 July 1915). In the relatively quiet second line of the Belgian front, he could steal time from the chores of war for reading Poincare's _Lecons sur les Hypotheses cosmogoniques_ and other monographs in physics. As the war dragged on, Lemaitre moved up the hierarchy of non commissioned officers to attain the top rank of master sergeant. He went through a period of training for a field commission as a first lieutenant in the artillery (20 March--27 September 1917). Alas, he challenged an instructor in the class--room on his erroneous solution to a problem in ballistics; he and his brother were expelled that day from the class, the instructor lodged a complaint, and the commanding officer entered a report stating that Lemaitre did not have the attitude expected from a candidate officer. Years after the incident, Lemaitre could still not dominate his resentment against the evaluation. A second period at the _Centre d'Instruction pour Sous--Lieutenants auxiliaires d'Artillerie_ was more successful, but it ended on 12 October 1918, that is a month before the Armistice, and so Lemaitre came out of the war without an officer's commission. On 29 November 1918, he was cited for bravery in the Belgian Army's Orders; he was awarded the _Croix de Guerre avec palmes_ on 28 February 1921 (a military distinction comparable to a Silver Star in the U.S. Army), the Yzer Medal (15 November 1922) and, much later (28 August 1946), the most coveted _Croix du Feu_.

After 53 months in the din of war, Lemaitre had made up his mind. He planned to realize his dream in two steps. Although he was still in the uniform, Lemaitre re-enrolled at the University of Louvain as soon as it reopened (21 January 1919); in less than six months, he presented the examinations for the degree of _candidat_ ("B. A.") in mathematics, the B. A. at the _Institut superieur de Philosophie_ which degree he obtained _cum laude_, and the first year of graduate studies toward a Ph. D. in mathematics which he passed _summa cum laude_. Having been discharged from the army in August 1919, without a pause he entered research and completed by July 1920 his doctoral dissertation on "The Approximation of Real Functions in Several Variables." Surprisingly enough, for he was not easy to please, his adviser, Charles de la Vallee Poussin (1866--1962), considered inviting Lemaitre to pursue an academic career of teaching and research in mathematical analysis. Lemaitre would have none of that. His first goal, a doctoral degree in science, having being reached, he shifted to the second stage of his personal plans. The 21 October 1920, Lemaitre entered the Maison Saint Rombaut, an extension of the major seminary of the Archdiocese of Malines, where adults were trained for the priesthood. The program of religious studies left him leisure for reviewing the literature published in special and general relativity. In his third year at the seminary, Lemaitre even found time to prepare three essays which he submitted in May 1923 to the state commission in charge of allocating scholarships for periods of study abroad. Lemaitre's second aspiration was fulfilled on 23 September 1923, when his spiritual mentor, Cardinal Desire Mercier, ordained him a priest in the clergy of his Archdiocese.

3. The first career: a visionary on the roads

"A Man there was, though some did count him mad.
"The more he cast away, the more he had.
John Bunyan.

With the scholarship granted by the Belgian government, Lemaitre could afford to apply for admission at Cambridge University as a research student in astronomy. Having been accepted to the Observatory by the director Arthur Eddington (1882--1944), he took his quarters in October 1923 for nine months at Saint Edmund's House, a residence for Catholic clergymen connected with the university. Beside attending the classes of Harold Jeffreys, Ernest Rutherford, Henri Baker and Arthur Eddington, he pursued his research on one of the topics of his scholarship essays, namely the concept of simultaneity in general relativity. Pleased with the results, Eddington added a foreword to the manuscript submitted to the *Philosophical Magazine*. Among the many friends Lemaitre made at Cambridge, most notable were Douglas Hartree (1897--1958), at the time preparing his doctoral dissertation under the direction of R. H. Fowler, William M. Smart then John Couch Adams astronomer and chief assistant at the Observatory, and, last but not least, Yusuke Hagihara (1897--1979) with whom Lemaitre cultivated a special interest in celestial mechanics.

The following academic year was spent at Cambridge in Massachusetts as a fellow of the C. R. B. Educational Foundation, a private institution funded by the Committee in Relief of Belgium out of the surplus of charities collected at the instigation of Herbert Hoover to assist the Belgian population during the German occupation. The fellowship tenure at the Harvard College Observatory was to be the most exhilarating period in Lemaitre's life. Harlow Shapley (1885--1972) had suggested that he work on the theory of variable stars after he became familiar with the observations. Lemaitre took the assignment with zest. He even made arrangements to spend the month of September 1924 in Ottawa at the Dominion Observatory; Francois Henroteau (1889--1951), ex-astronomer at the Royal Observatory of Belgium and an expert in variable stars, agreed to tutor him on the subject of Cepheids. Not that the summer of 1924 had been an idle season for Lemaitre. He had crossed the Atlantic in time for the Toronto meeting of the British Association for the Advancement of Science (6--13 August 1924); he had there some valuable discussions with Eddington on the Schwarzschild metric, and with Ludwik Silberstein (1872--1948) on his disputed conclusions from a linear relation he had derived between radial velocity and distance for galaxies in de Sitter's stationary model for the cosmos. The conversation with Silberstein stayed on Lemaitre's mind; eventually at the 133rd meeting of the American Physical Society (24--25 April 1925), he showed how, by reformulating de Sitter's solution to Einstein's equations in the manner suggested by Cornelius Lanczos (1893--1974), he could remove its spurious inhomogeneity, and how new coordinates separating space and time led to a relation velocity--distance not only linear but freed as well from the puzzling double sign introduced by Silberstein.

The meeting of the British Association overlapped with the International Mathematical Congress (11--16 August) also in Toronto. Lemaitre spent the balance of August in Montreal where he visited McGill University and attended several parties given at Beauharnais--la--Pointe by Miss Thibeaudeau, the most gracious guardian angel of European scientists setting foot for the first time in the New World.

While at the Harvard College Observatory, Lemaitre spared himself no effort extending his information and his contacts through meetings of various scientific societies. In that regard, the 33rd meeting of the American Astronomical Society in Washington during the Christmas season of 1924 marked a decisive turn in Lemaitre's career: listening to Henri Norris Russell reading Hubble's announcement that he had observed Cepheids in the galaxy Andromeda (1 January 1925), everyone in Corcoran Hall of George Washington University, and Curtiss, Lemaitre, Shapley and Stebbins more than anyone else, realized that sensational developments were imminent in cosmology.

Lemaitre was proud to have been among the founders of the Bond Astronomical Club organized by Shapley for the Boston area; he frequented the Harvard Mathematical Club; he joined the American Association of Variable Star Observers at its annual assembly at Harvard College Observatory (11 October 1924). He took part in "neighboring" meetings for astronomers at Yale University, and also, as a guest of the director Frederick Slocum (1873--1944), at the Van Vleck Observatory of the Wesleyan University in Middletown for the much publicized total eclipse of the sun on 24 January 1925.

Taking advantage of the Gordon--MacKay agreements, Lemaitre also registered as a graduate student in the Department of Physics at the Massachusetts Institute of Technology. To qualify as a candidate for the Ph. D. degree, he submitted in November 1924 to an examination in the theory of Fourier series, and went on taking the courses in electromagnetic theory and in quantum mechanics offered by a young bright mathematical physicist, Manuel Sandoval Vallarta (1898--1977), a Mexican who was destined to have the most pervasive influence on Lemaitre's first scientific career. Paul Heymans (1895--1960), a fellow Belgian who was specializing in photoelasticity although he turned later to economics and entered Belgium's politics, agreed to sponsor the doctoral dissertation. The memoir -- 37 pages long, never published -- dealt with paradoxes encountered by the Schwarzschild metric in a de Sitter universe; Eddington had suggested the topic. Not only does the dissertation constitute Lemaitre's very first step toward a theory of the expanding universe, it manifests as well mastery of a research style whereby invention of numerical procedures compensates for the impossibility of solving a problem by analytical methods.

On his way to the West Coast at the end of the academic year, Lemaitre stopped at the Yerkes Observatory (23 May 1925) where he happened to meet Leslie J. Comrie (1893--1950), a man whom Howard Aiken would salute as a genius in the art of scientific computing. Needless to say, the meeting was most significant in that it confirmed Lemaitre's proclivity toward numerical analysis and computer hacking in his teaching as well as his research. At the university of Chicago itself, Lemaitre conferred with Forest Ray Moulton (1872--1952) about Kant's and Laplace's nebular hypothesis, and with William Duncan Macmillan (1871--1948) about his just announced theory for continual creation of matter by dissipation of radiation while it traverses empty space. After a few weeks of vacation at Jasper, Banff and Lake Louise, Lemaitre passed through the Dominion Astrophysical Observatory in Victoria, the Lick Observatory, the Mount Wilson Observatory, and finally got an initiation to the physics of cosmic rays from Robert A. Millikan (1868--1953) at the California Institute of Technology, also a full briefing on the radial velocities of spiral galaxies from Vesto M. Slipher (1875--1969) at the Lowell Observatory in Flagstaff. No sooner had he returned to the family home in Brussels (8 July) than he hopped to Cambridge for the Second General Assembly of the International Astronomical Union (14--22 July 1925), so impatient was he to sort with Eddington his impressions and his fresh knowledge in astrophysics.

Soon after his return in Belgium, Lemaitre left the family home to take residence at the College du Saint Esprit in Louvain. Indeed he had been appointed associate professor in the Department of Mathematics at the University (July 1925). Eddington had taken the initiative of writing to Theophile De Donder (1872--1957) a very strong letter of support, even suggesting that Lemaitre be considered for an appointment at the Free University of Brussels, should the Catholic University of Louvain reject his candidacy.

The Recteur Magnifique ("provost"), Msgr Paulin Ladeuze, a wise scholar turned administrator, provided Lemaitre, under Spartan conditions, with inexhaustible leisure and a discriminatingly well--stocked Library. He personally saw to it that Lemaitre began his teaching career with a very light load, quite unusual a favor to the young faculty at the time. For several years, Lemaitre was in charge of creating a two--term course in relativity for the graduate students in mathematics and physics, of giving two one--term classes, respectively in the history of physics and mathematics, and on the methodology of mathematics in the secondary schools. He was also given the responsibility of a weekly session of exercises in analytical mechanics in support of the course taught by Charles de la Vallee Poussin to the undergraduates at the College of Engineering. Lemaitre who had embraced Comrie's doctrine of teaching applied mathematics by computational drills made the most of his graduate classes to raise the interest for scientific computing among his students and his colleagues.

This period of relative leisure as far as teaching and examinations were concerned lasted but a few years. Professor Maurice Alliaume (1882--1931), Lemaitre's most devoted colleague in the Department of Mathematics, died in a car accident one month into the academic year (24 October); Lemaitre was asked to step in immediately and take over the two graduate courses his friend had created in mathematical astronomy, one in spherical astronomy and the other in celestial mechanics. All of a sudden Lemaitre's teaching load was doubled; but it concerned graduate students for the most part, which left Lemaitre much at liberty to juggle with the academic schedules.

As for his research in the late twenties, Lemaitre focused primarily if not exclusively on de Sitter's cosmology. Unaware of the paper published in 1922 by Alexander Alexandrovich Friedmann (1888--1925), he first showed how to introduce a radius varying with time in de Sitter's metric (1925), then he deduced from the modified metric a linear relation between radial velocities and distances for galactic nebulae. Published in a relatively obscure periodical, the second paper went unnoticed, and Lemaitre felt somewhat dejected (1927). He tried to draw the attention of Einstein while he was attending the fifth Solvay Conference on Physics in Brussels (24--29 October 1927). Einstein was most abrupt: "Vos calculs sont corrects, mais votre physique est abominable" ("Your calculations are correct, but your physical insight is abominable"). Likewise Lemaitre failed with de Sitter at the Third General Assembly of the International Astronomical Union in Leiden

(5--13 July 1928). Basking as he was then in the vain glory of President of the Union ("hij was intens wolkenloos gelukkig" noted his wife), de Sitter had no time for an unassuming theorist without proper international credentials.

Meanwhile theorists and observers kept heaping criticisms on de Sitter's cosmology. Richard C. Tolman (1881--1948) and Howard P. Robertson (1903--1961), the latter in consultation with Hermann Weyl, entered the debate, blissfully unaware that they were retracing steps taken by Lemaitre several years earlier. Thus wooed, de Sitter appeared at a meeting of the Royal Astronomical Society (10 January 1930). Endorsing the careful observations made by Edwin P. Hubble at the Mount Wilson Observatory, he concurred, he said, with the majority in holding the linear relation between radial velocity and distance as a law of nature; he confessed though that he did not know how to assimilate it in his cosmology. As a comment on de Sitter's communication, Eddington indicated that he himself was working on the problem. By chance, Lemaitre picked up the February issue of The Observatory where the exchange between de Sitter and Eddington had been reported. He wrote at once to Eddington to remind him that he had already solved the problem, and also to ask him to send a reprint of his 1927 note to de Sitter (late March or early April 1930). This time Eddington paid attention to Lemaitre's contribution, dispatched a copy of it to de Sitter and Shapley, and reworked his communication to the Royal Society on "The Stability of Einstein's Universe" to make of it a critical review of Lemaitre's theory of the expanding universe (9 May 1930). By extraordinary favor the Royal Astronomical Society published an English translation of Lemaitre's note, with slight but telling amendments, in the Monthly Notices (March 1931). With Eddington as advocate, the theory gained rapid acceptance among most astronomers although Hubble, by training a lawyer as well as a physicist, never saw in it the "evidence beyond reasonable doubt" that recession velocities result from the expansion of space.

"O saisons, o chateaux,
"Quelle ame est sans defaut?
Arthur Rimbaud

Notwithstanding the sudden celebrity he was enjoying, Lemaitre was very conscious of the deficiencies of his theory. The current value of the Hubble constant led to a cosmological time scale of a couple of billion years, about a hundred times smaller than the geological and stellar time scales. Lemaitre proposed to stretch his time scale by imagining that, after a rapid expansion, the universe entered a period of "stagnation" when the repulsion due to the cosmological constant balanced the gravitational forces. The stagnation theory leads eventually to a Friedmann equation which Lemaitre found could be solved exactly by elliptic functions. Moreover, in Lemaitre's opinion, the stagnation process should explain why local condensations happen in the universe although all physical cosmologies since Einstein postulate the world to be both homogeneous and isotropic in the large. Stagnation in its double role as a retardant and as a condenser has been treated in detail by J. Wouters in his doctoral dissertation submitted in the schoolyear 1931--1932, the first Ph.D. thesis supervised by Lemaitre.

The clustering of galaxies became a challenge that devoured Lemaitre's research in cosmology. Time and again Shapley demanded that the theory of the expanding universe account for concentrations of nebulae he was charting close to the Milky Way. Lemaitre wanted foremost to satisfy the demand. Yet to the end of his life the solution eluded him. To be sure Lemaitre could not conceive of a solution that did not involve a cosmic repulsion. Hence his unmitigated conviction that the cosmological constant, however small it may be, must be held as an essential parameter in a physical cosmology. The importance he attached to the cosmological constant drew laconic objections, and even cold contempt, from Einstein. Although the constant was due to him, and he had used it to inject matter into a static universe, now that he had accepted the concept of universes with varying radii, Einstein felt sorry for having invented it. The controversy between the two men on this topic culminated in the magistral apology for the cosmological constant which Lemaitre presented to Einstein as a tribute for his seventieth birthday (1949) and in Einstein's summary rejection of it in a terse paragraph of five lines.

Eddington conceived the expansion as starting from a state of equilibrium represented by Einstein's model. Within that framework, for a popular conference on "The End of the World: from the Standpoint of Mathematical Physics" at the British Mathematical Association (5 January 1931), he offered a few thoughts on how the second law of thermodynamics could be extrapolated at both ends, in the future toward a state of complete disorganization, and in the past toward a beginning in time, which concept he was prompt to reject. Lemaitre found the text of Eddington's presidential address in the pages of Nature; it electrified his imagination. Aware by now of Friedmann's classification of cosmologies, Lemaitre had been toying for a while with models of the expanding universe starting from a radius equal to zero. In the use Eddington made of the second law of thermodynamics, he suddenly saw a way of giving physical meaning to the mathematical singularity. Interpreting entropy as a measure of fragmentation, Lemaitre read the second law of thermodynamics as meaning that the universe is relentlessly and irreversibly dividing itself into smaller and smaller pieces, or, going backward in time, that the universe descended from a state of supreme concentration, the Primeval Atom as Lemaitre dubbed it. By virtue of the identification between matter and space--time which is the essence of General Relativity, continued Lemaitre, the Primeval Atom cannot admit to being considered as an undivided fragment contained in a region of space, and the primordial explosion ("a day without yesterday") does not suffer to be assimilated to an instant ticked off an axis of coordinate time. Space--time itself originates from this physically indescribable event of which, for lack of a scientifically meaningful terminology, Lemaitre spoke in terms of fireworks, well aware though he was that there was no background sky against which the fireball burst out. Lemaitre's letter of 9 May 1931 to Nature is the charter of the Big Bang Theory.

Expounding these themes at the Centenary Meeting of the British Association for the Advancement of Science in the Great Hall of the University of London (24 October 1931), Lemaitre surmised that cosmic rays of high energy have their origin in the primordial Big Bang. In Arthur Compton's discovery that cosmic rays consisted of charged particles, Lemaitre saw material evidence of the "natural beginning" in a Big Bang.

Meanwhile Lemaitre applied for an Advanced Fellowship at the C.R.B. Educational Foundation in Brussels. He planned on spending a couple of months with Shapley at the Harvard College Observatory, on visiting Professor Henry Norris Russell (1877--1957), the director of the Fitzrandolph Observatory at Princeton University, and finally on working for another two months with Professor Tolman at the California Institute of Technology while at the same time meeting with Hubble and Humason at the Mount Wilson Observatory. He left Europe in August 1932 by the Canadian Pacific which brought him to Montreal. There he joined F.J.M. Stratton and his party from the Cambridge Observatory, and together they made the trip to Magog (in the province of Quebec), a locality on the totality path of the solar eclipse (31 August 1932). On the grounds of the Hermitage Country Club, they mixed with other groups coming from McGill University, the Leander McCormick Observtory at the University of Virginia, and the University of Utrecht. However the sky was wholly overcast, the clouds were the thickest along the central line of the eclipse zone, the expedition assembled at Magog saw nothing. Lemaitre who looked at the clouds "with his hands in his pockets" found the spectacle most ironic, the more so having heard there of Lyot's coronagraph.

The next day the various parties turned their steps in the direction of Cambridge in the Massachusetts where the International Astronomical Union was due to hold its Fourth General Assembly (2--9 September) Arrangements had been made for housing the participants and their guests in the dormitories of Radcliffe College. The event which in the minds of many was the climax of the week was the public lecture on "The Expanding Universe" by Sir Arthur Eddington in the Main Hall of the Walker Memorial at the Massachusetts Institute of Technology (7 September). For Lemaitre himself, Eddington's presidential address was an hour of triumph. Yet, at the discussion on "The Extra--Galactic Objects" held at the Harvard College Observatory (9--10 Sptember), the observers, among them Bok, Lindblad, Lundmark, Oort, J.S. Plackett, and last but not least, Shapley and Stebbins, closely questioned Eddington and Lemaitre in the final session on these strange views of theirs on a universe in expansion. Lemaitre in particular was called to defend his "fireworks theory of the beginning of things."

While in Cambridge, Lemaitre divided his time between the Department of Physics at the M. I. T. and the Harvard College Observatory. At Harvard, among other things, he discussed with Shapley ways of integrating observations of distant nebulae in his own model for an expanding universe; he also conversed with E. Opik about the conflict between the short time scale inherent to his model and the long time scales adopted by the current theories of stellar evolution. He did research mainly at the M.I.T. with Vallarta on his hypothesis for the cosmic rays. Granted that they were produced by the Big Bang, cosmic rays should enter the earth's upper atmosphere coming from the most remote outer space. If so, Liouville's central theorem of Hamiltonian mechanics guarantees, according to Lemaitre, that the distribution of cosmic rays of a given energy is invariant inside the arrival cone at any point on earth. It follows that the envelope of the cone is made of periodic orbits and orbits asymptotic to periodic orbits. Hence Lemaitre and Vallarta set themselves to the task of calculating these envelopes to recover the latitude effects detected by Compton. Considering that Bruno Rossi and his team at Arcetri had just predicted an east--west asymmetry in the distribution of the cosmic rays and had announced their intention of detecting and measuring it, Vallarta was anxious to establish priority. With Lemaitre, he rushed to the 180th meeting of the American Physical Society in Chicago (25--26 November) to present the approach taken at the M.I.T.; with the same haste they wrapped up a paper which they submitted to the Physical Review. No sooner was it published than it drew the harshest criticisms from Carl Stoermer (1874--1957). There ensued a bitter controversy which fortunately turned into a long competition where all parties concerned learned to appreciate their respective strengths and weaknesses.

The research program proposed by Lemaitre and Vallarta proved to be a very long undertaking, in fact a lifetime research for Vallarta, one in which several generations of students at the University of Louvain (L.P. Bouckaert, A. Descamps, O. Godart, R. de Vogelaere, L. Bossy, Tchang Yong--Li), the Massachusetts Institute of Technology (A. Banos, C. Graef, R. Albagi Hutner, S. Kusaka, E. J. Shremp), and the National University of Mexico (R. Gall, J. Lifshitz, H. Uribe) found topics for their doctoral dissertations and an intensive initiation to scientific computing. Fierce loyalties developed in the team work across the Ocean: in faraway Yunnan, throughout the war with Japan, the civil war and the political cataclysms of the 1960s, Professor Tchang Yong-Li (1913--1972) could not help but infuse his students with his fervent devotion for the man in Louvain who taught him in the most unpredictable ways how to enjoy playing the games of physics with computers. Indeed, piqued by the rude reception their program received from Carl Stoermer, encouraged however by the fact that, soon after them but independently, Enrico Fermi had also proposed to apply Liouville's central theorem, Lemaitre and Vallarta gave the problem the best of their imagination and of their computing skill. Both became very attentive to the efforts made at the M.I.T. by Vannevar Bush (1890--1974) in developing the analog computer called the Differential Analyzer.

However it was too early yet for harnessing the Differential Analyzer to the production of periodic and asymptotic orbits. In the second half of November 1932, Lemaitre left Cambridge for Princeton where he met with Professor Russell and, at the invitation of Percy H. Robertson, gave a seminar on his cosmology for the Departments of Mathematics and Physics. In the first week of December, he departed for Pasadena. At the California Institute of Technology, in the presence of Einstein and other "universe makers", he gave two sensational seminars, one on the theory of the expanding universe, and another one on the cosmic rays as the fossils of the Big Bang (12 January 1933). These lectures were even covered in the national daily press. Lemaitre also spent some time in conference with Edwin Hubble at the Administrative Building of the Mount Wilson Observatory on Santa Barbara Street. In his last week at the Atheneaum in Pasadena, he was interviewed by Duncan Aikman; the article "Lemaitre follows two paths to truth" which spread over two pages of the New York Times Magazine (19 February 1933) made him a public figure in the United States. On his way back to Belgium, Lemaitre took a couple of weeks sightseeing in the Tonto National Forest in the state of Arizona.

The Spring of 1933 in Belgium saw Lemaitre busy on two fronts outside his classes. On the one hand, he prepared a critical review of his mathematical methods in general relativity. The revision brought an original result, as a matter of fact a decisive step on the road that eventually led to the theory of the black holes. Lemaitre may indeed be given credit for having been the first in proving that, contrary to appearances, the gravitational radius of a star is not an essential singularity in the Schwarzschild metric.

On the other hand, with Theophile De Donder and L. Infeld, he obtained the support of the Francqui Foundation to organize a series of six seminars on spinors to be given by Albert Einstein. Having learnt that Adolf Hitler had been appointed Chancellor of the German Republic, Einstein upon his arrival in Antwerp from the California Institute of Technology, had gone to the German Embassy in Brussels to surrender his passport; he also resigned from his positions at the Prussian Academy of Sciences and at the university of Berlin. For a few months, he stayed with his family at a villa in De Haan on the Belgian seashore. Lemaitre paid him a visit there to obtain his agreement to the proposition by the Francqui Foundation. The organization of Einstein's lectures took a great deal of effort. For reasons of security, attendance was to be by invitation only, which restriction required that the seminars be held not in a university but in the rooms of the Fondation Universitaire in Brussels where access could be controlled. Lemaitre's efforts on behalf of Einstein drew upon him the grateful attention of the Royal Court in Belgium, in particular of Queen Elisabeth. Einstein gave three conferences in French on 3, 6, and 10 May, then conducted three seminars. Before the session of 13 May, at which De Donder presented some aspects of his research related to the theme of the seminar, Lemaitre had sounded Einstein on possible ways of simplifying proofs of the main results communicated by Einstein. At the end of De Donder's exposition, Einstein rose the curiosity of the audience by announcing that the next seminar would be given by Lemaitre "qui a des choses interessantes a nous dire" ("who has interesting things to tell us"). Lemaitre left the confernece room almost in a state of panic. He spent the whole weekend feverishly developing the ideas he had reviewed with Einstein. But, on Wednesday 17 May, he arrived ready for the seminar and suffered gleefully the enviable inconvenience of being interrupted several times by Einstein talking to himself but in a loud voice and exclaiming that it was "tres joli, tres, tres joli" ("very beautiful, very beautiful indeed").

For the first semester of the next schoolyear Lemaitre had accepted an appointment as "Guest Professor" in the Department of Physics at the Catholic University of America in Washington D.C. Leaving Belgium on 5 September 1933, he first went to Leicester where the British Association for the Advancement of Science had convened a short symposium on "The Expanding Universe" (6--13 September). The organizers had planned on a short session in which the pioneers (de Sitter, Eddington, Lemaitre, McCrea, McVittie, Milne) would assist the scientific public at large in surveying the difficult problems of the day in cosmology. In that regard the meeting was a failure. The experts however appreciated the high technical quality of the communications and, above all, enjoyed the clash of personalities. The duel about clusters of nebulae against the background of a homogeneous universe in expansion may well be summed up in Clausewitz's words: "Alles ist einfach (Lemaitre), aber das Einfache ist schwierig (McVittie)."

Without a pause, Lemaitre left for London and Southampton where he embarked on the Duchess of Bedford of the Canadian Pacific for New York. Besides a course on the "Astronomical Applications of the Theory of Relativity" to the graduate school, he was obligated to deliver three formal lectures to the general public in the Auditorium of the McMahon Hall. Lemaitre chose to develop his views on the "Time Scale" (14 December), the "Structure of Space"

((5 January 1934) and on "Cosmic Rays" (11 January). A presentation to the Washington Academy of Sciences (16 November) and to the U. S. National Academy of Sciences at its Autumn session in Boston (20 November), an exposition of his cosmogony in the presence of Cardinal Archbishop O'Connell of Boston at a Round Table of Catholic Scientists (28 December), a visit to Villanova College where he was presented the Mendel Medal for "outstanding services to science" (15 January 1934), these were the salient features of a period of prodigious activity in varied sectors, either research, teaching, scientific popularization, or public relations. Never in a hurry to return to Louvain, Lemaitre stopped in London and Cambridge, and then travelled to Newcastle. The Astronomer Royal for Scotland, Dr R. A. Sampson, had invited him on behalf of the Durham University Philosophical Society to deliver a lecture on the "Evolution in the Expanding Universe" at Armstrong College (12 February). He would turn up in Great Britain two months later at the meeting of the Royal Astronomical Society when its President, F.J.M. Stratton, presented the Gold Medal to Shapley and then opened a discussion on "The Expanding Universe" (11 May 1934). Lemaitre was called to defend his short cosmological time scale, a time long enough, he reiterated, for the stagnation process to trigger local condensations of the diffuse gas into nebulae. Yet Shapley was not convinced. After all, since this was his day at Burlington House, Shapley was allowed to leave the debate with the last word, in fact, with no more than a surmise that Opik's calculations of orbits for meteors might shed light on the problem.

Success breeds honors. In March 1934, Lemaitre, now a member of the Royal Academy of Belgium, became the second recipient, after the historian Henri Pirenne, of the Francqui Prize. Proposed by Charles de la Vallee Poussin and Count Alexandre de Hemptinne, the nomination had been seconded by Einstein (in a letter dated Le Coq 9 April 1933). It was examined, and approved by an international commission including Eddington and Paul Langevin professor at the College de France. With the award presented by King Leopold III (27 March 1934) came a check for 500,000 Belgian francs (equivalent in purchase power to 200,000 dollars of 1984). The Francqui Prize brought Lemaitre a most needed additional income, considering that, in Belgium at those days, full--time university professors were rather meagerly compensated, and also that clergymen teaching at the University of Louvain were only paid a third to a half of what the laymen received. The evening after the academic session honoring the hero in Louvain (17 April) was something between a riot and a ritual, one of the most memorable in the student history. Msgr Josef Van Roey Cardinal Archbishop of Malines made Lemaitre an honorary canon of his cathedral (27 July 1935). After Pope Pius XI reorganized the Academia dei Novi Lincei to make it the Pontificia Academia Scientiarum, Lemaitre was admitted into that prestigious international institution (28 October 1936). He was especially pleased to enter at the same time as one of his closest personal friends, Hugh Stott Taylor (1890--1974), David B. Jones Professor and chairman of the Department of Chemistry at Princeton University. The preceding year, Lemaitre had been awarded the J. Jansen Medal by the Societe astronomique de France.

For the next ten years, Lemaitre gave most of his research effort to the Stoermer problem dealing with a charged particle in the magnetic field of a dipole. Whatever were the circumstances, he would reach for the problem and push the calculations yet another step. He had agreed, for instance, to spend the first half of 1935 at the Institute for Advanced Study. He had been invited by Professor Oswald Veblen (1880--1960) who was most eager to establish in the fledgling Institute the policy of attracting mathematicians of the most varied specialties as temporary members of the School of Mathematics. Veblen himself had done extensive work on projective relativity theory and the geometry of four--dimensional spinors. While at Princeton, Lemaitre worked out for Dirac's equation an intrinsic representation of the group of symmetries in the algebra of quadri--quaternions or the so--called fourth order Clifford numbers. Compared to Eddington's befuddling treatment of the same problem by searching for a subalgebra in the Lie algebra $g\ell(16,R)$, Lemaitre's approach was a breakthrough no less than a precursor of van der Waerden's Spinor Analysis. Yet, at the time, Lemaitre was so much wrapped up in the Stoermer problem that he satisfied himself with dumping his Princeton results in a scientific journal of convenience where, as could be expected, they were lost. At the end of the Princeton semester, Lemaitre stopped in Montreal to receive an honorary degree from McGill University (30 May 1935). That summer the International Astronomical Union held its fifth general assembly in Paris (10--17 July). The controversy with Shapley about the time scale begun the year before at the Royal Society flared again at a mini--symposium of the Commission on Stellar Constitution; but, this time, Lemaitre was at liberty to express his skepticism in regard to the use Shapley intended to make of Opik's determinations for orbits of meteors. A busy agenda, in spite of the scorching heat of an exceptionally hot summer, where committee meetings competed with excursions, official receptions, and other social events, culminated in a banquet on the first platform of the Eiffel tower (14 July): the brilliant illuminations and fireworks of Bastille Day viewed from the second platform made a memorable spectacle.

Lemaitre's visit to the Institute was but a side episode in the relentless pursuit of his theory on the origin of the cosmic rays. In fact, while in residence at the Institute, he kept commuting between Princeton and Cambridge where he had joined Vallarta in running numerical integrations through Vannevar Bush's Differential Analyzer. In the summer of 1936, it was Vallarta's turn to visit Louvain for a year as Advanced Fellow of the C.R.B. Educational Foundation. Lemaitre and Vallarta had planned first on paying a visit to Carl Stoermer during the International Congress of Mathematicians which the latter was hosting in Oslo, his home town (13--18 July 1936). In his keynote address, Stoermer condescended to mention the research undertaken by Lemaitre and Vallarta; he also announced Professor Rosseland's plans for building a much extended Bush analyzer, and his intention of employing the machine for computing orbits of charges in the field of a magnetic dipole. Evidently the competition had not abated yet: Lemaitre and Vallarta left Oslo without having met Stoermer in private to settle the misunderstandings and coordinate their research.

From the close collaboration between Lemaitre and Vallarta in Louvain, there emerged an ambitious program for continuing systematically natural families of periodic orbits, for integrating numerically differential equations and the concomitant variational equations by Fourier series. The main difficulty resided in reducing the variational system to the Hill equation for the normal displacement and a quadrature for the tangential displacement, also in using the latter to obtain the variation caused by an infinitesimal change in the Jacobi constant in order to continue natural families in an analytical manner according to certain schemes proposed, altogether offhandedly, by Henri Poincare. In that area, credit must be given to Lemaitre for having been the first one with his student Louis Bouckaert in normalizing numerically a Hamiltonian system at an equilibrium, and the first one with the collaboration of another of his student, Odon Godart, in generalizing Hill's alogorithm for determining numerically periodic so-lutions for systems of differential equations with periodic coefficients.

Well ahead of his time, Lemaitre was busy creating a mix of algorithms equivalent to what one would learn later to call the Galerkin method and the Fast Fourier Transform, so busy in fact that he never cared to establish his priorities in the area of scientific computing. Spurred by the success Leslie J. Comrie had met in automating H. M. Nautical Almanac Office at Greenwich Observatory, Lemaitre created a Laboratory of Scientific Calculations in Louvain. By now computing had become a passion for Lemaitre. His enthusiasm was infectious: for Charles Manneback, Lemaitre devised an efficient technique for determining the quadratic potential of an ethylene molecule from the vibration frequencies measured in the Raman spectrum.

In February 1938, Lemaitre boarded the Normandie for New York on his way to the University of Notre Dame in Indiana where he had accepted a visiting professorship. This was the result of a suggestion made by Professor Arthur Haas (1884--1941) to Vallarta who passed it to Lemaitre (2 June 1937). By the standards of the time, the offer was generous: a salary of $3500, living expenses at Corby Hall, and $400 in travel allowances. Father John F. O'Hara, the president of Notre Dame, was eager to upgrade his departments; Lemaitre's arrival coincided with the appointment of Emil Artin, most anxious to escape the political oppression at the University of Hamburg, with an arrangement with Eugene Guth who was starting an academic career in theoretical physics at the University of Vienna, and with the visit of Kurt Godel (1906--1978) on leave, also, from the University of Vienna. Lemaitre's departure caused a sensation in the American scientific circles; Edmund Bartnett in the New York Sun (2 September 1937) compared it to Einstein's exile from Berlin to Princeton. To be sure, Lemaitre never entertained the prospect of migrating to the United States.

Karl Menger, the Chairman of the Department of Mathematics and a reputed geometer, had scheduled the second Notre Dame Symposium on mathematics to begin the week after Lemaitre disembarked in New York (3 February 1938). Next to Marshall H. Stone, Garrett Birkhoff, Oystein Ore, Adrian Albert and Emil Artin, in short all the tenors in "The Algebra of Geometry", Lemaitre talked of "The Algebraic Details of the Relativity Theory of Protons and Electrons" proposed by Eddington (12 February). His semester course on cosmology was attended by the graduate students and the faculty in the Departments of Mathematics and Physics. The most significant event in that period was the Notre Dame Symposium on the Physics of the Universe and the Nature of Primordial Particles (2--3 May). The particle physicists, Gregory Breit, Arthur Haas and Eugen Guth, tossed ideas about approaches to the physics of what they understood was Lemaitre's Primeval Atom, whereas Lemaitre, more interested in Shapley's observations of faint galaxies and also of hierarchies of clusters of galaxies, kept on developing purely gravitational models to account for the kind of density fluctuations measured by Shapley and Hubble. His friend Vallarta who had also been invited gave a progress report on the Stoermer problem. Both he and Lemaitre failed to perceive that the communications of Arthur H. Compton and Carl D. Anderson at the symposium itself indicated that cosmic rays could not be the much sought after remnants of the Primeval Atom. The Notre Dame symposium showed Lemaitre missing the cues about his Big Bang Theory on the verge of taking a critical turn.

Both indefatigable travellers, Lemaitre and Vallarta made arrangements for touring eastern Canada together in the month of June. As was expected from Lemaitre, they stopped in Montreal to pay their hommage to Miss Thibeaudeau, and they met for long hours of discussions with Francois Henroteau at the Dominion Observatory in Ottawa. The summer vacations in Louvain were interrupted by the Sixth General Assembly of the International Astronomical Union in Stockholm (3--10 August 1938). His last conversation with Eddington took place in the ferry--boat from Malmo to Copenhagen. It was for Lemaitre a most memorable occasion. He had tried once more to overcome Eddington's hesitations toward the Big Bang Theory, and he expected Eddington to retort that no scientific hypothesis is admissible unless it is confirmed by experiments or observations. Much to Lemaitre's surprise Eddington, in a somewhat confidential tone, declared that, to the contrary, "he could not trust an experimental result unless it was confirmed by theory."

At the University of Louvain, Lemaitre resumed his teaching; he now occupied the Chair of Mechanics that Maurice Biot had vacated when he accepted a permanent appointment at Columbia University. Repeated alerts on the German border disrupted progressively scientific research in Belgium. Young professors, research assistants, graduate students kept getting in and out of the uniform; foreign students hastened to repatriate. Yet everyone in place pretended to act as if life continued on its normal course. On 12 March 1939, H.C. Plummer, president of the Royal Astronomical Society, announced that the Council had elected Lemaitre as an Associate of the Society. Cardinal Eugenio Pacelli elected Pope on 2 March insisted on holding an extraordinary session of the Pontifical Academy of which he had been a charter member. In spite of the drole de guerre ("the phoney war") which closed the borders between France and Germany, Lemaitre travelled to Rome and pronounced the eulogy of Lord Rutherford at the solemn session presided by H.H. Pius XII (3 December 1939).

4. Second intermission: a lonely teacher

In May--June 1940, fleeing the second German invasion, Lemaitre ran away from Belgium with his parents and other relatives, but was stopped in the Pas de Calais by the panzer divisions investing the Dunkirk beaches. He returned to Louvain where he faced a very bleak situation. For the second time in twenty-six years the University Library had been burnt down to ground level; a number of his colleagues who had been drafted in the Belgian army, had been taken to Germany as prisoners of war. Nevertheless the new Recteur Magnifique, Msgr Honore Van Waeyenbergh who had succeeded Msgr Paulin Ladeuze, reopened the University on 8 July, the students completed the last term of the current year, and a new academic year opened on 12 November. All international contacts were cut off, and so was funding in support of scientific research. Recruitment of faculty was next to impossible. A course left vacant, either because the instructor died or was retired or arrested, was routinely re--attributed to a colleague. Thus, when it became apparent that Canon De Strijcker could not return to the University after Belgium's capitulation, his course in philosophy for the freshmen at the College of Engineering was re--assigned to Canon Lemaitre; he kept that charge from 1940 until 1945. After Germany invaded the Soviet Union, material conditions in occupied Belgium deteriorated at an accelerated pace. As the war went on, foreign occupation turned to repression and even outright oppression. When the Kommandantur closed the University of Brussels, its students were immediately admitted to the University of Louvain; two years later, the Recteur of Louvain was arrested for refusing the enemy the roster of matriculated students; the Kommandantur claimed it needed it to track down freshmen hiding from forced labor in Germany. With the winter of 1943--1944, rationing of food and fuel fell to levels of starvation.

In the middle of that utter misery, Lemaitre saw it as his duty to help maintain a minimum of scientific activity among his colleagues in occupied Belgium. Personally, he spent his free time meditating on the classics of mechanics, especially Poincare's _Methodes nouvelles de la Mecanique celeste_. Much to the surprise of everyone who remembered the compulsive traveller he had been, Lemaitre involved himself with the affairs of the _Societe belge d'Astronomie, de Meteorologie et de Physique du Globe_; he was elected its president in 1943. As a token of the nation's appreciation for the moral leadership gallantly assumed in the most depressing circumstances, Prince--Regent Charles bestowed on Lemaitre the insignia of Commander of the Order of Leopold II, the highest honors the King of the Belgians confers to a citizen who is not a member of the armed forces (23 April 1947).

In the early morning of 12 May 1944, the U.S. Air Force bombed Louvain; it planned to destroy the railway depot, instead it ravaged the university district. Lemaitre's apartment on the third floor of the Debelva pastry--shop was blown out by a direct hit. Rescued from the debris, Lemaitre was taken to the University hospital where he was treated for shock and multiple contusions. With Louvain in ruins, the school closed that day. For a while, until the city was liberated and the communications with Brussels re--established, Lemaitre camped in the attics above his office at the Premonstratensian College. More destruction occurred when the Germans rearguard blew up all bridges, small and large, a few hours before the English troops entered the town (4 September).

5. The second career: a craftsman at the bench

> Il y a trois categories de personnes
> qui aiment les plaisirs du gadget:
> les physiciens par obligation,
> les gens d'affaires par ostentation
> et les theoriciens par compensation.

The war years had exaceted a heavy toll on Lemaitre. His physical vigor had lost its youthful punch. Once again he was tied by family obligations. He felt that his mother, now a widow, could not be left to live alone. Since he himself was homeless, he returned to live with her in the family house in Brussels. Until her death in 1956, he will commute by train and by taxi -- Lemaitre never drove a car -- between Brussels and Louvain two or three days a week. His teaching duties were not onerous, at any rate by the standards of the time in Belgium. He crammed his undergraduate classes in the late morning hours, then walked to the Hotel Majestic for a long lunch with batchelor colleagues, as if they were sitting at the high table of a Cambridge college. After a nap in his office at the Institute of Physics, he would confer with his research assistant on the progress of a calculation or the correction of galley proofs, and then came the time for tea at a table in the middle of his computer laboratory to which visitors and junior colleagues in the Departments of Mathematics and Physics were welcome. The afternoon usually ended with a graduate class. His teaching work done for the day, he walked back to the railway station, except every other Wednesday when he would preside after dinner over the Cercle mathematique. He was sentimentally attached to that informal seminar run by the engineering students because it had been founded by Charles de la Vallee Poussin in the early years of his tenure at the University.

For two weeks each summer, he took his mother to Switzerland, often to the Lake of Bienne for vacations among the habitues, a closely knit ring of Belgian and Swiss friends and acquaintances. After his mother died (25 March 1956), Lemaitre sold the family house, and took an apartment in Louvain, first on President Hoover Square in the hub of the University district, then on King Albert Street in the quiet shadow of Saint Peter's Church. All the same he kept to his routine of classes two days a week, and of work at home interrupted by discussions with his assistants and graduate students at his office in the Institute of Physics at the Premonstratensian College.

Lemaitre had no special taste for a prominent place in the councils of the University or any other national institution. Which is to say that he never volunteered for administrative chores, although he acquitted himself of whatever he was elected to assume, and this rarely happened, much to his satisfaction. Thus he served as Dean of the Sciences Faculty (1948--1950), and for a period as director of the Sciences Class at the Royal Academy (1949--1950).

His academic standing brought him into the ambit of an affluent and cultivated society in Brussels interested in the arts and music. His youngest brother, Maurice, Chief Engineer for the nationalized Belgian railways, played the alto in the quatuor Queen Elisabeth; his sister-in-law organized at home concerts of chamber music which Lemaitre attended frequently. His relaxations were modest. He enjoyed playing piano at home. His hobby was photography, chiefly for the producing of unusual prints of ordinary scenes under unexpected lights. For a time, he studied Chinese rather assiduously. Lemaitre found his chief human happiness in the families of his brothers among his many nephews and nieces. Above all he loved travelling. He was delighted to be appointed to the Belgo--Italian commission for cultural exchanges. He took the habit of extending mission trips to Italy into vacation tours through various regions of Italy, the Ligurian coast one year, Naples and Capri another year, or Florence and Bologna, Ravenna and Venice. On several occasions, he arranged his trip to stop a couple of days in Assisi and visit the tomb of Saint Francis.

Thus, at the age of 50, Lemaitre settled in his second scientific career. It was to be the life of a professor in academic semi--retirement, and also that of a research scientist intensely preparing himself backstage for applying the latest developments in computer technology to the progress of non linear dynamics in general and of celestial mechanics in particular. Apparently he felt no regret for having withdrawn from the limelight. The scientific world to which he returned in 1945 had changed radically, and he was well aware of it. While the German occupation had kept him stagnating in Belgium, his friends in Great Britain and the United States had moved in step with the technical and scientific evolution determined by the war effort. Vallarta had resigned from the M.I.T. to go back to his native Mexico where he assumed the leadership in promoting scientific research and peaceful applications of nuclear energy. Hugh Stott Taylor had assumed the role of Dean of the Graduate School at Princeton. Leslie J. Comrie was now managing his own company, the celebrated Scientific Computing Service Limited. In Japan, Hagihara had his hands full with reconstructing research and teaching in astronomy. Lemaitre had lost his patron in Louvain, Msgr Ladeuze. The new Recteur, a man of heroic modesty as he proved it during the occupation, was a builder and a wizard at raising money and suggesting donations, but was not a scholar by any standards. Beyond their sacerdotal persuasion and their dedication to the University, there was nothing these two men shared on a personal basis.

With a caution that had been so unusual with him in the past fifteen years, Lemaitre undertook to renew his international contacts. His friends and colleagues in the United States were eager to re--establish communication. Lemaitre was not forgotten; on 19 April 1945, the American Philosophical Soceity elected him a foreign member. At a meeting of the Societe helvetique des Sciences naturelles in Fribourg (1--3 September 1945), Lemaitre was introduced to Fernand Gonseth at the ETH ("Federal Institute of Technology") in Zurich, a professor of mathematics who had studied very closely relativity and cosmology, but was now established as a philosopher of considerable notoriety in the fields of logic and scientific methodology. Gonseth offered to publish a collection of Lemaitre's popular lectures and conferences on the Big Bang Theory. Lemaitre however insisted that a mathematical summary be added as an appendix. The volume appeared in 1946 under the title "L'Hypothese de l'Atome Primitif. Essai de Cosmogonie" with a long preface by Gonseth. It was immediately translated into Spanish; the English translation with a prologue by Henri N. Russell followed in 1950. With the meeting of the Societe helvetique came the opportunity of renewing acquaintance with the nuclear physicist Paul Scherrer at the Department of Physics of the ETH, to whom Lemaitre had been introduced by Vallarta and Manneback. It was then arranged that Scherrer would visit Belgium and advise the

Belgian government on how to proceed in order to establish a national competence in nuclear engineering.

The publication of Lemaitre's popular papers on his Big Bang Theory drew the attention of the French speaking public. It resulted in a number of invitations to speak on the topic to general audiences in Belgium and abroad. Lemaitre made a special effort in preparing a major address he had agreed to deliver at the Palais de la Decouverte in Paris (13 May 1947). The text of that conference became, one might say, the manifest of his theory. He read it with resounding success at the first postwar meeting of the Pontifical Academy (4--19 February 1948). At the same meeting, he presented a delightful treatment of the elliptic geometry by quaternions; the paper reveals first hand knowledge of the XIXth century sources. Such a display of erudition is unusual in Lemaitre's publications. One owes it undoubtedly to the years of confinement in Louvain among the classical works which his negligence in returning books borrowed from the Library had involuntarily salvaged from the Holocaust of May 1940. Finally the Seventh General Assembly of the International Astronomical Union in Zurich (11--18 August 1948) gave Lemaitre the full opportunity of renewing friendships with colleagues of his American years. Yet he was not interested in resuming his first career.

Lemaitre's second stage in scientific life exhibits elements of culmination and of decline. It was then that the mature -- perhaps the aging -- Lemaitre set out to retrench himself in Louvain.

In 1946 Prime Minister Eamon De Valera, desirous of the scientific achievement of his country and mindful perhaps of his youthful aspirations as a student of astrophysics under Whittaker, had appropriated a budget for rehabilitating Dunsink Observatory as a national institution. Thanks to the vigorous leadership of Professor Hermann Bruck, the first phase of the restoration was completed within two years. At Bruck's invitation, Lemaitre flew to Dublin to visit the renovated Dunsink (14--24 March 1950). He toured the Institute of Advanced Studies, also the Departments of Physics and Mathematics at the National University of Ireland in Dublin; he was driven to Armagh Observatory where an old acquaintance of Harvard days, Emil Opik, was cutting for himself a reputation in estimating the age of the universe from galactic and extragalactic statistics. Lemaitre was enthralled by what he had seen and heard; his hosts were delighted with their guest. Nevertheless, when there came a letter from Erwin Schrodinger offering a visiting professorship at the School of Physics in the Institute for Advanced Studies (9 May 1951), Lemaitre declined. After the Van Allen belt around the earth had been detected by satellites, the European division of the U.S. Air Force Office for Scientific Research sent delegates to Louvain to invite Lemaitre's collaboration in recruiting a team for analyzing the data in relation to the Stoermer problem. Once again Lemaitre declined.

Honors nonetheless kept coming to the doorstep of Lemaitre's retreat in Louvain. A bronze portrait head, executed by Mr Charles Leplae, was commissioned by the Belgian Ministry of Public Instruction (August 1951) and placed in the Aedes Academiae, the seat of the Royal Academy in Brussels (8 January 1955). Lemaitre was awarded the Decennial Prize for Applied Mathematics by the Royal Academy of Belgium (19 September 1950), the Eddington Medal by the Royal Astronomical Society (13 February 1953). He received the honorary D.Sc of the University of Dublin (1954). H.H. Pope John XXIII appointed him President of the Pontifical Academy of Sciences for a period of four years (27 March 1960). Prior to that nomination, as prescribed by the etiquette at the Vatican Court, Lemaitre was bestowed the honorary title of prelat domestique de Sa Saintete ("prelate in the Pope's Household"), by virtue of which title he was henceforth addressed formally as Monsignor (19 March). The presidential mandate was renewed in 1964 for another period of four years. The Accademia Nazionale (delle Scienze detta) dei XL elected Lemaitre as one of its twelve soci stranieri or foreign members (17 March 1961). Lemaitre was given the seat first attributed in 1786 to the Prussian meteorologist Franz Carl Achard, and successively occupied by the medical genius Jons Berzelius, Urbain Le Verrier, the geologist David Owen, the mathematician James Sylvester, John W. Strutt third Baron Rayleigh, Lord Rutherford, and Max Von Laue.

Only after the death of his mother did Lemaitre return to the U. S. A., and only for short visits. In 1961, as the guest of the International Business Corporation of Belgium, he attended the Eleventh General Assembly of the International Astronomical Union at the University of California at Berkeley. On that occasion he was invited to make a communication to the conference on "The Instability of Systems of Galaxies" at the University of California at Santa Barbara (8--9 August); he also appeared at the IAU Symposium on "Problems of Extra--Galactic Research" where he contented himself with chairing a session of papers pertaining to cosmology (10--12 August). By that time however, his research interests had drifted away from cosmology into celestial mechanics. During the working session of Commission VII (16 August), at the instigation of its chairman, Professor Dirk Brouwer of Yale University, he made a report on his regularization of the problem of three bodies. It was brilliant; the effect it made on the public was marred however by the following communication he insisted on giving about his versions for numerical integration by finite differences. Lemaitre returned to Berkeley in the summer of 1962 as the guest of the Space Science Laboratories. He wrote there his last scientific paper, a most elegant treatment of some problems left unsolved by Elie Cartan regarding the problem of three bodies. While in Berkeley, Lemaitre shared an office with Professor A. van Wijngaarden director of the Mathematisch Centrum at the University of Amsterdam and one of the authors of ALGOL. van Wijngaarden converted Lemaitre to the use of programming languages more advanced than assembler language. So much so that, in the academic year 1962--1963 at Louvain, Lemaitre turned himself into an instructor in computer sciences and opened a course in Algol.

In cosmology, Lemaitre's postwar attitude was one of deferment: he was waiting for observations to confirm the Big Bang Theory. For the past fifteen years, since his memorable address to the British Association in 1931, he had been put on the defensive. The Big Bang Theory had been held in suspicion by most astronomers, not the least by Einstein, if only for the reason that it was proposed by a Catholic priest and seconded by a devout Quaker, hence highly suspect of concordism. In that regard, the personal opinion of Pope Pius XII came to Lemaitre as a most embarrassing surprise. The Pontifical Academy of Sciences was meeting for a week long seminar on "The Problem of the Microseisms". The session was to begin on the morning of 22 November 1951 with an audience given by the Pope in the Great Hall of the Consistory. There, in the presence of several cardinals and of the Italian Minister for Education, the Pope pronounced his famous speech "Un Ora". After a brief review of the traditional Catholic teaching regarding the creation, the Pope entered a long and detailed exposition of the physical cosmology to lead to his conclusion that the initial singularity postulated by the Big Bang Theory could be made the antecedent of the scholastic syllogism concluding to the Catholic concept of creation. Astronomers present at the ceremony, Lemaitre among them, had recognized in the pontifical address arguments developed by their fellow member, Sir Edmund Whittaker (1873--1956), either in his Riddell lectures at Durham University (1942) on the "Beginning and end of the world" or in the Donnellan lectures on "Space and Spirit" delivered in 1947 at the University of Dublin. A direct quotation from "Space and Spirit" explicitly acknowledged in the official text published the following day by the Osservatore Romano confirmed the general opinion. The Pope's speech did not go unnoticed. Excerpts of it were quoted jokingly by George Gamow as if the pontifical declaration had made an "unquestionable truth" out of his theory on the role of turbulence in the expansion of the universe (15 April 1952).

Needless to say, regarding the philosphical and theological implications of the Big Bang, Lemaitre and Whittaker held diametrically opposite views. Always on the alert lest he be drawn into a religious controversy, Lemaitre had never reacted to Whittaker's apologetics. Nor did he ever comment on the opinions Fred Hoyle had expressed in his sixth lecture to the B.B.C. on "The Nature of the Universe" about alleged incompatibilities between the theory of the continuous creation and what Hoyle regarded as the Judeo-Christian tradition rooted in a fundamentalist interpretation of the Genesis.

Fortunately Lemaitre had made a friend of Father Daniel O'Connell, a Jesuit director of the Specola Vaticana and a man who was gradually emerging as a most trusted scientific adviser to the Vatican Curia. Lemaitre had accepted the invitation by the Union of South Africa to join Sir Lawrence Bragg, Professors E.J. Brouwer and Jan H. Oort at the Science Congress commemorating the fiftieth anniversary of the South African Society for the Advancement of Science (7--12 July 1952). On his way to Cape Town, he stopped in Rome to consult with O'Connell and dignitaries of the Vatican Curia, in particular Msgr dell'Acqua and Cardinal Tisserand, about the pontifical address planned for the Eighth General Assembly of the International Astronomical Union (4 --13 September 1952). At the reception in Castel Gandolfo, H.H. Pope Pius XII exalted the progress accomplished in observational astronomy, but made no allusion to the modern cosmology and the Big Bang Theory save for a poetic allusion to "les processus cosmiques qui se sont deroules au premier matin de la creation"(7 September). Lemaitre felt relieved. Father O'Connell had well exercised his legendary discretion in obtaining that the Pope, by his silence, vindicate Lemaitre's integrity as a physicist.

Lemaitre wanted his Hypothesis of the Primeval Atom to be judged solely as a physical theory, and exclusively on grounds of both mathematical consistency and adequation with observations and experiments. After the two pontifical addresses, he felt it necessary to explain himself to an areopage of physicists. The opportunity arose at the XIth Solvay Conference in Physics dealing with "The Structure and The Evolution of the Universe" (Brussels, 9--13 June 1958). In his communication, in essence a full account of his Primeval Atom Hypothesis, after he had explained, rather well, what he meant by the natural beginning of the universe, he faced the issue squarely. "I do not pretend, digressed Lemaitre, that such a singularity is inescapable in Friedmann's theory, but I simply point out how it fits with the quantum outlook as a natural beginning of multiplicity and of space--time." Then he added:

> "As far as I can see, such a theory remains entirely outside any metaphysical or religious question.
> "It leaves the materialist free to deny any transcendental Being. He may keep, for the bottom of space--time, the same attitude of mind he has been able to adopt for events occurring in non--singular places of space--time.
> "For the believer, it removes any attempt to familiarity with God, as were Laplace's chiquenaude or Jeans' finger. It is consonant with Isaias speaking of the Hidden God, hidden even in the beginning of creature."

At the Notre Dame Symposium of 1938, Lemaitre had virtually committed himself to solving the problem of the clustering of nebulae. From thereon it had been the sole topic of his research in cosmology. In his view, due to the instability at the equilibrium that is the Einstein model, amidst a universe expanding as a whole, there were individual regions that failed to expand. Soon, at these points, the exchange of nebulae became an exchange between the cluster itself and the neighbouring field with the latter expanding more and more to become finally the general field of nebulae. Can the continuous relay of nebulae between the cluster and the external field explain why the local condensation persists in spite of the general expansion? Put in so general terms, the problem looked impenetrable. To make it simpler, Lemaitre limited himself to small clusters in an expanding universe close to the Einstein equilibrium; even so reduced to its bare minimum, the problem remained a challenge. For the spherical models of clusters, which he built with the collaboration of R. Vander Borght, Lemaitre could produce simple solutions that were both static and isotropic, but the resulting clusters had too large a radius and presented no marked condensation in the central region (1948). At the next level of complexity, dropping the requirement that the solutions be static, Lemaitre still tried to keep the problem somewhat manageable by imposing that spatial densities in the clusters be determined only by the radial motion of the nebulae (1951). These so--called quasi--isotropic models turned out to be inconsistent with the boundary conditions (1958). Finally, dropping this pseudo--simplification, Lemaitre and his assistant Andree Bartholome embarked on a large computing program to produce at least some classes of particular solutions by separating the variables. At that stage the clustering of nebulae became for Lemaitre and his assistant a problem that they carried from machine to machine as a benchmark to test how far the available computing power would develop the solutions.

For the Stoermer problem, Lemaitre hit an interesting idea. He proposed to develop in literal form by Poincare's technique a canonical transformation that would eliminate the terms periodic in the latitude. He carried the reduction by hand to the fourth order. Having satisfied himself that the calculation could be pursued in a consistent manner, he went looking for ways of executing algebraic calculations automatically by computer. On that score, considering the state of the technology at the time, he was reaching rather far into the future. Yet he was not day--dreaming. He followed closely the work done by Wallace Eckert at Columbia University first in automating Brown's Lunar Tables, then in reconstructing by computer the solution developed by Hill and Brown for the main problem in Lunar Theory. At the same time he was studying closely Delaunay's masterpiece looking ahead for the time when he would reproduce Delaunay's operations by machine. Past 1950 however, the only activity on record regarding the cosmic rays was the session on "The problem of cosmic radiation in the intergalactic space", the first manifestation he organizsed as president of the Pontifical Academy of Sciences (1--6 October 1962). Thorough screening of the invitations and a jovial but firm stirring of the debates made it a success.

Through his solitary readings of the classics during the war, Lemaitre had come across the famous Lecons sur les Invariants integraux ("Lectures on the integral invariants") given in 1921--1922 by Elie Cartan at the Sorbonne. Cartan's complete reduction of several particular cases in the problem of three bodies intrigued Lemaitre . At first he strived to reconstruct Cartan's solution in the simplest manner possible, exclusively by elementary geometric transformations (1950). Having succeeded for three particles of equal masses in a fixed plane, he extended his technique by inventing a new set of symmetric coordinates to cover gradually more general cases, removing in turn the restriction on the masses and the condition that they be located in a fixed plane. Considering the enormous amount of research that had been done on this question for two centuries, the reduction effected by Lemaitre is simply amazing in its originality and in its simplicity. This is not all. Somewhat to his surprise Lemaitre discovered that his reduction by elementary geometry led to a representation transforming the moving binary collisions into fixed singularities which he could then regularize by conformal mappings, a procedure somewhat analogous to Levi--Civita's transformation in the planar restricted problem of three bodies. He was most interested in exploiting his representation in order to explore numerically the qualitative structures of the phase space in the two special cases which Jean Chazy had studied qualitatively by Tauberian arguments, the collinear configuration in which the three mass points move on a fixed line, and the isosceles configuration in which the triangle of the three particles remains isosceles at all times. A considerable number of numerical explorations were made in these two problems, but few of them have been published. For Lemaitre waited for the moment when all these preliminary results would be reconstituted systematically by automatic prospection on a large computer.

Old enough to have experienced the antiquated ways of computing, either numerically or algebraically, by hand with tables of logarithms, Lemaitre in his sixties was young enough to have entered fully into the spirit of the revolution in scientific computing that was taking place in the 1950s. Moreover he kept himself acquainted with almost every one of the astronomers engaged in mastering computers for the benefit of mathematical astronomy. When electronic machines became available commercially, he kept before himself the purpose of bringing the equipment to Louvain. The President of the University tried to represent that all the resources were committed to reconstructing the campus devastated by the war and to meeting the maddeningly rapid expansion in the student population. Lemaitre would have none of that: he lent the Recteur the money to buy the first electronic computer installed at the University, a Burroughs E101 (1957). On that machine he learned the basics of machine language and computer organization. Later, on an IBM 1620, then on an Elliot 801, he became proficient in assembler languages. In many respects he had become a computer hacker; but he also had a vision of the revolution computers were making in mathematical research. Almost single--handedly he instigated the creation of a computing center in his university. At this point hovever, friction developed with administrative authority, and Lemaitre withdrew from further initiative on the academic scene.

The friction arose as a result of the linguistic conflicts in Belgium. The 1960s were indeed sad years for the French section of the University of Louvain. The "language laws" voted by the Flemish majority in the Belgian Parliament had resolved that only Dutch should be used for teaching in schools situated in Flanders (the northern part of the country), and French only in schools located in Wallonie (the southern provinces). From a strict constructionist viewpoint, the law did not apply to the French section of the University although it was established in the Flemish city of Leuven. Yet the Flemish majorities in the Parliament, in the press, and in the streets to be sure, never ceased to proclaim their will to make Louvain the Katholieke Universiteit te Leuven, a school totally and exclusively Flemish. Since the university was a private institution, the decision of abolishing the French section fell on its Flemish faculty. For a while after the vote of the linguistic laws, two universities lived side by side in Leuven. Most of the French speaking professors grouped themselves in an association aimed at negotiating compromises with the Flemish majority. Lemaitre was elected its president, and he accepted the charge (1962). That made him the prime target of the campaign waged in the Flemish press for the abolition of the French section. "Walen buiten" ("Oust the Walloons") was the battle--cry calling the Flemish students to rioting in the streets of the university district. A couple of times the rioters broke the windows in Lemaitre's apartment.

Before the French problem in Leuven reached the final solution, Lemaitre passed the age limit. By way of retiring him from the academic service, the University promoted him Emeritus Professor at the Faculty of Sciences, which promotion safeguarded his privileges (July 1964). An office at the Institute of Nuclear Physics in the Arenberg park, access to the computer and the library, the employment of a research assistant: he had everything to make him happy and contented, yet life had become wearisome to him. People who saw him at the XIIIth Solvay Conference in Physics on "The Structure and Evolution of Galaxies" in Brussels the next September hardly realized how sick he was and worn out. Yet he had to keep busy, and he did hack it on the computer -- in an aimless fashion. His health declined steadily; he suffered a heart attack in 1965 from which he did not fully recover. In January 1966, he received news of his election to the Academia Neocastrum. To his bedside at the Hopital Saint Pierre of the School of Medecine, Professor Godart brought him the issue dated 1 July 1965 of the Astrophysical Journal which contained the now famous parallel letters, one signed by R.H. Dicke, P.J.E. Peebles, P.G. Roll, and D.T. Wilkinson, the other by A.A. Penzias and R.W. Wilson. At last there came the evidence from observations about the Big Bang.

After a long illness Monsignor Georges Lemaitre died Monday 20 June 1966. At the solemn service in the University parish church attended by the recteurs of the Flemish and French universities of Leuven/Louvain, the French faculty, members of the Belgian ministerial Cabinet, and representatives of the diplomatic corps, the eulogy was pronounced, in keeping with the academic etiquette, by the Dean of the Sciences Faculty, Albert Bruylants (24 June); in the afternoon, Lemaitre was buried in the family plot in the cemetery of Marcinelle, a suburb of Charleroi.

"Ergo vivida vis animi pervicit et extra
"processit longe flammantia moenia mundi
"atque omne immensum peragravit mente animoque
"unde refert nobis victor quid possit oriri
"quid nequeat.
Aurelius Lucretius Carus

P.S. Many people helped in the preparation of this notice, especially Jacques Absil, Luc Arnould, P. Darcy Barnett, Emile Boulpaep, Rolf Brahde, Andre Berger, Louis Bouckaert, Lucien Bossy, C. Bruneel, Charles Courtoy, Richard Cushman, Andree Deprit--Bartholome, Raoul Deprit, Rene de Vogelaere, Jules Deutsch, Albert d'Haenens, Howland Fowler, Dionigi Galletto, Robert Graas, Robert Howland, Xu Jia, Patricia Jordan, Charles Lamb, Monique Lavendhomme, Gilbert Lemaitre, Pierre Lemaitre, Marguerite Manneback, Edouard Massaux, Karl Menger, Marcel Michel, Paul Paquet, Elizabeth Pessek, Alfredo Ramirez--Araiza, Joseph Schnaubelt, Paul Smeyers, Miles Standish, Pierre Triest, Roderick and Marjorie Webster, Chen Zhong--Xuan, and Anthony Zito.

THE SCIENTIFIC WORK OF GEORGES LEMAITRE (*)

O. Godart

Institut d'Astronomie et de Géophysique G. Lemaître,
Université Catholique de Louvain-la-Neuve, Belgium

The scientific interests of Georges Lemaître were rather broad, mainly centered on applied mathematics, mechanics, physics and astronomy.

His first university studies before the first world war (1914-1918) were in mechanical engineering at Catholic University of Louvain. Although he changes afterward, mechanics kept a very important place in Lemaître's investigations. It is thus quite natural that he became interested specially in the famous problem of the three body motion. By using symmetrical coordinates, referred to the instantaneous principal axes, he was able to construct what one calls nowadays the "regularization of double collisions" (1952). This partly explains why he was so fond of graphical and mechanical aids to calculations. In 1933, he brought in, for the University, three great electrical calculating machines (Mercedes) and, in 1956 the first electronic computer (Burroughs E101). Already in 1935 at M.I.T., he had enjoyed the use of the Bush machine. Although not in charge of the computing center of the Louvain University, but convinced of the important possibilities of electronic computers, he preserved up to the end of his life a great inclination to all problems connected to computers, mainly those concerning languages and programming. In this context, he invented new digits based on the binary system which enable to build up automation in computations, making unnecessary to memorize the multiplication table (1954). This became for him a great diver-

(*) The bibliography of Lemaître to which the reader is invited to refer is published in the next section.

A. Berger (ed.), The Big Bang and Georges Lemaitre, 393–397.

sion. He was in fact a prominent calculator in algebra and arithmetic, numerical analysis and computations were not only tools for his researches but matters for investigations. He published several interesting works on harmonic analysis, rational iteration, and integration of systems of differential equations (1955).

But the main theme of his research was relativistic cosmology (Godart and Heller, 1984). Independently of his University lessons, very early after the first world war, he became acquainted with the work of Einstein. Still a student, in the seminary of Malines (1922), he wrote a monography about the theory of Relativity with comments that would have been worth publishing (Lemaître unpublished, 1922). His doctoral thesis on: "L'Approximation des Fonctions de Plusieurs Variables Réelles" and other related works were awarded by a research fellowship. He used that opportunity to deepen his knowledge in relativistic cosmology and its astronomical background as a student of Eddington (Cambridge, England, 1924) and as a research fellow of the Harvard Observatory with Shapley (1925). Soon, he realized the instability of the Einstein solution of the General Relativity for an homogeneous Universe and, although few data were then available, he made the connection between the redshifts of galaxies and the expansion of the Universe.

A dynamical cosmological model was then proposed in 1927 : "Un univers homogène de masse constante et de rayon variable rendant compte de la vitesse radiale des nébuleuses extra-galactiques". At that time, Friedman (1924) had already proposed solutions of the General Relativity showing an expansion of the Universe (1924). Lemaître had no knowledge of these publications. Moreover up to then, relativistic cosmology was rather a gravitational and geometrical branch of Science. Lemaître felt that it was indispensible to introduce astronomical and physical considerations in the macroscopic picture of the Universe.

What caused the expansion of a Universe initially macroscopically homogeneous and at equilibrium ? : the formation of condensations proposed Lemaître. Referring to Jean's theory (1918) of gravitational instability, he studied the consequences of the growth of small singularities and concluded to the expansion of the "neutral zones" between such condensations (1931). Rather unsatisfied by the extension in an infinite past of a Universe previously static, he computed the "Friedmann" solutions with positive cosmological constant. These were unpublished but as a technical collaborator, I have had the pleasure to help him in these calculations (Heller, 1979).

What were the arguments that led him to choose the now called "Lemaître cosmological model" admitting a singular beginning, an initial expansion damped by a coasting period near the static scale factor of Einstein model, followed by another indefinite expansion ? There is usually a philosophical background of new ideas. For instance, Einstein (1916) was greatly influenced by Mach philosophy inferring that particle's inertia is due to some interaction of that particle with all the other masses of the Universe. In the case of Lemaître, Godart and Heller (1978) discovered in an unpublished manuscript written around 1922 that "as the genesis suggested it, the Universe had begun by light". However Lemaître was too carefull a scientist to build his theory on what was no more than an intuitive opinion; a scientific basis was necessary. Thermodynamics envisaged from the point of view of quantum theory, including the splitting of energy in ever increasing quanta, gave him in 1931 the idea of what he called later the Primeval Atom Hypothesis, forefather of the actual Big Bang Theory.

Although exposed in broad lines at the 1931 Royal Society meeting on "the question of the relation of the Physical Universe to life and mind" (Godart and Heller, 1979), it took a structural form in his paper : "l'Univers en expansion" published in 1933, date which was chosen as a reference for this colloquium. The revolutionnary idea of a singular beginning of the Universe was not well accepted at that time. Meeting Einstein in Pasadena in 1933, Einstein objected to the isotropy of his solution, probably responsible of the singularity (Godart and Heller, 1979). But in his 1933 paper, Lemaître showed that models slightly anisotropic will also have a singularity. Actually, we know mainly from the works of Hawking (1973) that singularities are features of solutions of General Relativity. By the time of his 1933 paper, he had read the work of Friedman and, in fact, he proposed then for a coarse description of the Universe, imbricated Friedman solutions inside a macroscopic space in expansion with regions assimilated to cluster of galaxies in equilibrium, containing fluctuations collapsing eventually to form protogalaxies. The shortness of the Hubble time with respect to the geophysical age of the Earth convinced him with the existence of an appreciable coasting period in the expansion and it was his principal argument in favour of the cosmological constant. In that context, calculations on the formation of galaxies were started including pressure term and were published much later (Godart, 1968). Also the question of cluster of galaxies was reexamined later in order to take into account the great speed of individual galaxies envisaging a continual exchange between galaxies in the cluster and field galaxies (1948).

The physical picture of the remnants of the primeval atom remains rather vague. Besides a powerful primeval radiation, Lemaître thought that there would form early, besides a gaseous phase, a primary population of big stars. In fact, he felt rather uneasy about the lack of knowledge concerning hyperdense matter and the uncertainties of elementary particles theory and he abstained researches on such subjects. However, he was strongly convinced, already since 1931, that cosmic radiation would give the clues of the primeval splitting and since 1935 he was engaged with collaborators, and in particular with professor Vallarta of M.I.T., in researches concerning the distribution of primary cosmic radiation received at the surface of the Earth. This project suited his taste and ability for mechanical and numerical problems. Instead of computing innumerable trajectories, as it was done by Stormer, he studied the structure of the dynamical problems, calculating singular period orbits and their assymptotes limiting the directions of cosmic radiation reaching the Earth. The sequels of these studies brought interesting results in analytical mechanics and numerical analysis. However, from the point of view of cosmology, it appears more and more that cosmic radiation could be explained by astrophysical processes.

The isolation due to the war, some personal difficulties to restart international scientific relations and the lack of physical proof of a "primordial" radiation and also of the cosmological constant indispensible for the long age of the Universe envisaged in Lemaître's cosmology, appeared to render absolete his cosmological hypothesis. Moreover, the success of the competing steady state theory (1950) brought in its development new ideas in physical cosmology such as star's evolution and elements nucleosynthesis. Lemaître scientific activities were then mainly oriented to other fields (celestial mechanics, numerical analysis, history of science). From time to time he tried to modernize his cosmology theory. In particular, he made for the XI Solvay Congress (Brussels, 1958) a remarkable report "The hypothesis of primeval atom and the problem of cluster of galaxies" (1958) (Godart and Turek, 1982). The discovery of micro-wave radiation came too late (1965) to give a new impulse in his cosmological researches. He was very ill but I had the pleasure to inform him that the proof of the initial firework, main object of his life researches, had been discovered (Dicke et al., 1965).

The reader wanting to know more about the work of Lemaître in cosmology is invited to refer to his 1945 book "The Primeval Atom", unhappily out of print. A new French edition, with comments and his Solvay paper, has been published in 1972, "L'Hypothèse de l'Atome Primitif", Editions Culture et Civilisation, Bruxelles, 1972.

REFERENCES

Dicke, R.H., Peebles, P.J.E., Roll, P.C. and Wilkinson, D.T., 1965. Cosmic black body radiation, Astrophysical Journal 142, p. 414.

Einstein, A., 1916. "Die Grundlage der allgemeinen Relativitäts Theorie", Ann. Phys. 49, pp. 769-822.

Friedman, A.F., 1924. Zeitschrift für Physic 21, pp. 326-332.

Godart, O., 1968, Cosmologie et Observations. Annales de la Société Scientifique de Bruxelles, T82, I et III.

Godart, O., Heller, M., 1978, Les relations entre la Science et la Foi chez Georges Lemaître, Commentarii Pontifica Academia Scientiarum, vol. III n°21, 12pp.

Godart, O., Heller, M., 1979, Einstein-Lemaître rencontre d'idées, Revue des Questions Scientifiques, T.150 n°1, pp. 23-43.

Godart, O., Turek, J., 1982. Le développement de l'atome primitif, Revue des Questions Scientifiques, T 153 n°3.

Godart, O. and M. Heller, 1984. "Lemaître Cosmology", ed. Pachart Publishing House, Tucson, in press.

Hawking, S.W., Ellis, G.F.R., 1973. The large-scale structure of space time, chap. 8, Univ. Cambridge Press, pp. 286-289.

Heller, M., 1979, Questions of Infallible Oracle, Springer Verlag (ed.), Lecture Notes in Physics n°109, Physics of the Expanding Universe, pp. 199-210.

Jeans J., 1918. Cosmogony and Stellar Dynamics, Cambridge University Press.

Lemaître, G., 1922. Archives. Unpublished manuscripts kept at the Institut d'Astronomie et de Géophysique G. Lemaître.

BIBLIOGRAPHY OF G. LEMAITRE

The following list of publications by Georges Lemaître has been compiled by Esqs. O. Godart, A. Deprit, P. Pâquet, L. Bossy and Mrs L. Moens. This is an up-date and corrected version of the following bibliographies published earlier :

1937. Annuario pontificia Acad. Sci. 1, pp. 489-493.
1961. Annuario gen., Acad. nazion. XL, pp. 615-620.
1974. Poggendorf : in "Biografish - Literarisches Handwörterbuch der Exacten Natuurwissenschaften - Band VII - b - Teil 5.
1984. Godart O. et Heller M. in "Lemaître Cosmology", Pachart Publ. House, Tucson, USA.

Those interested in a more detailed account of the personality of G. Lemaître, are invited to read the Addresses delivered during the Opening Ceremony of this symposium and published in Revue des Questions Scientifiques, fasc. 2, 1984.

1923. Sur une propriété des Hamiltoniens d'un multiplicateur. Académie Royale de Belgique, Bulletin de la Classe des Sciences, 5e série, tome IX, pp. 380-388.
1924. The motion of a rigid body according to the relativity principle. Philosophical Magazine 48, pp. 164-176.
1925. Note on De Sitter's Universe. The Physical Review 25, pp. 903.
1925. Note on De Sitter's Universe. Journal of Mathematics and Physics 4, pp. 37-41; Publ. du Labo. d'Astronomie et de Géodésie, Univ. Cath. de Louvain 2, n°17, pp. 37-41.
1925. Note on the theory of pulsating stars. Harvard College Observatory Circular 282, 6pp.
1926. La théorie de la relativité et l'expérience. La Revue des Questions Scientifiques, 4e série, tome IX, pp. 346-374.
1927. A propos d'une note du P. Schaffers sur la relativité. Annales de la Société Scientifique de Bruxelles 47A, pp. 15-21.

1927. Un univers homogène de masse constante et de rayon croissant, rendant compte de la vitesse radiale des nébuleuses extra-galactiques. Annales de la Société Scientifique de Bruxelles 47A, pp. 49-59; Publ. du Labo. d'Astronomie et de Géodésie, Univ. Cath. de Louvain 4, n°48, pp. 101-111.

1927. Le mouvement varié d'un solide d'après la théorie de la relativité. Annales de la Société Scientifique de Bruxelles 47A, pp. 103-109; Publ. du Labo. d'Astronomie et de Géodésie, Univ. Cath. de Louvain 4, n°54, pp. 239-245.

1929. La grandeur de l'espace. La Revue des Questions Scientifiques, 4e série, tome XV, pp. 189-216; Publ. du Labo. d'Astronomie et de Géodésie, Univ. Cath. de Louvain 6, n°58, pp. 9-36.

1930. On the random motion of material particles in the expanding universe. Explanation of a paradox. Bulletin of the Astronomical Institutes of The Netherlands 5 n°200, pp. 273-274.

1930. L'hypothèse de Millikan-Cameron dans un univers de rayon variable. Congrès national des Sciences, Bruxelles, pp.3.

1931. L'indétermination de la loi de Coulomb. Annales de la Société Scientifique de Bruxelles 51B, lère partie, pp. 12-16.

1931. Sur l'interprétation d'Eddington de l'équation de Dirac. Annales de la Société Scientifique de Bruxelles 51B, lère partie, pp. 89-99.

1931. The beginning of the World from the point of view of quantum theory. Nature 127, pp. 706.

1931. A homogeneous universe of constant mass and increasing radius accounting for the radial velocity of extra-galactic nebulae. Monthly Notices of the Royal Astronomical Society 91, pp. 483-490.

1931. The expanding universe. Monthly Notices of the Royal Astronomical Society 91, pp. 490-501.

1931. Contributions to a British Association Discussion on the Evolution of the Universe. Nature 128, pp. 704-706.

1931. L'expansion de l'espace. La Revue des Questions Scientifiques, 4e série, tome XX, pp. 391-410.

1933. L'Univers en expansion. Annales de la Société Scientifique de Bruxelles 53A, pp. 51-85; Publ. du Labo. d'Astronomie et de Géodésie, Univ. Cath. de Louvain 9, n°86, pp. 172-205.

1933. (with M.S. Vallarta). On Compton's latitude effect of cosmic radiation. The Physical Review 3 n°2, pp. 87-91.

1933. The uncertainty of the electromagnetic field of a particle. The Physical Review 43 n°2, 15 janvier 1933.

1933. Condensations sphériques dans l'univers en expansion. Comptes Rendus de l'Académie des Sciences de Paris 196, pp. 903-904.

1933. La formation des nébuleuses dans l'univers en expansion. Comptes Rendus de l'Académie des Sciences de Paris 196, pp. 1085-1087.

1933. The cosmical significance of the clusters of nebulae, Rep. Annu. Meeting, Brit. Assoc. Advancement Sci. 103, pp. 448.

1934. Evolution of the expanding universe. Proceedings of the National Academy of Sciences USA 20, pp. 12-17.

1934. Evolution in the expanding universe, Nature 133, pp. 654.

1934. Manifestation en l'honneur de M. l'Abbé Georges Lemaître, Lauréat du Prix Francqui 1934, Union des Ingénieurs sortis des Ecoles spéciales de Louvain, Bull. Techn., numéro spécial, 1934, pp. 17-56.

1934. Réponse de M. le Professeur Abbé Lemaître. Union des Ingénieurs sortis des Ecoles spéciales de Louvain. ler Bulletin technique, numéro spécial, pp. 39-43.

1934. Contributions à la théorie des effets de latitude et d'asymétrie des rayons cosmiques. I.-Trajectoires infiniment voisines de l'équateur. Annales de la Société Scientifique de Bruxelles 54A, pp. 162-173; Publ. du Labo. d'Astronomie et de Géodésie, Univ. Cath. de Louvain 10, n°88, pp. 21-33.

1934. Contributions à la théorie des effets de latitude et d'asymétrie des rayons cosmiques. III.-Trajectoires périodiques. Annales de la Société Scientifique de Bruxelles 54A, pp. 194-207; Publ. du Labo. d'Astronomie et de Géodésie, Univ. Cath. de Louvain 10, n°90, pp. 55-68.

1934. L'univers en expansion. Académie Royale de Belgique, Bulletin de la Classe des Sciences 20, pp. 1182-1200; Publ. du Labo. d'Astronomie et de Géodésie, Univ. Cath. de Louvain 10, n°87, pp. 1-19.

1935. (with M.S. Vallarta and L. Bouckaert). On the North-South asymmetry of cosmic radiation. The Physical Review 47, pp. 434-436.

1935. L'univers en expansion. La Revue des Questions Scientifiques 27, pp. 357-375.

1935. L'expansion de l'univers. Société Astronomique de France, L'Astronomie 49, pp. 153-168.

1935. La expansion del universo, Ibérica 22, pp. 3-32, 78-80, 204-207.

1936. (with M.S. Vallarta). On the geomagnetic analysis of cosmic radiation. The Physical Review 49, pp. 719-726.

1936. (with M.S. Vallarta). On the allowed cone of cosmic radiation. The Physical Review 50, pp. 493-504.

1936. (with M.S. Vallarta). Contributions à la théorie des effets de latitude et d'asymétrie des rayons cosmiques. IV. Calcul d'une famille d'orbites asymptotiques. Annales de la Société Scientifique de Bruxelles 56A, pp. 102-130; Publ. du Labo. d'Astronomie et de Géodésie, Univ. Cath. de Louvain 10, n°91, pp. 69-97.

1936. La culture catholique et les sciences positives, Actes du VIe congrès catholique de Malines, vol. 5, pp. 65-70.

1937. (with J.M. Delfosse, J.C. Jungers, Y.L. Tchang, C. Manneback). Raman spectrum of monodeuteroethylene, Nature 139, pp. 1111.

1937. Longitude effect and the asymmetry of cosmic radiation. Nature 140, pp. 23-24.

1937. (with C. Manneback and Y.L. Tchang). Calcul des fréquences et modes de vibration de la molécule de monodeutéroéthylène. Annales de la Société Scientifique de Bruxelles 57A, pp. 120-128.

1937. Sur l'interprétation d'Eddington de l'équation de Dirac. Annales de la Société Scientifique de Bruxelles 57A, pp. 165-172.

1938. Les rayons cosmiques et le champ magnétique terrestre. Réunion internationale de physique - chimie - biologie, Congrès du Palais de la Découverte, oct. 1937, Actualités Scientifiques et Industrielles n°720, Paris, Hermann et Cie, pp. 165-172.

1938. (with O. Godart). Généralisation de la méthode de Hill. Académie Royale de Belgique, Bulletin de la Classe des Sciences 24, pp. 19-23.

1938. (with O. Godart). Contributions à la théorie des effets de latitude et d'asymétrie des rayons cosmiques. V.-Détermination des exposants caractéristiques des trajectoires périodiques, Annales de la Société Scientifique de Bruxelles 58A, pp. 27-41.

1942. L'itération rationnelle, Académie Royale de Belgique, Bulletin de la Classe des Sciences, 5e série, 28, pp. 347-354; Publ. du Labo. d'Astronomie et de Géodésie, Univ. Cath. de Louvain 11, n°97.

1942. Intégration d'une équation différentielle par itération rationnelle, Académie Royale de Belgique, Bulletin de la Classe des Sciences, 5e série, 28, pp. 815-825.

1942. (with F. Moreau). Eléments d'astronomie et de géodésie. Ciel et Terre 58, pp. 378.

1943. Champ magnétique et rayons cosmiques. Ciel et Terre 59, pp. 1-16; Publ. du Labo. d'Astronomie et de Géodésie, Univ. Cath. de Louvain 11, n°98.

1944. Fortschritte der Astronomie, Band II : Heckman, Otto, Theorie der Kosmologie, Ciel et Terre 60, pp. 140.

1945. (with L. Bossy). Sur un cas limite du problème de Stoermer, Académie Royale de Belgique, Bulletin de la Classe des Sciences, 5e série, 31, pp. 357-364.

1945. Les variables canoniques dans le mouvement Képlérien, Académie Royale de Belgique, Bulletin de la Classe des Sciences, 5e série, 31, pp. 365-371.

1945. L'hypothèse de l'atome primitif, Actes de la Société Helvétique des Sciences Naturelles, Fribourg, pp. 77-96.

1945. Hypothèses cosmogoniques, Ciel et Terre 61, pp. 61-71; Publ. du Labo. d'Astronomie et de Géodésie, Univ. Cath. de Louvain 11, n°101.

1945. A propos de la théorie de Milne-Haldane, Ciel et Terre 61, pp. 213-218.

1945. Cinquantenaire de la Société Belge d'Astronomie, de Météorologie et de Physique du Globe, discours du Président le 27 octobre 1945, Ciel et Terre Mémorial, pp. 3-6.

1946. Un modèle de nébuleuse, Annales de la Société Scientifique de Bruxelles 60, pp. 103-114.

1947. Calcul des intégrales elliptiques, Académie Royale de Belgique, Bulletin de la Classe des Sciences, 5e série, 33, pp. 200-211; Publ. du Labo. d'Astronomie et de Géodésie, Univ. Cath. de Louvain 12, n°108.

1947. Interpolation dans la méthode de Runge-Kutta, Annales de la Société Scientifique de Bruxelles 61A, pp. 106-111.

1948. Quaternions et espace elliptique, Acta Pontificiae Academiae Scientiarum 12, pp. 57-78.

1948. Modèles mécaniques d'amas de nébuleuses, Académie Royale de Belgique, Bulletin de la Classe des Sciences, 5e série, 34, pp. 551-565; Publ. du Labo. d'Astronomie et de Géodésie, Univ. Cath. de Louvain 12, n°112.

1948. (with R. van der Borght). Modèles de nébuleuses à vitesses radiales, Académie Royale de Belgique, Bulletin de la Classe des Sciences, 5e série, 34, pp. 955-965; Publ. du Labo. d'Astronomie et de Géodésie, Univ. Cath. de Louvain 12, n°115.

1948. L'hypothèse de l'atome primitif, La Revue des Questions Scientifiques, (e série, 9, pp. 321-339; Publ. du Labo. d'Astronomie et de Géodésie, Univ. Cath. de Louvain 12, n°110.

1948. L'hypothèse de l'atome primitif, Acta Pontificiae Academiae Scientiarum 12, pp. 25-40.

1948. L'hypothèse de l'atome primitif, Université de Paris, Conférence du Palais de la Découverte, 13 mai 1947, Maison Poulet-Malassis, Alençon, 15p.

1949. Cosmological application of relativity, Reviews of Modern Physics 21, pp. 357-366.

1949. The cosmological constant. Albert Einstein : Philosopher - Scientist (ed. P.A. Schilpp). Evanston : The Library of Living Philosophers, pp. 439-456.

1949. Application des méthodes de la mécanique céleste au problème de Stoermer, Annales de la Société Scientifique de Bruxelles 63A, pp. 83-97; Publ. du Labo. d'Astronomie et de Géodésie, Univ. Cath. de Louvain 13, n°115.

1949. Rayons cosmiques et cosmologie, Gazette Astronomique, (Anvers : Société d'Astronomie) 31, pp. 1-14; Publ. du Labo. d'Astronomie et de Géodésie, Univ. Cath. de Louvain 12, n°113.

1949. L'énigme de l'hydrogène, Académie Royale de Belgique, Bulletin de la Classe des Sciences, 5e série, 35, pp. 1158-1163.

1950. Application des méthodes de la mécanique céleste au problème de Stoermer, Annales de la Société Scientifique de Bruxelles 64A, pp. 76-82; Publ. du Labo. d'Astronomie et de Géodésie, Univ. Cath. de Louvain 13, n°116.

1950. Laplace et la mécanique céleste, Société Astronomique de France, L'Astronomie, 64, pp. 23-31.

1951. Modèles mécaniques d'amas de nébuleuses, Académie Royale de Belgique, Bulletin de la Classe des Sciences, 5e série, 37, pp. 291-306; Publ. du Labo. d'Astronomie et de Géodésie, Univ. Cath. de Louvain 13, n°118.

1952. Coordonnées symétriques dans le problème des trois corps, Académie Royale de Belgique, Bulletin de la Classe des Sciences, 5e série, 38, pp. 582-592; Publ. du Labo. d'Astronomie et de Géodésie, Univ. Cath. de Louvain 13, n°120.

1952. Coordonnées symétriques dans le problème des trois corps (suite), Académie Royale de Belgique, Bulletin de la Classe des Sciences, 5e série, 38, pp. 1218-1234; Publ. du Labo. d'Astronomie et de Géodésie, Univ. Cath. de Louvain 13, n°121.

1952. Clusters of nebulae in an expanding universe, Monthly Notes of the Astronomical Society of South Africa 11, pp. 110-117.

1952. The clusters of nebulae in the expanding universe, South African Journal of Science/Suid Afrikaanse Joernaal van Wetenskap 49, pp. 80-86.

1954. Comment calculer ? Académie Royale de Belgique, Bulletin de la Classe des Sciences, 5e série, 40, pp. 683-691.

1954. Régularisation dans le problème des trois corps, Académie Royale de Belgique, Bulletin de la Classe des Sciences, 5e série, 40, pp. 759-767; Publ. du Labo. d'Astronomie et de Géodésie, Univ. Cath. de Louvain 13, n°122.

1955. Intégration par analyse harmonique, Annales de la Société Scientifique de Bruxelles 70, pp. 117-123; Publ. du Labo. d'Astronomie et de Géodésie, Univ. Cath. de Louvain 13, n°123.

1955. Regularization of the three body problem, Vistas in Astronomy 1, (A. Beer) London, New York, 1955; J. Atm. and Terrestrial Phys., Spec. Suppl. 3, 1955, pp. 207-215.

1955. L'oeuvre scientifique d'Albert Einstein, La Revue des Questions Scientifiques, 5e série, 16, pp. 475-487.

1955. Die Kosmologische Konstante, in : Albert Einstein, Kohnhammer Verlag, 1955, pp. 312-327 - German translation of "The cosmological constant" (1949).

1955. Réunion intime en l'honneur du Chanoine G. Lemaître à

l'occasion de la remise de son buste à la Classe, Bull. Cl. Sc. Acad. Royale de Belgique : adresse de M. Ch. Manneback, pp. 73-76; réponse du Chanoine Lemaître, pp. 77.

1955. Pourquoi de nouveaux chiffres ? La Revue des Questions Scientifiques, 5e série, 16, pp. 379-398.

1956. Le calcul élémentaire, Académie Royale de Belgique, Bulletin de la Classe des Sciences 42, pp. 1140-1145.

1958. Instability in the expanding universe and its astronomical implications. in : Semaine d'étude sur le problème des Populations stellaires, Rome 20-28 mai 1957, Pont. Acad. Scient. Scripta Varia, 16, 1958, pp. 475-488.

1958. The primeval atom hypothesis and the problem of the clusters of galaxies. in : La Structure et l'Evolution de l'Univers, Rapports et Discussions du Onzième Conseil de Physique, Institut International de Physique Solvay, Université de Bruxelles, Stoops (Ed.), pp. 1-31.

1958. (with A. Bartholomé). Contribution au problème des amas de galaxies, Annales de la Société Scientifique de Bruxelles 72, pp. 97-102.

1958. Rencontres avec A. Einstein, La Revue des Questions Scientifiques, 5e série, 19, pp. 129-132.

1960. L'étrangeté de l'Univers, La Revue Générale Belge 95, pp. 1-14.

1960. L'étrangeté de l'Univers, La Scuola in Azione Notiziario n°15, Varese : Multa Paucis, pp. 3-22.

1961. Remarques sur certaines méthodes d'intégration des systèmes d'équations différentielles. Colloque sur l'analyse numérique (Centre Belge de Recherches Mathématiques), Paris : Gauthier-Villars, pp. 9-23.

1961. Exchange of galaxies between clusters and field, Astron. J. 66, pp. 603-606.

1962. Charles-Jean de la Vallée Poussin, Pontificiae Academiae Scientiarum Commentarii 1, pp. 1-4.

1962. Visite à la chambre de Monseigneur Georges Lemaître, Chamber Primitif, Essai de Cosmogonie, Editions CuCanada, n°11-12, pp. 5-8.

1967. L'expansion de l'Univers, Réponses à radio-Canada le 15 avril 1966, La Revue des Questions Scientifiques 28, pp. 153-162.

1967. Le principe de continuité d'après Jean-Victor Poncelet, La Revue des Questions Scientifiques, 5e série, 28, pp. 381-395.

1978. L'Univers, problème accessible à la science humaine, Revue d'Histoire Scientifique 31, pp. 345-359; the text of Lemaître's conference at the Catholic Institute in Paris, found in Lemaître's archive at Louvain-la-Neuve and published by O. Godart and M.

Heller with an introduction on "Un travail inconnu de Georges Lemaître".

Obituary

1945. Sir Arthur Stanley Eddington (1882-1944), Ciel et Terre 61, pp. 114-115.
1948. M. le Chanoine René De Muynck, Professeur émérite à la Faculté des Sciences, Discours prononcé aux funérailles célébrées à Louvain, le19 juin 1948.
1949. Lord E. Rutherford of Nelson. Discorso commemorativo pronunciato alla augusta presenza del S. Padre Pio XII nella solenne tornata inaugurale del IV anno academico il 3 dicembre 1939, 13, pp. 93-96.
1966. Ch. Manneback, Bull. Cl. Sci., Acad. Roy. Belgique 52, pp. 1034-1039.; Rev. Questions Sci. 27, pp. 453-461.
1967. A.V. Douglas, J. Roy. Astron. Soc. Canada 61, pp. 77-80.

Books

1946. L'hypothèse de l'Atome Primitif - Essai de Cosmogonie, préface de F. Gonseth, Ed. du Griffon, Neuchâtel, 203p.
1946. Cosmogonia-Hipotesis del atomo primitivo, Ibero-Americana, Buenos Aires 1946, 189p. - Spanish translation of the book "L'hypothèse de l'atome primitif".
1949. Rayons Cosmiques et Cosmologie, Louvain, E. Nauwelaerts, 37pp.
1950. L'Univers, (Causeries de l'Université radiophonique internationale, Paris 1950), Nauwelaerts, Louvain, 73pp.
1950. The Primeval Atom - An Essay on Cosmogony, Van Nostrand Comp., Toronto - New-York - London, 1950. English translation of "L'hypothèse de l'atome primitif" (1946) by B.H. and S.A. Korff, 9+186pp.
1954. Calculons sans fatigue, Louvain : E. Nauwelaerts, 41pp.
1961. Un nouveau système de chiffres et autres essais, Scuola di Studi Superiori sugli Idrocarburi dell'Ente Nazionale Idrocarburi, Quaderno n°10, Varese : Multa Paucis, 57pp.
1963. The three-body problem, Space Science Laboratories, Technical Report n°49, Berkeley, University of California, 40pp.
1965. Code machine symbolique pour calculatrice NCR-Elliott 802, Laboratoire de Recherches Numériques, Labo. d'Astronomie et de Géodésie, 25pp.
1965. Emploi des routines classiques dans le code machine symbolique pour calculatrice NCR-Elliott 802,

Laboratoire de Recherches Numériques, Université de Louvain, Labo. d'Astronomie et de Géodésie, 29pp.

1972. L'hypothèse de l'Atome Primitif, Essai de Cosmogonie, Editions Culture et Civilisation, Bruxelles 1972, 203+99pp.; photographical re-edition of the book (1946) with the French version of the paper delivered by Lemaître at the Solvay Conference in Brussels (1958), the article by O. Godart "Georges Lemaître et son oeuvre", and the bibliography of Lemaître's works.

AUTHOR INDEX

SUBJECT INDEX

GPSR Compliance

The European Union's (EU) General Product Safety Regulation (GPSR) is a set of rules that requires consumer products to be safe and our obligations to ensure this.

If you have any concerns about our products, you can contact us on ProductSafety@springernature.com

In case Publisher is established outside the EU, the EU authorized representative is:

Springer Nature Customer Service Center GmbH
Europaplatz 3
69115 Heidelberg, Germany

Batch number: 10815974

Printed by Printforce, the Netherlands